# VCE Units 3 & 4

# MATHEMATICAL METHODS

SUE GARNER

2023–2027 STUDY DESIGN • 2023–2027 STUDY DESIGN •

A+

- + topic summaries
- + graded exam practice questions with worked solutions
- + study and exam preparation advice

STUDY NOTES

**A+ VCE Mathematical Methods Study Notes**
**3rd Edition**
**Sue Garner**
**ISBN 9780170465366**

Publisher: Cathy Beswick-Davison
Associate publisher: Naomi Campanale
Project editor: Tanya Smith
Editor: Geoffrey Brent
Series text design: Nikita Bansal
Series cover design: Nikita Bansal
Series designer: Cengage Creative Studio
Artwork: MPS Limited
Production controller: Karen Young
Typeset by: Nikki M Group Pty Ltd



**Acknowledgements**
Selected VCE examination questions and extracts from the VCE Study Designs are copyright Victorian Curriculum and Assessment Authority (VCAA), reproduced by permission. VCE ® is a registered trademark of the VCAA. The VCAA does not endorse this product and makes no warranties regarding the correctness or accuracy of this study resource. To the extent permitted by law, the VCAA excludes all liability for any loss or damage suffered or incurred as a result of accessing, using or relying on the content. Current VCE Study Designs, past VCE exams and related content can be accessed directly at www.vcaa.vic.edu.au.

Casio ClassPad: Images used with permission by Shriro Australia Pty Ltd

For product information and technology assistance,
in Australia call **1300 790 853**;
in New Zealand call **0800 449 725**

For permission to use material from this text or product, please email
**aust.permissions@cengage.com**

ISBN 978 0 17 046536 6

**Cengage Learning Australia**
Level 5, 80 Dorcas Street
Southbank VIC 3006 Australia

**Cengage Learning New Zealand**
Unit 4B Rosedale Office Park
331 Rosedale Road, Albany, North Shore 0632, NZ

For learning solutions, visit **cengage.com.au**

Printed in China by 1010 Printing International Limited.
1 2 3 4 5 6 7 26 25 24 23

# CONTENTS

## AREA OF STUDY 3: CALCULUS



CHAPTER 4

## UNIT 4

## AREA OF STUDY 4: DATA ANALYSIS, PROBABILITY AND STATISTICS



CHAPTER 5

## SOLUTIONS

# HOW TO USE THIS BOOK

The *A+ VCE Mathematical Methods* resources are designed to be used year-round to prepare you for your VCE Mathematical Methods exam. *A+ VCE Mathematical Methods Study Notes* includes topic summaries of all key knowledge in the VCE Mathematical Methods Study Design. The first four chapters of this book address the four Areas of Study of the course. The following gives you a brief overview of each chapter and the features included in this resource.

## Concept maps

The concept map at the beginning of each chapter provides a visual summary of each Area of Study.

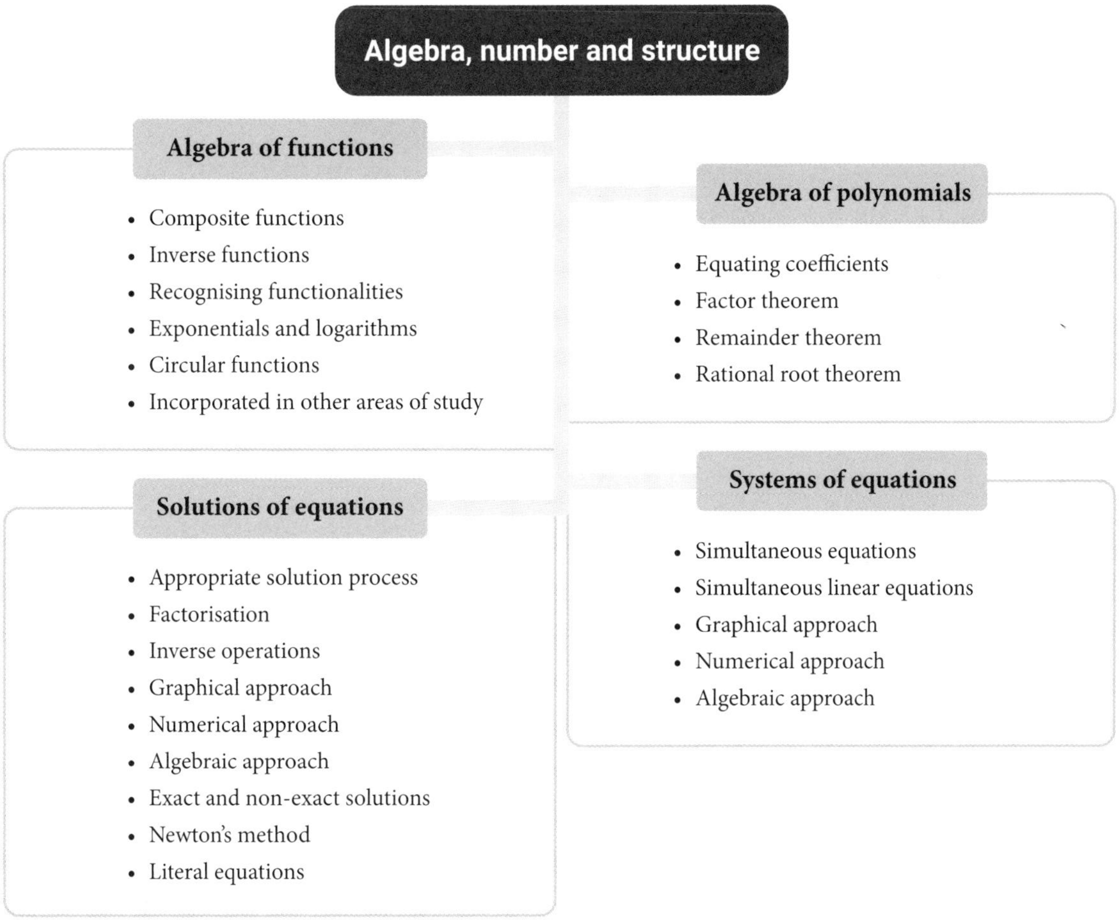

**Algebra, number and structure**

**Algebra of functions**

- Composite functions
- Inverse functions
- Recognising functionalities
- Exponentials and logarithms
- Circular functions
- Incorporated in other areas of study

**Algebra of polynomials**

- Equating coefficients
- Factor theorem
- Remainder theorem
- Rational root theorem

**Solutions of equations**

- Appropriate solution process
- Factorisation
- Inverse operations
- Graphical approach
- Numerical approach
- Algebraic approach
- Exact and non-exact solutions
- Newton's method
- Literal equations

**Systems of equations**

- Simultaneous equations
- Simultaneous linear equations
- Graphical approach
- Numerical approach
- Algebraic approach

## Exam practice

Exam practice questions appear at the end of each chapter to test you on what you have just reviewed in the chapter. These are written in the same style as the questions in the actual VCE Mathematical Methods Exams 1 and 2. There are also official past VCAA exam questions in each chapter.

### 1 Short-answer questions (Exam 1: Technology free)

There is approximately 20 short-answer questions in each chapter. These questions require you to apply your knowledge across single or multiple concepts. Mark allocations have been provided for each question.

### 2 Multiple-choice questions (Exam 2: Technology active)

There are approximately 50 multiple-choice questions in each chapter.

### 3 Extended-answer questions (Exam 2: Technology active)

There are a wide range of extended-answer questions in each chapter. These questions are longer and often consist of parts that require you to apply your knowledge across multiple concepts. Mark allocations have been provided for each question.

# Solutions

The last chapter of the book provides the solutions to all exam practice questions. Solutions to the extended-answer questions have been written to reflect a high-scoring response and include explanations of what makes an effective answer.

## Explanations

Where relevant, the solutions section includes explanations of multiple-choice answers, both correct and incorrect. Short-answer and extended-answer solutions outline what a high-scoring response looks like and signpost potential mistakes. The VCAA extended-answer questions state the success percentage with each question or question part, where available.

**Question 3**

Set up the equation $y = a\sqrt{2-x} + 2$ and substitute $(0, -4)$.

$-4 = a\sqrt{2} + 2$

$-6 = a\sqrt{2}$

$a = -\frac{6}{\sqrt{2}} = -3\sqrt{2}$

$\therefore y = -3\sqrt{2}\sqrt{2-x} + 2$

Equation is $y = -3\sqrt{4-2x} + 2$.

**39 A**

$y = \cos(x)$ is transformed to $y = -\cos\left(4x - \frac{1}{2}\right)$ by

- a reflection in the $x$-axis
- dilation by a factor of $\frac{1}{4}$ from the $y$-axis
- a horizontal translation of $\frac{\pi}{8}$ units to the right.

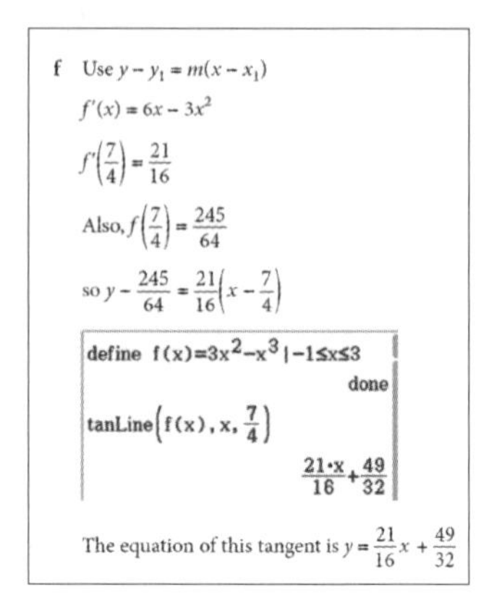
f Use $y - y_1 = m(x - x_1)$

$f'(x) = 6x - 3x^2$

$f'\left(\frac{7}{4}\right) = \frac{21}{16}$

Also, $f\left(\frac{7}{4}\right) = \frac{245}{64}$

so $y - \frac{245}{64} = \frac{21}{16}\left(x - \frac{7}{4}\right)$

```
define f(x)=3x^2-x^3 | -1≤x≤3
                                done
tanLine(f(x), x, 7/4)
                          21·x/16 + 49/32
```

The equation of this tangent is $y = \frac{21}{16}x + \frac{49}{32}$

# Icons

The icons below occur in the summaries and exam practice sections of each chapter to provide additional tips and support.

**Notes**
Remember that a function is defined by its rule and domain.

Hint and note boxes appear throughout the summaries to provide additional tips and support.

©VCAA 2020 2BQ3

This icon appears with official past VCAA exam questions.

These icons indicate the question's level of difficulty: easy, medium or hard.

# A+ DIGITAL

Just scan the QR code or type the URL into your browser to access:

- A+ Flashcards: revise key terms and concepts online
- Revision summaries of all concepts from each extended-answer question.

Note: You will need to create a free NelsonNet account.

https://get.ga/aplus-vce-maths-methods

# PREPARING FOR EXAMS

Exam preparation is a year-long process. It is important to keep on top of the theory and consolidate often, rather than leaving study to the last minute. You should aim to have the theory learnt and your notes complete so that by the time you reach the pre-exam study period, the revision you do is structured, efficient and meaningful.

## Study tips

To stay motivated, try to make the studying experience as comfortable as possible. Have a dedicated study space that is well lit and quiet. Create and stick to a study timetable, take regular breaks, reward yourself with social outings or treats.

### Revision techniques

Here is a useful technique to help with your revision: **'STIC'** - a framework for structuring your learning so that you study less, but learn more!

| | |
|---|---|
| **Spaced repetition** | This technique helps to move information from your short-term memory to your long-term memory by spacing out the time between revision and recall flash card sessions. As the time between retrieving information is slowly extended, the brain processes and stores the information for longer periods. |
| **Testing** | Testing is necessary for learning and is a proven method for exam success. If you test yourself continually before you learn all the content, your brain becomes primed to retain important information when you learn it. As part of this process, engage with the marking criteria provided to help decide on the areas where improvement is needed. |
| **Interleaving** | This is a revision technique that sounds counterintuitive but is very effective for retaining information. Most students tend to revise a single topic in a session, and then move onto another topic in the next session. With interleaving, you choose three topics (1, 2, 3) and spend 20 to 30 minutes on each topic. You may choose to study 1–2–3 or 2–1–3 or 3–1–2, 'interleaving' the topics and repeating the study pattern over a long period of time. This strategy is most helpful if the topics are from the same subject and are closely related. |
| **Chunking** | An important strategy is breaking down large topics into smaller, more manageable 'chunks' or categories. Essentially, you can think of this as a branching diagram or mind map where the key theory or idea has many branches coming off it that get smaller and smaller. By breaking down the topics into these chunks, you will be able to revise the topic systematically. |

These strategies take cognitive effort, but that is what makes them much more effective than re-reading notes or trying to cram information into your short-term memory the night before the exam!

# Time management

It is important to manage your time carefully throughout the year. Make sure you are getting enough sleep and the right nutrition, and that you are exercising and socialising to maintain a healthy balance.

To help you stay on target, plan your study timetable. Here is one way to do this:

1 Assess your current study time and social time. How much will you dedicate to each?

2 List all your commitments and deadlines, including sport, work, assignments etc.

3 Prioritise the list and reassess your time to ensure you can meet all your commitments.

4 Decide on a format, whether it be weekly or monthly, and schedule in a study routine.

5 Keep your timetable where you can see it.

6 Be consistent.

Studies suggest that one-hour blocks with a 10-minute break are most effective for studying. You will also have free periods during the school day that you can use for study, note-taking, practice questions and past papers, meeting with your teachers and group study sessions. Studying does not have to take hours if it is done effectively. Use your timetable to schedule short study sessions often.

# The exams

The examinations will contribute 60 per cent to your study score. Examination 1 (technology free) will contribute 20 per cent and Examination 2 (technology active) will contribute 40 per cent to your study score.

## Examination 1

This examination comprises 9 or 10 short-answer questions that may have one part or several linked parts in the one question. Examination 1 is a one-hour technology free examination.

The examination is designed to assess students' knowledge of mathematical facts, concepts, techniques and analysis that can be achieved without notes or a calculator.

## Examination 2

This examination comprises 20 multiple-choice questions covering all areas of study and then 4, 5, or 6 extended-answer questions that will have multiple parts linked in the one question, and sometimes across a few topics. Examination 2 is a two-hour technology active examination.

Any approved CAS technology is permitted and assumed in all the questions, and a scientific calculatior is permitted as well, if the student would like this.

Examination 2 is designed to assess students' ability to select and apply mathematical facts, concepts, models and techniques to solve multiple-choice questions and extended-answer questions in a range of contexts.

## Conditions

The examinations will be completed under the following conditions:

### Examination 1

- Duration: 1 hour
- 15 minutes reading time
- The same formula sheet will be provided for both Examinations 1 and 2.
- No technology, student notes or reference texts are permitted.

### Examination 2

- Duration: 2 hours
- 15 minutes reading time
- The same formula sheet will be provided for both Examinations 1 and 2.
- Student access to an approved technology with numerical, graphical, symbolic, financial and statistical functionality will be assumed.
- One bound reference text (which may be annotated) or lecture pad may be brought into the examination. This *A+ Study Notes* may be the bound reference text.

The following strategies will help you prepare for the exams.

## Practise using past papers

To help prepare, download past papers from the VCAA website and attempt as many as you can in the lead-up to the exam. Make a plan as to when you will start these and in what order you will complete them. Some students like to leave the most recent ones until the week before the exams. These will show you the types of questions to expect and provide practice in writing answers. As you head towards the exam period, it is a good idea to make the trial exams as much like the real exam as possible (conditions, time constraints, materials etc.). Some students like to begin the process of attempting past papers early in the year with the answers beside them in order to build confidence.

Use trial papers, school-assessed coursework and comments from your teacher to pinpoint weaknesses, and work on improving these areas. Do not just tick or cross your answers; look at the suggested answers and try to work out why your answer was different. What misunderstandings do your answers show? Are there gaps in your knowledge? Read the examiners' reports to find out the common mistakes students make.

Make sure you understand the material, rather than trying to rote learn information. Most questions are aimed at your understanding of concepts and your ability to apply your knowledge to new situations.

Clearly the more past papers you can complete the more ready you will be for the range of questions you will be asked, so make a decision as to the time of the year you will begin your past paper revision. For instance, are you going to start with the 2016 paper in March and complete one past paper every fortnight? Remember there are 2 papers for every year.

In recent years VCAA exams have also been written for the Northern Hemisphere exams. These papers are labelled NHT and are an additional resource to utilise in your exam preparation.

## The day of the exam

The night before your exam, try to get a good rest and avoid cramming, as this will only increase stress levels. On the day of the exam, arrive at the venue early and bring everything you need with you. If you rush to the exam, your stress levels will increase thereby lowering your ability to do well. Furthermore, if you are late you may not be allowed to enter the examination room. Do not worry too much about exam jitters. A certain amount of stress can help you concentrate and achieve an optimum level of performance. If, however, you are feeling very nervous, breathe deeply and slowly. Breathe in for a count of 6 seconds and out for 6 seconds until you begin to feel calm. Know yourself if you want to gather with your peers and teacher and talk Maths as you wait to enter the exam room. And also know yourself if you would like to discuss your answers to Examination 1 before you attempt Examination 2 the next day. Students vary widely in their approach to this and any choice you make is your own choice without judgement.

## Important information from the Study Design

Your school-based assessment will address the skills required for working mathematically, which contributes to 40% of your study score.

## Exam 1

Exam 1, worth 40 marks, consists of a question and answer book with space to write answers. Read each question carefully and underline key words. If you are given a graph or diagram, make sure you understand what the graph is telling you. You may make notes on the diagrams or graphs.

As exams are now scanned for external assessing, take care to ensure your handwriting is clear and legible, and do not overwrite a number or word. If you make a mistake, cross out clearly with one line and write below or beside. Attempt all questions: do not leave gaps! Marks are not deducted for incorrect answers, and you might get some marks if you make an educated guess. You will definitely not get any marks if you leave a question blank!

You will see from past papers that the early questions in Exam 1 are quite similar over the years. Practise these so that you can be sure of scoring full marks on at least the first few pages.

You have one hour to complete 40 marks. This means you have an average of 1.5 minutes per mark.

## Exam 2

Exam 2 consists of a question and answer book with space to write answers.

Section A, worth 20 marks, includes 20 multiple-choice questions, where the answer must be recorded on the answer sheet provided. A correct answer scores 1, and an incorrect answer scores 0. There is no deduction for an incorrect answer, so attempt every question. Read all the possible solutions. One possible technique is to eliminate any clearly wrong answers.

Section B, worth 60 marks, includes the extended-answer questions. The space provided is possibly an indication of the detail required in the answer. Most questions will be broken down into several parts, and each part will be testing new information, so read the entire question carefully to ensure you do not repeat yourself. Use correct mathematical terminology. If you make a mistake, cross out any errors. If you cross out your answer, mark on your paper clearly if you have used another space on the page to answer the question.

Make sure your handwriting is clear and legible. Attempt all questions; marks are not deducted for incorrect answers, and you might get some marks if you make an educated guess. You will definitely not get any marks if you leave a question blank!

Do not be put off if you do not recognise an example or context; questions will always be about the concepts that you have covered. In fact, top performing students are expected to apply learnt knowledge to an unfamiliar context.

You have 2 hours to complete 80 marks. This means you have an average of 1.5 minutes per mark.

## Reading time

Use your time wisely! Do not use the reading time to try and figure out the answers to any of the questions until you have read the whole paper. Plan your approach so that when you begin writing you know which section, and ideally which question, you are going to start with. Remember that you are permitted to look at your notes in Examination 2 reading time, but not CAS and no writing is allowed.

In Exam 2 make the decision beforehand if you are going to start Section A or Section B first. If you start Exam 2 with the extended-answer questions, make sure you leave 45 minutes for the multiple-choice section. This is something to discuss with your teacher.

# ABOUT THE AUTHOR

**Sue Garner** has taught VCE Mathematics at Ballarat Grammar School for over 30 years and lectured tertiary students in Mathematics Education. Sue has been a VCAA examination assessor for both Mathematical Methods and Specialist Mathematics and works for the VCAA in the role of Curriculum Manager for VCE Coursework. Sue has expertise in CAS technology and regularly presents workshops to teachers on the senior Mathematics classroom.

# UNIT 3

# Chapter 1
# Area of Study 1: Functions, relations and graphs

## Content summary notes

**Functions, relations and graphs**

**Key features of functions**

- Relations and functions
- Domain, co-domain and range
- Axis intercepts
- Stationary points
- Points of inflexion
- Asymptotic behaviour
- Symmetry

**Elementary functions**

- Power functions
- Exponential and logarithmic functions
- Circular functions
- Polynomial functions

**Transformations of the plane**

- Functions approach
- Mapping approach

**Behaviour of elementary functions**

- Inverse functions
- Sum, difference and product
- Composite functions
- Piecewise (hybrid) functions

**Modelling contexts**

- Theoretical investigations
- Modelling investigations

The area of study: Functions, relations and graphs, covers transformations of the plane and the behaviour of some elementary functions of a single real variable, including key features of their graphs such as axis intercepts, stationary points, points of inflection, domain (including maximal, implied or natural domain), co-domain and range, asymptotic behaviour and symmetry. The behaviour of functions and their graphs is to be explored in a variety of modelling contexts and theoretical investigations.

This area of study includes:

- graphs of polynomial functions and their key features
- graphs of the following functions:
  - power functions, $y = x^n, n \in Q$
  - exponential functions, $y = a^x, a \in R^+$, in particular $y = e^x$
  - logarithmic functions, $y = \log_e(x)$ and $y = \log_{10}(x)$
  - circular functions, $y = \sin(x)$, $y = \cos(x)$, $y = \tan(x)$ and their key features
- transformation from $y = f(x)$ to $y = Af(n(x + b)) + c$, where $A, n, b$ and $c \in R$, $A, n \neq 0$, and $f$ is one of the functions specified above, and the inverse transformation
- the relation between the graph of an original function and the graph of a corresponding transformed function (including families of transformed functions for a single transformation parameter)

- graphs of sum, difference, product and composite functions involving functions of the types specified above (not including composite functions that result in reciprocal or quotient functions)
- modelling of practical situations using polynomial, power, circular, exponential and logarithmic functions, simple transformation and combinations of these functions, including simple piecewise (hybrid) functions.

## 1.1 Relations and functions

A **relation** is a numeric, algebraic or graphical relationship between the set of ordered pairs $(x, y)$, for example, $x^2 + y^2 = 1$. This relation is, in fact, a circle where ordered pairs on the circle will include the points clockwise $(1, 0)$, $(0, -1)$, $(-1, 0)$ and $(0, 1)$.

A special type of relation is a **function** such that no two ordered pairs have the same first element. To the right, the labelled points $(0, -1)$ and $(0, 1)$ have the same first element with a different second element, so the relation $x^2 + y^2 = 1$ is NOT a function.

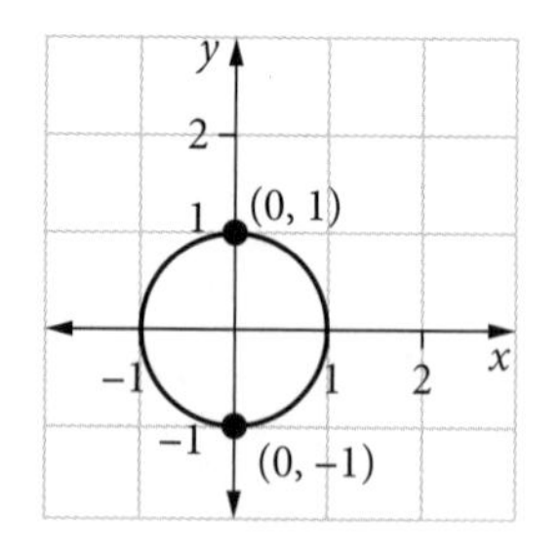


In a function, there is only one unique $y$ value for every $x$ value.

The easiest way to test if a relation is a function is to show vertical lines through the graph: if any vertical line cuts no more than once, then it is a function. If any horizontal line also cuts the function no more than once, then you have a **one-to-one function**. A one-to-one function is important for the existence of inverse functions.

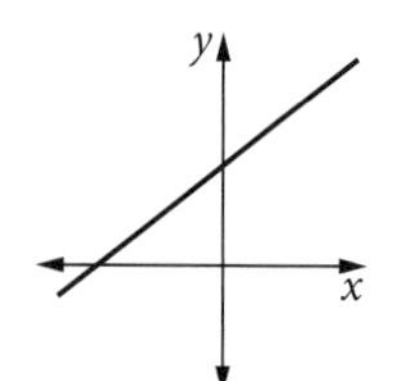

Function, one-to-one

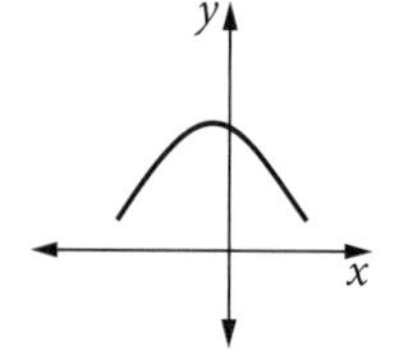

Function, many-to-one

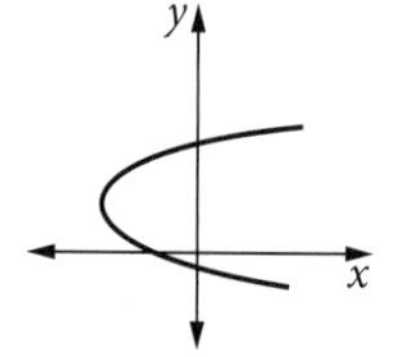

Not a function, one-to-many

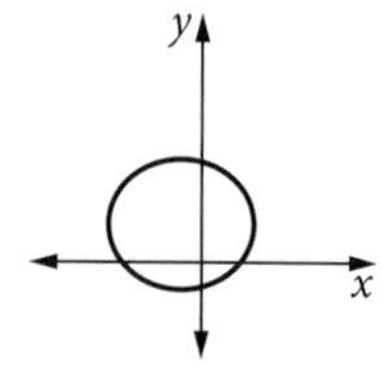

Not a function, many-to-many

The **independent variable** is the $x$ value. The **dependent variable** is the $y$ value.

## 1.2 Domain, co-domain and range

For every function, the **domain** is significant. A function is defined by its **rule and domain**.

Always take note of the domain (**maximal**, implied or natural domain). If not stated, the implied domain is the largest possible domain for which the rule is defined.

Using the mapping notation form $f: [0, \infty) \to R$, where $f(x) = x^2$, the domain is $[0, \infty)$ and the rule is $f(x) = x^2$.

These two, the rule and domain, define the function.

In this form the $R$ is a **co-domain**, where $R$ is all real numbers; in other words, a co-domain consists of all possible $y$ values that might be obtained via the domain and function.

The **range**, also called the **image**, represents the $y$ values that are actually obtained via the domain and function.

### Example 1

Consider the function $f: [1, 5] \to R$, where $f(x) = 3x$.

The implied domain for a line is all real numbers, but in this case, the **domain** is restricted to $[1, 5]$ and the rule is a **linear function** $y = 3x$.

The range is $[3, 15]$.

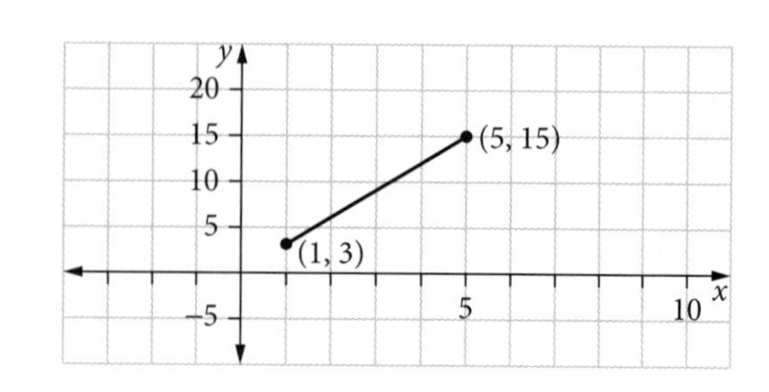

### Example 2

Consider the function $g: R\backslash\{2\} \rightarrow R$, where $g(x) = \frac{1}{x-2}$.

The **implied, or naturally occurring, domain** for this **hyperbola** is $R\backslash\{2\}$, the rule is $y = \frac{1}{x-2}$.

The range is $R\backslash\{0\}$.

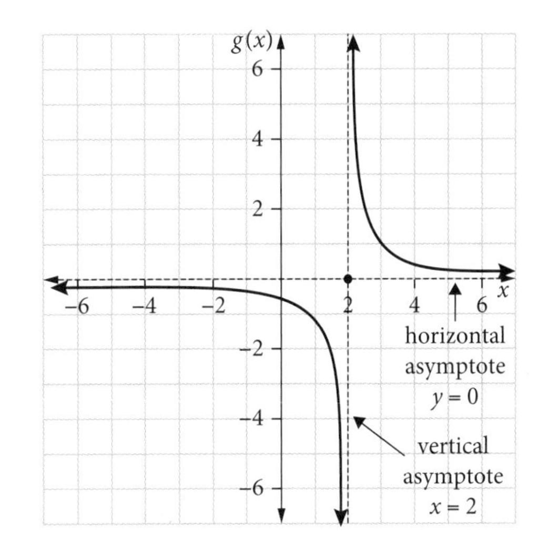


## 1.3 Graphs of power functions $y = x^n, n \in Q$

This section will cover

- graphs of the following functions: power functions $y = x^n$, $n \in Q$.

VCE Mathematics Study Design 2023–2027 p. 98, © VCAA 2022

The graphs of **power functions** include:

- linear $y = x$
- quadratic $y = x^2$
- cubic $y = x^3$
- quartic $y = x^4$ … and so on
- hyperbola or reciprocal $y = \frac{1}{x}$
- **truncus** $y = \frac{1}{x^2}$ …
- rational power $y = x^{\frac{1}{2}} = \sqrt{x}, y = x^{\frac{1}{3}}, y = x^{\frac{2}{3}}, y = x^{\frac{5}{2}}$ …

### Example 3

The graphs of power functions $y = x^n$ for different values of $n \in Q$ are shown below.

$y = x^n$ where $n = \frac{1}{2}$

**Square root** $y = \sqrt{x}$

domain: $x \geq 0$ or $R^+ \cup \{0\}$

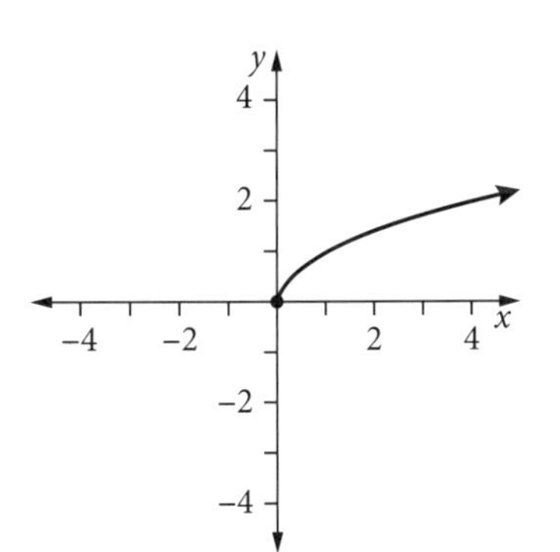


$y = x^n$ where $n = -1$

**Hyperbola** $y = \frac{1}{x}$

domain: $x \neq 0$ or $R\backslash\{0\}$

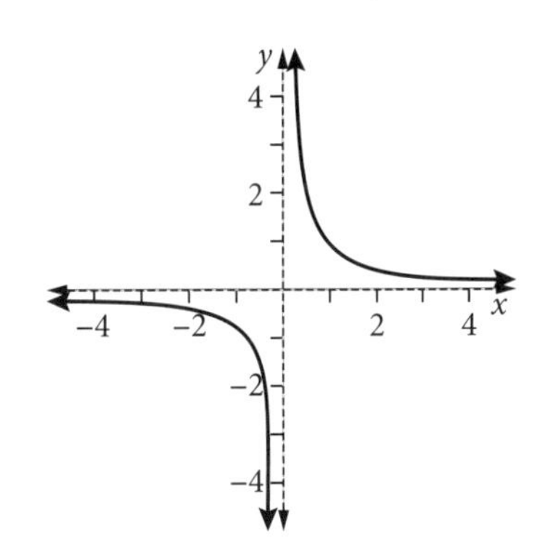


$n = -2$

**Truncus** $y = \frac{1}{x^2}$

domain: $x \neq 0$ or $R\backslash\{0\}$

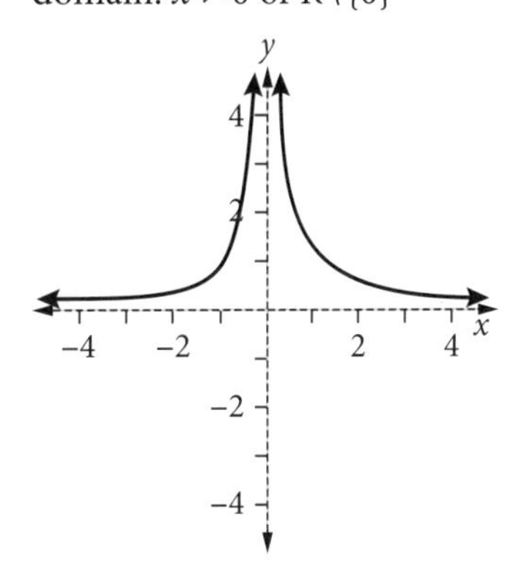

$n = \frac{1}{3}$

$y = x^{\frac{1}{3}}$

domain: $R$

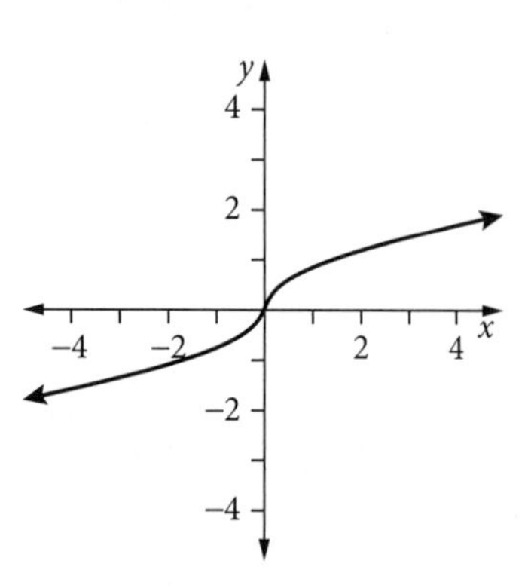


$n = \frac{2}{3}$

$y = x^{\frac{2}{3}}$

domain: $R$

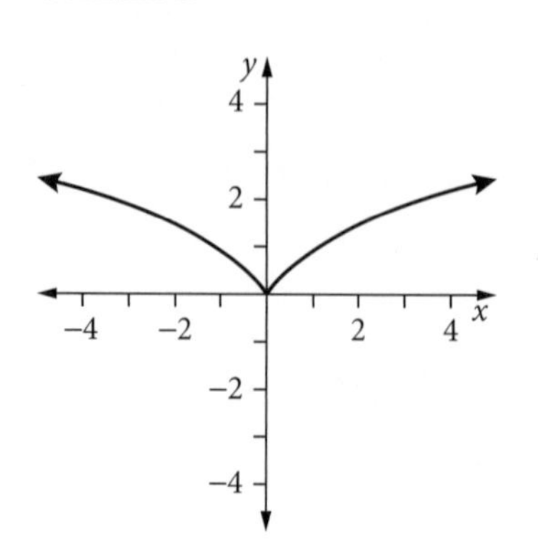


$n = \frac{3}{2}$

$y = x^{\frac{3}{2}}$

domain: $x \geq 0$ or $R^+ \cup \{0\}$

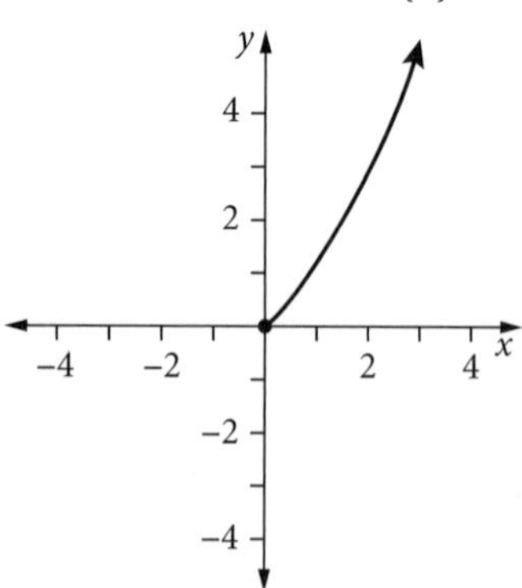


For reciprocal powers, note the domain and concavity formed by odd or even numbers on the denominator and if the numerator is > (greater than) the denominator or < (less than) the denominator.

## 1.3.1 Interval notation

In the previous graph of $y = x^{\frac{3}{2}}$, the domain is written as $x \geq 0$ or $R^+ \cup \{0\}$.

An efficient way to write domain and range uses **interval notation**.

In interval notation, $x \geq 0$ is written as $[0, \infty)$ including zero and not including $\infty$.

When we use $\infty$, we always use the round bracket ( or ).

| Set notation | Interval notation |
|---|---|
| $x \geq a$ | $[a, \infty)$ |
| $x > a$ | $(a, \infty)$ |
| $x \leq a$ | $(-\infty, a]$ |
| $x < a$ | $(-\infty, a)$ |
| $a < x < b$ | $(a, b)$ |
| $a \leq x \leq b$ | $[a, b]$ |
| $a < x \leq b$ | $(a, b]$ |
| $a \leq x < b$ | $[a, b)$ |

**Example 4**

The graphs of power functions $y = x^{\frac{p}{q}}$ for different values of $p$ and $q$ are shown below.

$y = x^{\frac{3}{4}}$

$q$ is even and $q > p$

domain: $[0, \infty)$

range: $[0, \infty)$

concave down

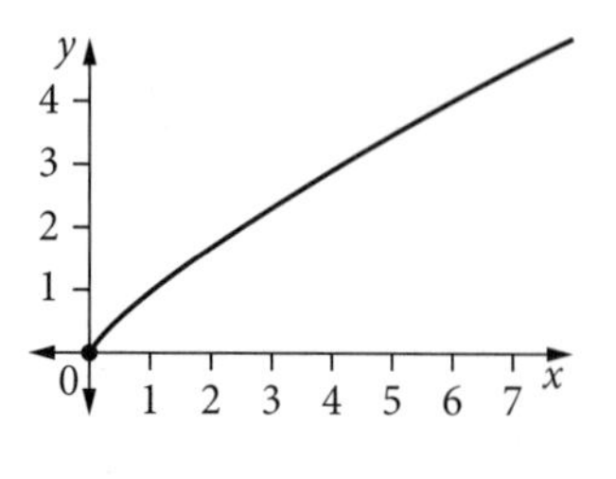


$y = x^{\frac{5}{3}}$

$q$ is odd and $q < p$

domain: $R$

range: $R$

moves from concave down to concave up

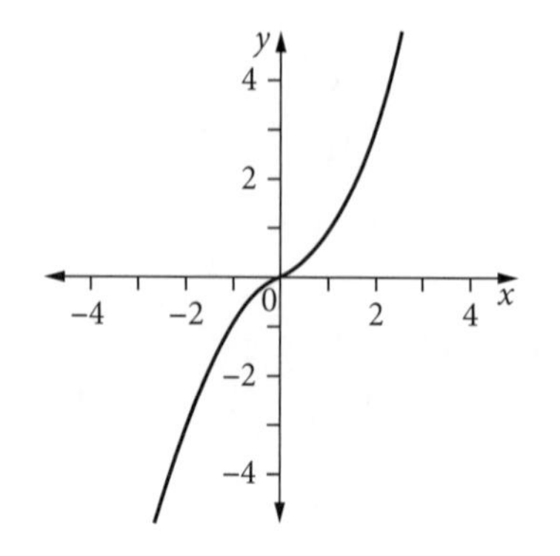


$y = x^{\frac{2}{3}}$

$q$ is odd and $q > p$

domain: $R$

range: $[0, \infty)$

concave down

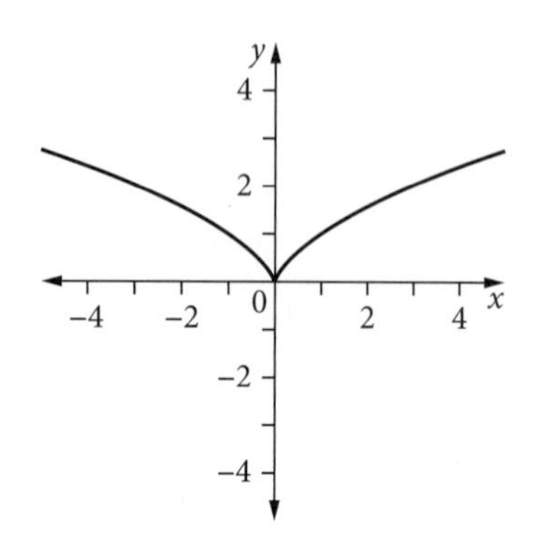

# 1.4 Graphs of exponential and logarithmic functions

This section will cover

- graphs of the following functions: exponential functions, $y = a^x$, $a \in R^+$, in particular $y = e^x$; logarithmic functions, $y = \log_e(x)$ and $y = \log_{10}(x)$.

VCE Mathematics Study Design 2023–2027 p. 98, © VCAA 2022



## 1.4.1 Exponential functions

**Exponential functions** are of the form $f: R \to R, f(x) = a^x$, where $a \in R$.

Using the special **base** $e$, the exponential function is:

$f: R \to R, f(x) = e^x$, where $e$ is Euler's number.

$e$ is a transcendental irrational, where $e \approx 2.71828...$

**Hint**
CAS will label $\log_e(x)$ as $\ln(x)$, meaning log natural. You may also use ln in your working.

(One way of defining $e$ is by using the power series, $e = \sum_{n=0}^{\infty} \frac{1}{n!} = \frac{1}{0!} + \frac{1}{1!} + \frac{1}{2!} + \frac{1}{3!} ...$)

### Example 5

The graph of $y = e^x$ has the domain $R$ and range $R^+$, with a horizontal **asymptote** at $y = 0$, the $x$-axis. The $y$-intercept is at $(0, 1)$.

The graph of $y = 10^x$ has the domain $R$ and range $R^+$, with an asymptote at $y = 0$, the $x$-axis. The $y$-intercept is at $(0, 1)$.

Both graphs represent **exponential growth.**

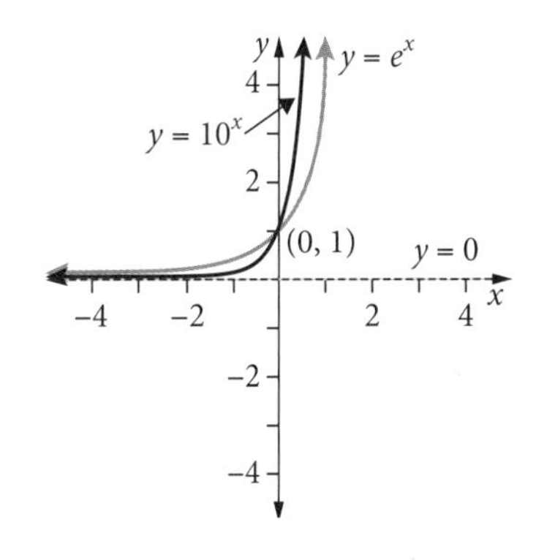


If the base $a$ is less than 1, this can be written as $y = (a^{-1})^{-x}$. For example, $y = 0.1^x$ can be written as $y = 10^{-x}$.

The graph of $y = 10^{-x}$ is a **reflection** of the graph of $y = 10^x$ in the $y$-axis and represents **exponential decay**.

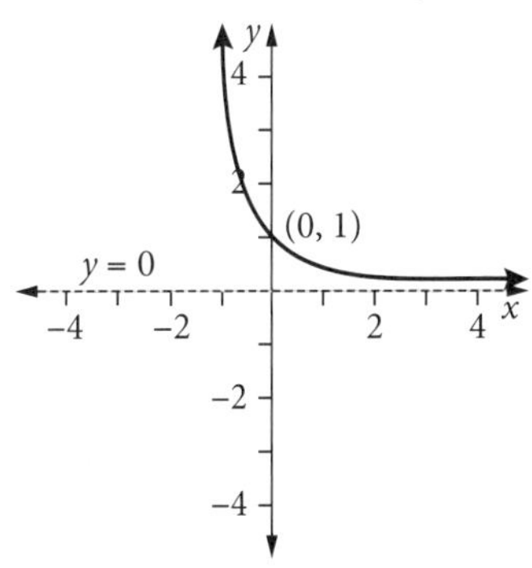


Transformations of dilations, reflections and translations can be applied to all exponential functions. Note the mnemonic DrT (dilation, reflection, translation) for the default order of these transformations (refer to section 1.7 for further information on transformations).

9780170465366

**Example 6**

Sketch the graph of $y = -3e^{x} + 2$, labelling key features.

**Solution**

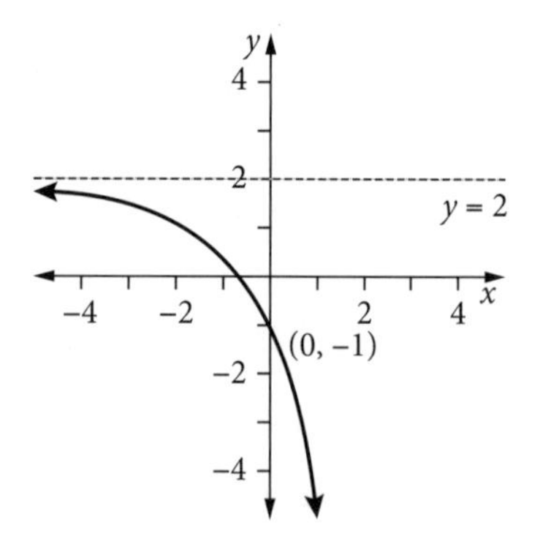


This graph is the transformed graph of $y = e^{x}$ and has:

- a dilation from the $x$-axis of 3 units
- a reflection over the $x$-axis
- a translation 2 units in the positive direction of the $y$-axis.
- domain $R$, range $(-\infty, 2)$
- horizontal asymptote at $y = 2$
- $x$-intercept $\left(\log_e\left(\frac{2}{3}\right), 0\right)$
- $y$-intercept $(0, -1)$

## 1.4.2 Logarithmic functions

**Logarithmic functions** are of the form $f: R^{+} \rightarrow R, f(x) = \log_a(x)$, where $a \in Z^{+}$.

**Hint**

You can use either the notation $\log_e(x)$ or $\ln(x)$ in the exams.

Using base $e$, the logarithmic function is: $f: R^{+} \rightarrow R, f(x) = \log_e(x)$, where $e$ is Euler's number.

$\log_e(x)$ is written as $\ln(x)$, standing for 'log natural'.

The graph of $y = \log_e(x)$ has the domain $R^{+}$, and range $R$, with an asymptote at $x = 0$, the $y$-axis. The $x$-intercept is at $(1, 0)$.

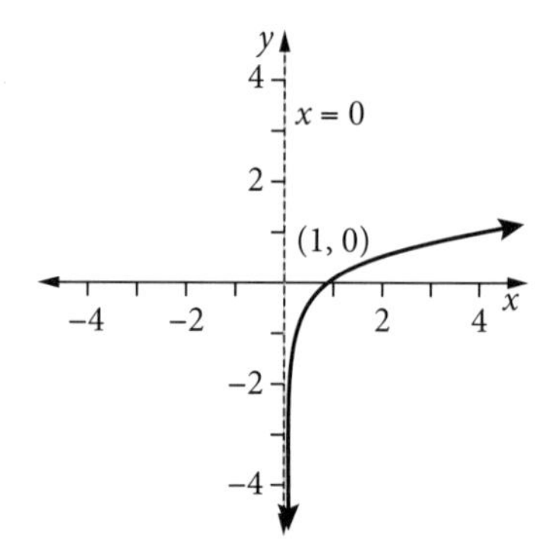


The graph of $y = \log_{10}(x)$ has the domain $R^{+}$, and range $R$, with an asymptote at $x = 0$, the $y$-axis. The $x$-intercept is at $(1, 0)$.

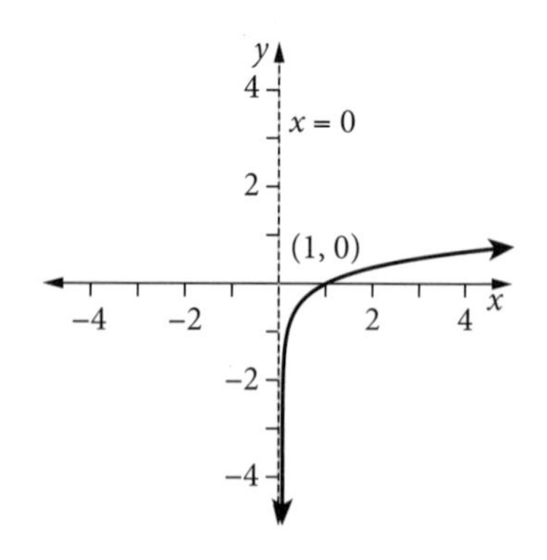

Transformations of dilations, reflections and translations can be applied to all logarithmic functions.

---

**Example 7**

Sketch the graph of $y = -3\log_e(x-2)$, labelling key features.

**Solution**

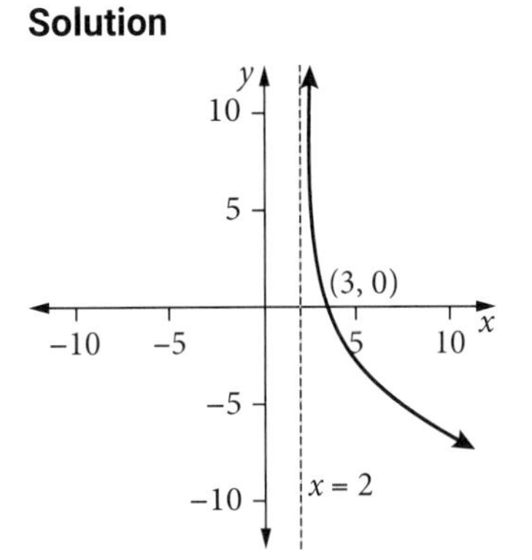


This graph is the transformed graph of $y = \log_e(x)$ and has:

- a dilation from the $x$-axis of 3 units
- a reflection over the $x$-axis
- a translation of 2 units in the positive direction of the $x$-axis.
- domain $(2, \infty)$, range $R$
- vertical asymptote at $x = 2$
- $x$-intercept $(3, 0)$
- no $y$-intercept.

**Hint**

Note that the gap between the vertical asymptote and the $x$-intercept for $y = \log_e(x)$ is 1 unit. Use this as a guide for the dilation from the $y$-axis. For example, the graph of $y = \log_e(2x)$ dilates the graph of $y = \log_e(x)$ from the $y$-axis by a factor of $\frac{1}{2}$, transforming this 'gap' from 1 to $\frac{1}{2}$.

## 1.4.3 Inverse relationship of exponential and logarithmic functions

$y = a^x$ can be written in logarithmic form.

$x$ is the power or **logarithm**, so $x = \log_a(y)$ is an equivalent expression to $y = a^x$; that is, $y = a^x \Leftrightarrow x = \log_a(y)$.

Logarithmic and exponential functions are **inverses** of each other.

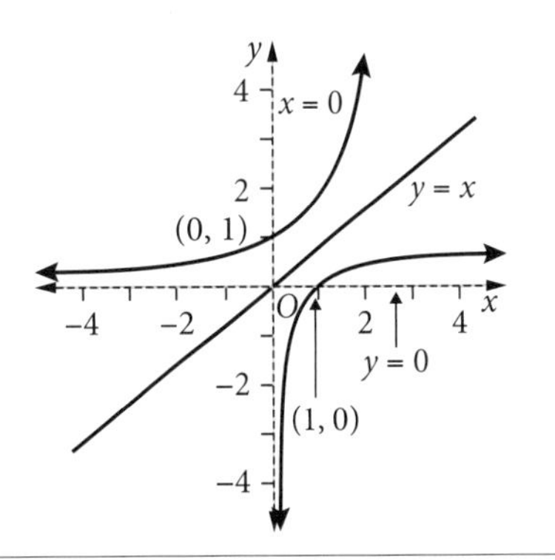


Their graphs

- are reflections over the line $y = x$
- swap the $x$ and $y$ intercepts
- swap the asymptotes
- swap the domain and range

The inverse of function $f$ is $f^{-1}$.

---

**Example 8**

$f(x) = e^x$

This is a one-to-one function where domain $f = R$ and range $f = (0, \infty)$.

Interchange $x$ and $y$ where $y = f(x)$:

$x = e^y$

$x = e^y \Leftrightarrow y = \log_e(x)$

Thus, $f^{-1}: (0, \infty) \to R, f^{-1}(x) = \log_e(x)$

domain: $f^{-1} = R^+$

range: $f^{-1} = R$

**Hint**

You need to name the inverse of function $f$ as $f^{-1}$.

# 1.5 Circular functions $y = \sin(x)$, $y = \cos(x)$ and $y = \tan(x)$

This section will cover

- graphs of the following functions: circular functions $y = \sin(x)$, $y = \cos(x)$ and $y = \tan(x)$ and their key features.



## 1.5.1 Sketching the graphs of sin, cos and tan

- Calculate the **period** of the graph (how long it is before the graph starts to repeat)
- Calculate the **amplitude** of the graph. Note that amplitude is always positive.
- If $y = a\sin(nx)$ or $y = a\cos(nx)$:
  - period $= \frac{2\pi}{n}$
  - amplitude $= a$
- If $y = a\tan(nx)$:
  - period $= \frac{\pi}{n}$
  - amplitude is meaningless for tan as the graph vertically approaches $\infty$ or $-\infty$.
  - vertical asymptotes at $x = \frac{\pi}{2n} \pm \frac{\pi}{n}$

### 1.5.1.1 Graph of $y = \sin(x)$

period $= \frac{2\pi}{1} = 2\pi$

amplitude $= 1$

**Hint**

Note that the graphs of sin $(x)$ and cos $(x)$ are the same graph $\frac{\pi}{2}$ apart.

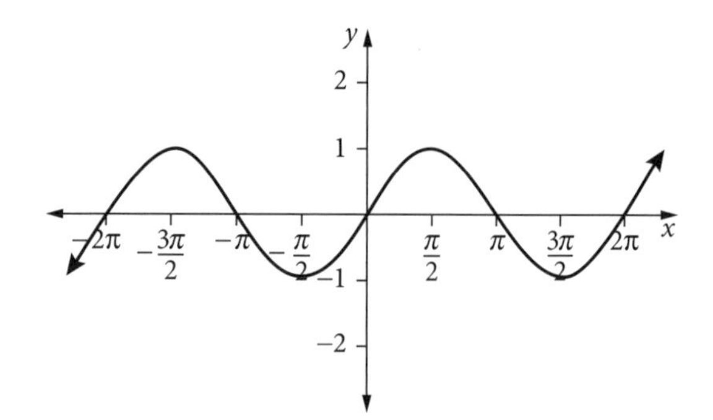


### 1.5.1.2 Graph of $y = \cos(x)$

period $= \frac{2\pi}{1} = 2\pi$

amplitude $= 1$

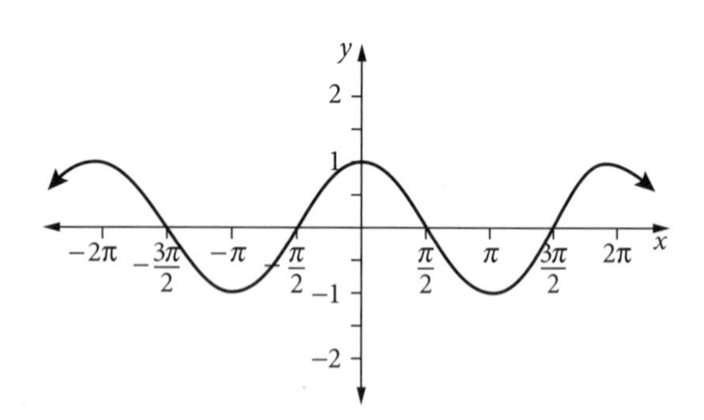


### 1.5.1.3 Graph of $y = \tan(x)$

period $= \frac{\pi}{1} = \pi$

vertical asymptotes at $x = \pm\frac{\pi}{2}, \pm\frac{3\pi}{2}, \pm\frac{5\pi}{2} \ldots$

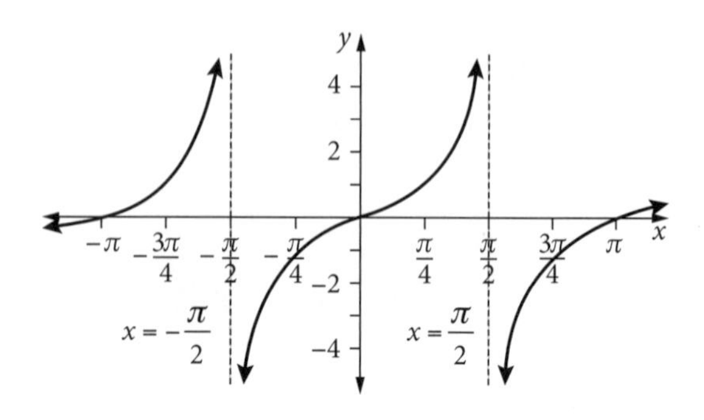

### Example 9

Sketch the graph of $y = -\frac{1}{2}\sin(3x) + \frac{1}{2}$ for $x \in \left[0, \frac{2\pi}{3}\right]$.

- period $= \frac{2\pi}{3}$
- amplitude $= \frac{1}{2}$
- reflected over the $x$-axis
- translation by $\frac{1}{2}$ units in the positive direction of the $y$-axis.

**Hint**
The period and amplitude of trigonometric graphs are, respectively, the horizontal and vertical dilations.

#### Solution

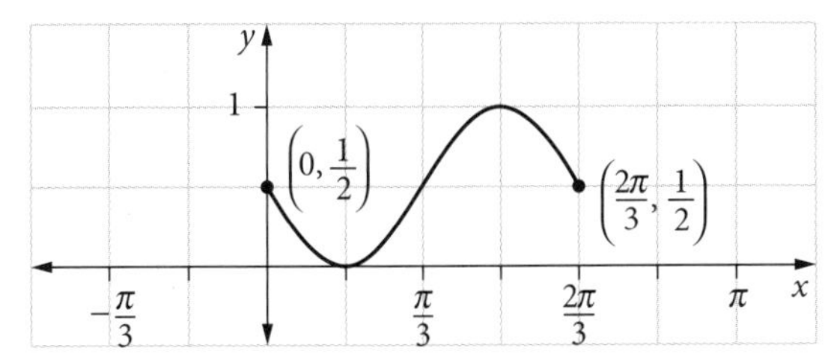


### Example 10

Sketch the graph of $y = 3\tan(2x) - 1$ for $x \in \left[-\frac{\pi}{2}, \pi\right]$.

- period $= \frac{\pi}{2}$
- dilated by factor of 3 from the $x$-axis
- dilated by factor of $\frac{1}{2}$ from the $y$-axis
- translation by 1 unit in the negative direction of the $y$-axis.
- asymptotes of $x = -\frac{\pi}{4}$, $x = \frac{\pi}{4}$, $x = \frac{3\pi}{4}$

#### Solution

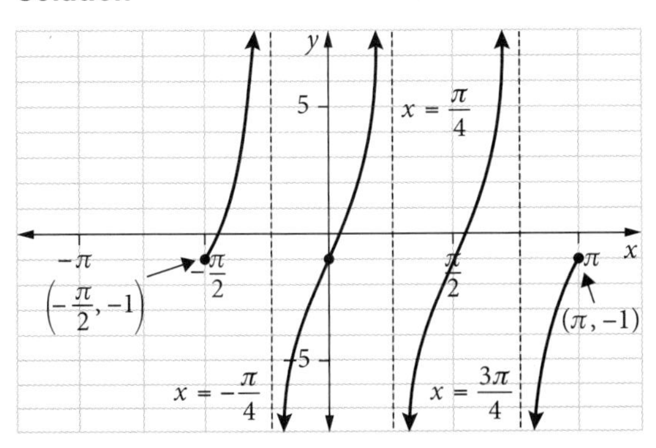

# 1.6 Polynomial functions

This section will cover

- graphs of polynomial functions and their key features.

VCE Mathematics Study Design 2023–2027 p. 98, © VCAA 2022

## 1.6.1 Linear functions $y = x^n$, where $n = 1$

The gradient of a straight line is $m = \frac{y_2 - y_1}{x_2 - x_1}$, where $(x_1, y_1)$ and $(x_2, y_2)$ are two points on the line.

The equation of a straight line can be found using any of the forms:

- $y = mx + c$ (given gradient and $y$-intercept)
- $\frac{y - y_1}{x - x_1} = \frac{y_2 - y_1}{x_2 - x_1}$ (given two points)
- $y - y_1 = m(x - x_1)$ (given gradient and one point)
- $\frac{x}{a} + \frac{y}{b} = 1$ (given $x$- and $y$-intercepts)
  - The form $y - y_1 = m(x - x_1)$ is often most useful as it can be found from any two points, or one point and the gradient.
- $m = \tan(\theta)$ finds the gradient, when given the angle that a line makes with the positive direction of the $x$-axis.
- $m_1 = m_2$ occurs when two lines are parallel and their gradients are equal.

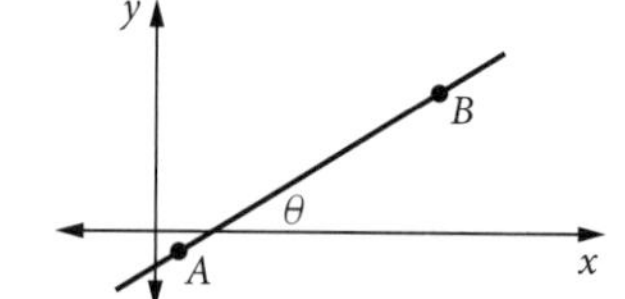


- $m_1 \times m_2 = -1$ or $m_2 = -\frac{1}{m_1}$ occurs when two lines are perpendicular; the gradient of one line is the negative reciprocal of the gradient of the other.
- $d(AB) = \sqrt{(x_2 - x_1)^2 + (y_2 - y_1)^2}$ finds the distance between two points $A(x_1, y_1)$ and $B(x_2, y_2)$.
- Midpoint $= \left(\frac{x_1 + x_2}{2}, \frac{y_1 + y_2}{2}\right)$ finds the midpoint of a line joining two points $(x_1, y_1)$ and $(x_2, y_2)$.

---

**Example 11**

Find the equation of the line joining the two points $(2, 3)$ and $(4, 7)$.

**Hint**

Rather than using the gradient formula $m = \frac{y_2 - y_1}{x_2 - x_1}$, it is often easier to plot the points and use $m = \frac{\text{rise}}{\text{run}}$.

Plot the points and find the gradient between them.

gradient $= \frac{\text{rise}}{\text{run}} = \frac{4}{2} = 2$

**Solution**

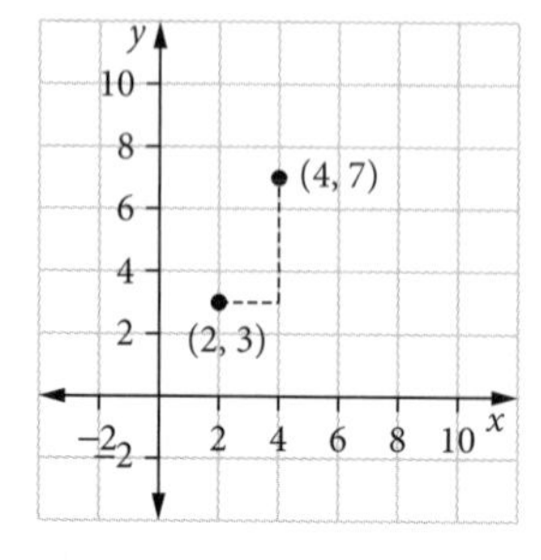


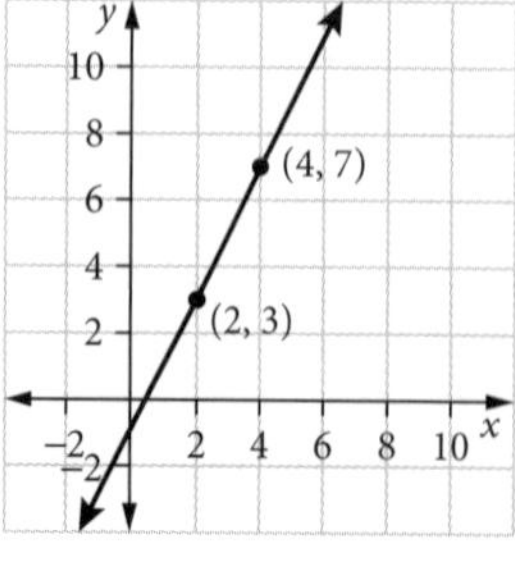


Use $y - y_1 = m(x - x_1)$ with gradient 2 and either point; in this case $(2, 3)$.

$y - 3 = 2(x - 2)$

$y = 2x - 1$

## 1.6.2 Linear regression

A **linear model** is defined by two sets of coordinates. We can use linear regression on CAS to find the equation of the line, or use simultaneous equations either on CAS or by hand.

---

### Example 12

Find the equation of the straight line joining points $(1.5, 2)$, and $(3.5, 5)$.

### Solution

- Using linear regression:

  Equation is $y = \frac{3}{2}x - \frac{1}{4}$.

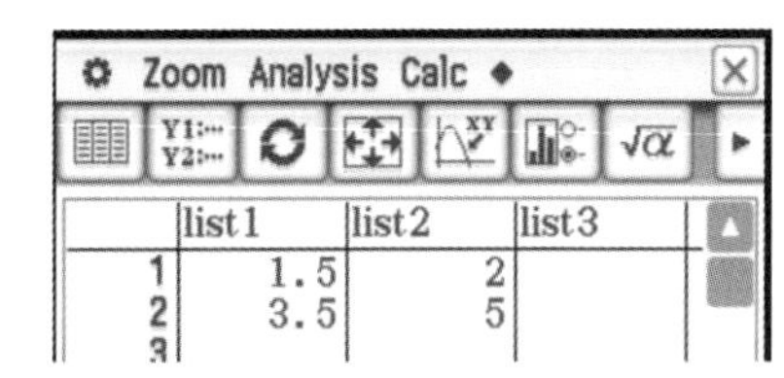


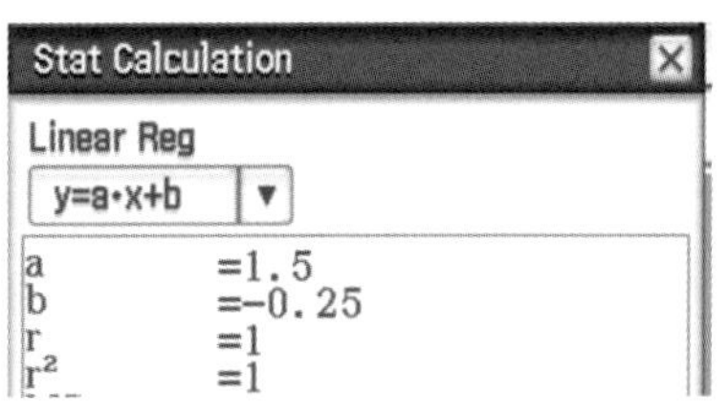


- Using simultaneous equations on CAS:

  Equation is $y = \frac{3}{2}x - \frac{1}{4}$.

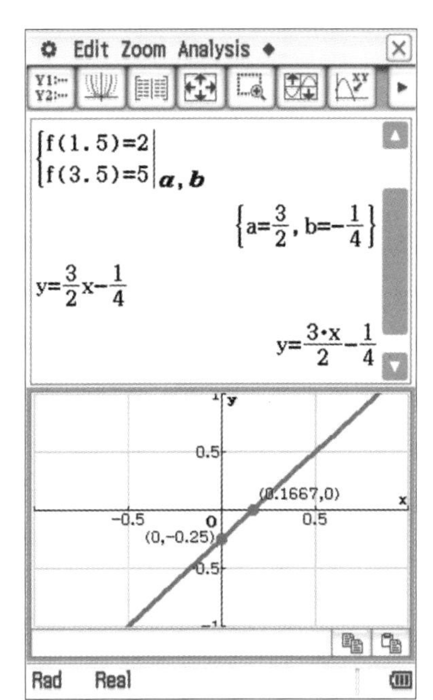


- Using a by-hand method:

  Plot the two points on a graph and determine the gradient using $m = \frac{\text{rise}}{\text{run}}$.

$$m = \frac{5-2}{3.5-1.5} = \frac{3}{2}$$

  Use $y - y_1 = m(x - x_1)$ with gradient $\frac{3}{2}$ and either point; in this case, $(1.5, 2)$.

$$y - 2 = \frac{3}{2}\left(x - \frac{3}{2}\right)$$
$$y = \frac{3}{2}x - \frac{1}{4}$$

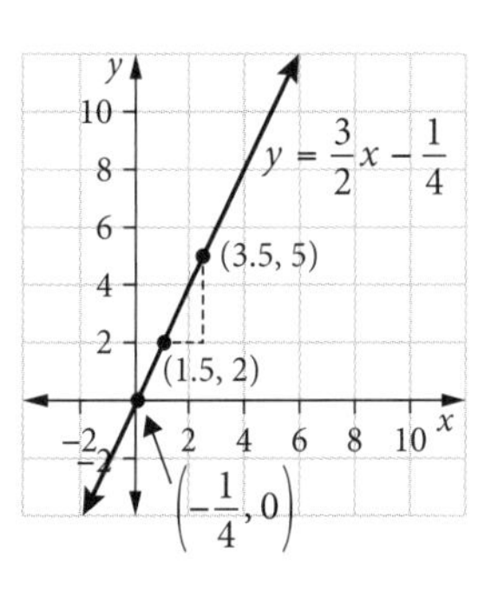


  Equation is $y = \frac{3}{2}x - \frac{1}{4}$.

## 1.6.3 Quadratic functions $y = x^n$, where $n = 2$

- The **general form** of a quadratic is $y = ax^2 + bx + c$.
- Turning point form: $y = a(x - h)^2 + k$ where the vertex has coordinates $(h, k)$.
- Factorised form: $y = a(x - b)(x - c)$ where the $x$-intercepts are $x = b$, $x = c$.

  CAS commands **factor** and **rfactor** (to find **roots**) can help you factorise quadratics.

```
factor(x^2+2x-15)
                              (x+5)·(x-3)
rFactor(3x^2-4x-10)
   3·(x+√34/3-2/3)·(x-√34/3-2/3)
```

- The turning point (vertex) of the **parabola** can be found using $x = \frac{-b}{2a}$.
- The **axis of symmetry** of the parabola is $x = \frac{-b}{2a}$.
- The **quadratic formula** to find $x$-intercepts is $x = \frac{-b \pm \sqrt{b^2 - 4ac}}{2a}$.

  Note the importance of the expression under the square root sign.
- The **discriminant** $\Delta = b^2 - 4ac$ gives you information about the number and types of **roots** in the graph, or solutions to the equation.
  - $\Delta = 0$: one real $x$-axis intercept, touching the $x$-axis.
  - $\Delta < 0$: no real $x$-axis intercepts.
  - $\Delta > 0$: two real $x$-axis intercepts.
  - When $a, b, c$ are integers: If $\Delta$ is a perfect square, the two roots are rational; if not, the two roots are real, but irrational (containing surds)

---

**Example 13**

Find the number and nature of the roots of the parabola with the rule $y = 2x^2 + 3x - 5$.

**Solution**

Using $\Delta = b^2 - 4ac$

$\Delta = 3^2 - 4(2)(-5)$

$\Delta = 49$

$\Delta > 0$ and is also a perfect square so there are **two rational roots.**

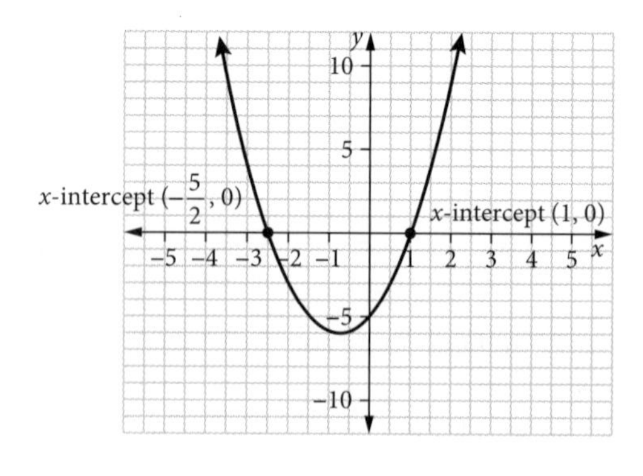


Factorising:

$y = 2x^2 + 3x - 5 = (2x + 5)(x - 1)$

Letting $y = 0$, solve the equation $2x^2 + 3x - 5 = 0$.

$x = -\frac{5}{2}, x = 1$

## 1.6.4 Determining the quadratic equation

Depending on the information given, we can use any of the following forms to find the rule for the required parabola:

- $y = ax^2 + bx + c$ (given the coefficients of terms)
- $y = a(x - h)^2 + k$ (given the turning point and another point)
- $y = a(x - b)(x - c)$ (given the $x$-intercepts)

### Example 14

Determine the rule for the parabola that has a vertex at the point $(1,3)$ and $y$-intercept $(0,5)$.

#### Solution

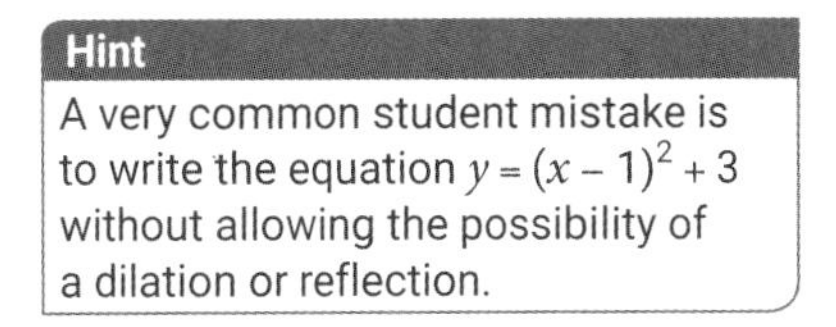
**Hint**
A very common student mistake is to write the equation $y = (x-1)^2 + 3$ without allowing the possibility of a dilation or reflection.

Select the turning point form:

$$y = a(x-h)^2 + k$$
$$y = a(x-1)^2 + 3$$

Use the other information given, in this case the point $(0,5)$.

$$y = a(x-1)^2 + 3$$
$$(0,5) \Leftrightarrow 5 = a(0-1)^2 + 3$$

So $a = 2$

Rule for the parabola is $y = 2(x-1)^2 + 3$.

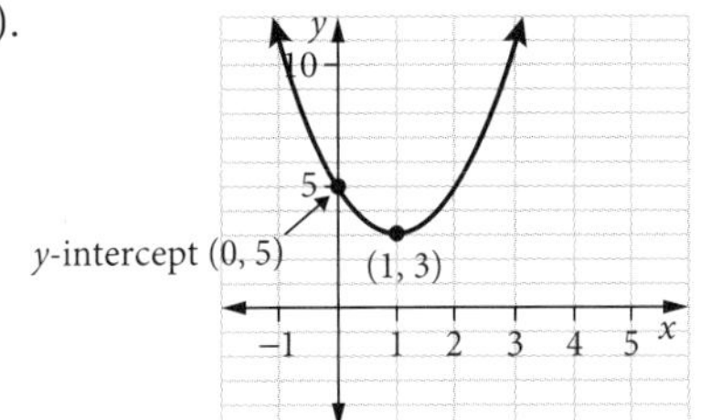


## 1.6.5 Quadratic regression

A quadratic model is defined by three sets of coordinates. We can use quadratic regression on CAS to find the equation of the parabola, or use simultaneous equations either on CAS or by hand.

### Example 15

Find the equation of the parabola joining points $(1,4)$, $(3,8)$ and $(-3,2)$.

#### Solution

- Using quadratic regression:

  Equation is $y = \frac{1}{4}x^2 + x + \frac{11}{4}$

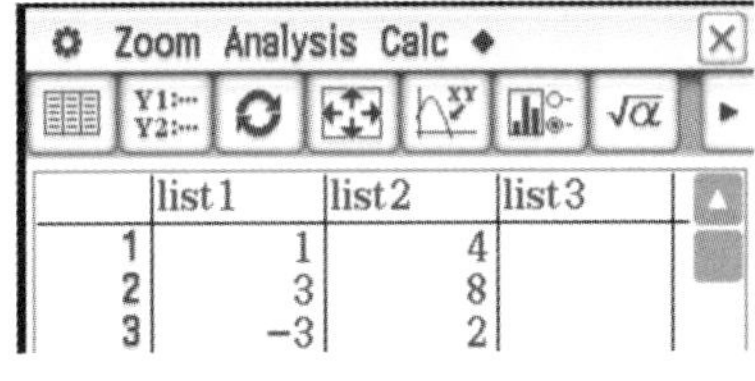


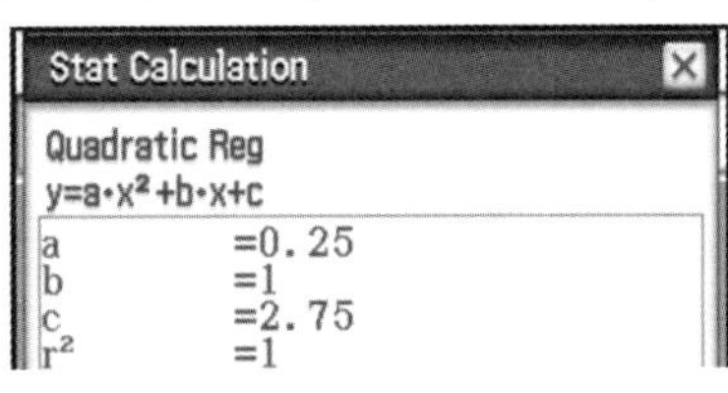


- Using simultaneous equations on CAS:

  Equation is $y = \frac{1}{4}x^2 + x + \frac{11}{4}$.

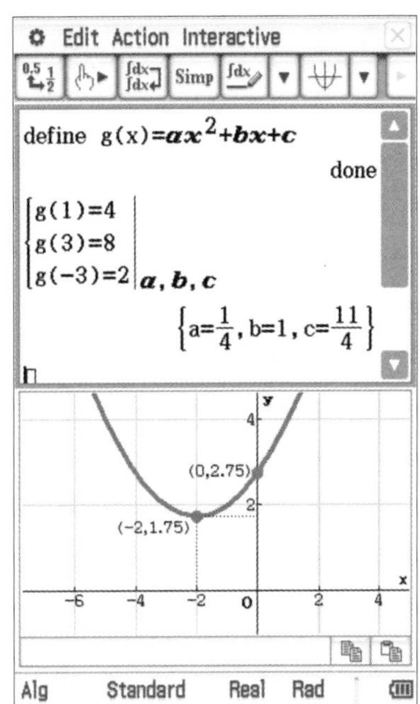

### 1.6.6 Cubic functions $y = x^n$, where $n = 3$

- The general form of a cubic is $y = ax^3 + bx^2 + cx + d$.
- Point of inflection form: $y = a(x - h)^3 + k$ where the stationary point of inflection has coordinates $(h, k)$.
- Factorised form: $y = a(x - b)(x - c)(x - d)$ where the $x$-intercepts are $x = b$, $x = c$, $x = d$.
- Repeated factor form: $y = a(x - b)(x - c)^2$ where the $x$-intercepts are $x = b$, $x = c$, touching the $x$-axis at $x = c$.

  Use the Factor Theorem (see later) or the factor or **rfactor** command on CAS to factorise cubics.

Cubic functions have the following shapes:

Positive cubics

Negative cubics

A cubic model is defined by four sets of coordinates. When more points are given, use CAS (Cubic Reg) to find the equation of the cubic, or use algebraic methods: simultaneous equations, point of inflection form or factorised form.

---

**Example 16**

Find the equation of the cubic function joining points (1, 0), (2, 0), (6, 3) and (10, 0).

**Solution**

- Using cubic regression:

  Equation is $y = -0.0375x^3 + 0.4875x^2 - 1.2x + 0.75$.

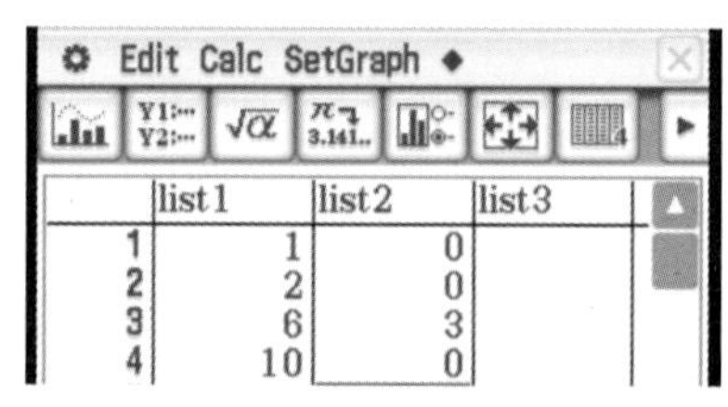


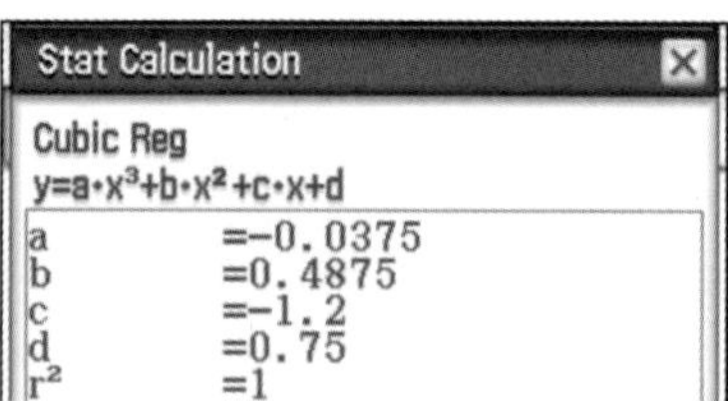


- Using simultaneous equations on CAS:

  Equation is $y = \frac{-3x^3}{80} + \frac{39x^2}{80} - \frac{6x}{5} + \frac{3}{4}$.

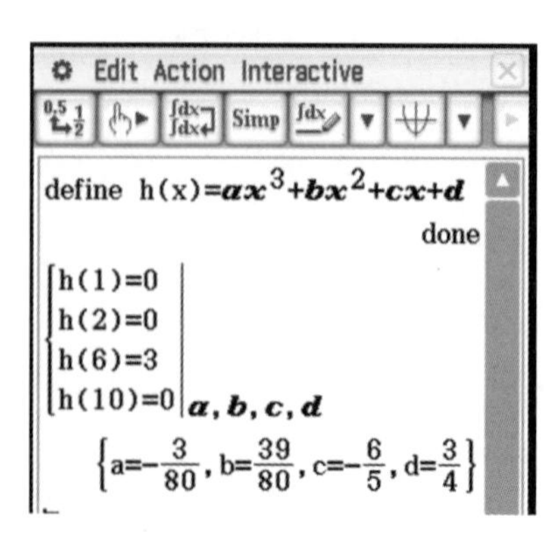

- Using a by-hand method:

  Given the points $(1,0)$, $(2,0)$, $(6,3)$ and $(10,0)$, we recognise that we are given three $x$-intercepts:

  Factorised form: $y = a(x-1)(x-2)(x-10)$

  Substitute $(6,3)$ to find $a$:

  $(6,3) \Leftrightarrow 3 = a(6-1)(6-2)(6-10)$ gives $a = -\frac{3}{80}$.

$y = -\frac{3}{80}(x-1)(x-2)(x-10)$

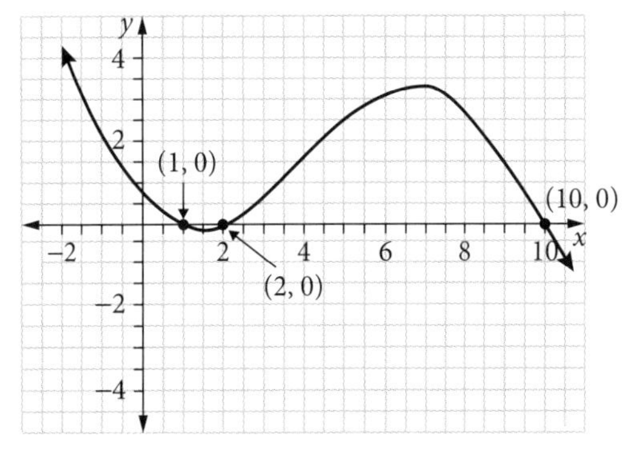


Graph of the cubic $y = \frac{-3x^3}{80} + \frac{39x^2}{80} - \frac{6x}{5} + \frac{3}{4}$



## 1.6.7 Quartic functions $y = x^n$, where $n = 4$

- The general form of a quartic is $y = ax^4 + bx^3 + cx^2 + dx + e$.
- Vertex form: $y = a(x-h)^4 + k$ where the vertex has coordinates $(h, k)$.
- Factorised form: $y = a(x-b)(x-c)(x-d)(x-e)$ where the $x$-intercepts are $x = b$, $x = c$, $x = d$, $x = e$.
- Repeated factor form: $y = a(x-b)^2(x-c)(x-d)$ or $y = a(x-b)^3(x-c)$ or $a(x-b)^2(x-c)^2$.

Quartic functions have the following shapes:

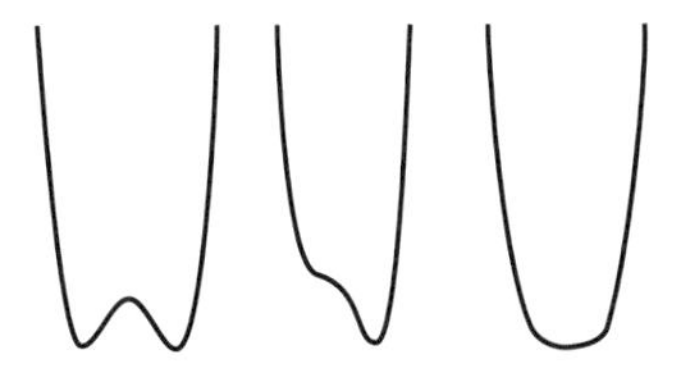

Positive quartics

Negative quartics

# 1.7 Transformations

This section will cover

- transformation from $y = f(x)$ to $y = Af(n(x+b)) + c$, where $A$, $n$, $b$ and $c \in R$.
- the relation between the graph of an original function and the graph of a corresponding transformed function (including families of transformed functions for a single transformation parameter).

VCE Mathematics Study Design 2023–2027 p. 99, © VCAA 2022

**Transformations** can be carried out on any of the functions previously studied.

The three types of transformations studied here are:

- **dilation**
- **reflection**
- **translation**

If a combination of transformations is required, and the order is not specifically stated, the default order is dilation, reflection, translation (DrT).

## 1.7.1 Functions approach

A transformation from $y = f(x)$ to $y = Af(n(x - b)) + c$ can be interpreted as follows.

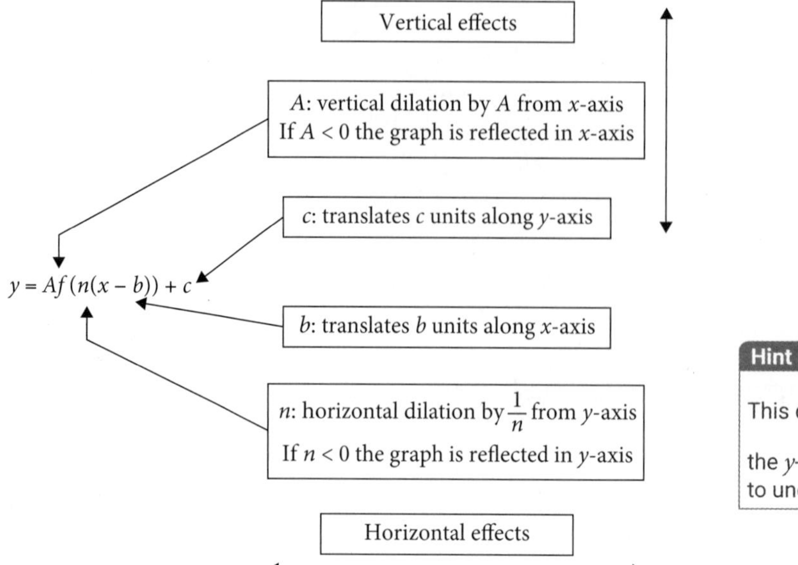


**Hint**

This dilation by a factor of $\frac{1}{n}$ from the $y$-axis can be the most difficult to understand.

Vertical dilation can also be expressed as

- a dilation from the $x$-axis
- a dilation parallel to the $y$-axis
- a dilation in the $y$ direction.

Horizontal dilation can also be expressed as

- a dilation from the $y$-axis
- a dilation parallel to the $x$-axis
- a dilation in the $x$ direction.

### 1.7.1.1 Dilation

- $y = Af(x)$ means $y = f(x)$ has been dilated (stretched) by a factor of $A$ from the $x$-axis.
- $y = f(nx)$ means $y = f(x)$ has been dilated by a factor of $\frac{1}{n}$ from the $y$-axis.

### 1.7.1.2 Reflection

- $y = -f(x)$ means $y = f(x)$ has been reflected in the $x$-axis.
- $y = f(-x)$ means $y = f(x)$ has been reflected in the $y$-axis.

### 1.7.1.3 Translation

- $y = f(x - b)$ means $y = f(x)$ has been translated '$b$' units to the right (in the positive direction of the $x$-axis).
- $y = f(x) + c$ means $y = f(x)$ has been translated '$c$' units up (in the positive direction of the $y$-axis).

**Example 17**

Using the functions method, find the rule for the image of $y = \sqrt{x}$ under the transformations, in order,

- a dilation of 2 units from the $x$-axis
- a dilation of 3 units from the $y$-axis
- a reflection in the $x$-axis
- a translation of 5 units in the positive direction of the $y$-axis.

Step 1: $y_1 = 2\sqrt{x}$

Step 2: $y_2 = 2\sqrt{\frac{x}{3}}$

Step 3: $y_3 = -2\sqrt{\frac{x}{3}}$

Step 4: $y_4 = -2\sqrt{\frac{x}{3}} + 5$

Image equation $y = -2\sqrt{\frac{x}{3}} + 5$

## 1.7.2 Mapping approach

- Dilate each point by a factor of $A$ from the $x$-axis means each $y$ value is multiplied by $A$.

  $(x, y) \to (x, Ay)$

- Dilate each point by a factor of $\frac{1}{n}$ from the $y$-axis means each $x$ value is multiplied by $\frac{1}{n}$.

  $(x, y) \to (\frac{x}{n}, y)$

- Reflect in the $x$-axis; $y = -f(x)$ changes the sign of every $y$ value.

  $(x, y) \to (x, -y)$

- Reflect in the $y$-axis; $y = f(-x)$ changes the sign of every $x$ value.

  $(x, y) \to (-x, y)$

- Translate by $b$ units in the positive direction of the $x$-axis adds $b$ to each $x$ value.

  $(x, y) \to (x + b, y)$

- Translate by $c$ units in the positive direction of the $y$-axis adds $c$ to each $y$ value.

  $(x, y) \to (x, y + c)$

**Example 18**

Using **mapping notation**, find the rule of the image of $y = \frac{1}{x^2}$ when

- dilated by a factor of 2 from the $x$-axis
- dilated by a factor of 4 from the $y$-axis
- reflected in the $x$-axis
- translated by 5 units in the positive direction of the $x$-axis.

**Solution**

$(x, y) \rightarrow (x', y')$, where $(x', y')$ are the coordinates of the image of $(x, y)$, so

- $(x, y) \rightarrow (x, 2y)$
- then $(x, 2y) \rightarrow (4x, 2y)$
- then $(4x, 2y) \rightarrow (4x, -2y)$
- then $(4x, -2y) \rightarrow (4x + 5, -2y)$.

Final mapping $(4x + 5, -2y) = (x', y')$ giving $4x + 5 = x'$ and $-2y = y'$.

Rearranging to $x = \frac{x' - 5}{4}$ and $y = -\frac{y'}{2}$.

Substitute in $y = \frac{1}{x^2}$: $\quad -\frac{y'}{2} = \frac{1}{\left(\frac{x' - 5}{4}\right)^2}$

$$\therefore y' = \frac{-2 \times 16}{(x' - 5)^2}$$

The image rule is $\quad y = \frac{-32}{(x - 5)^2}$.

Comparing the graphs:

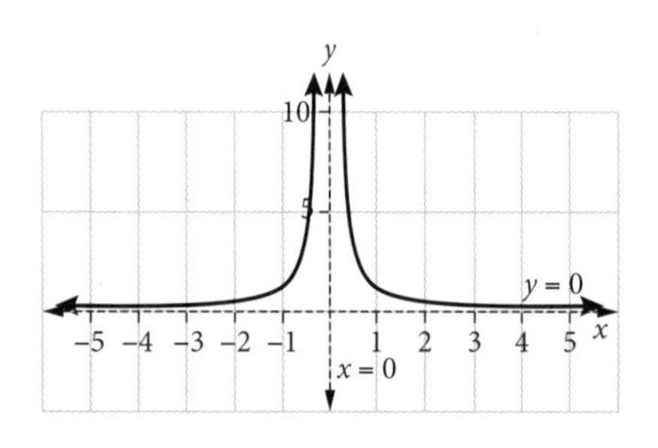


$y = \frac{1}{x^2}$

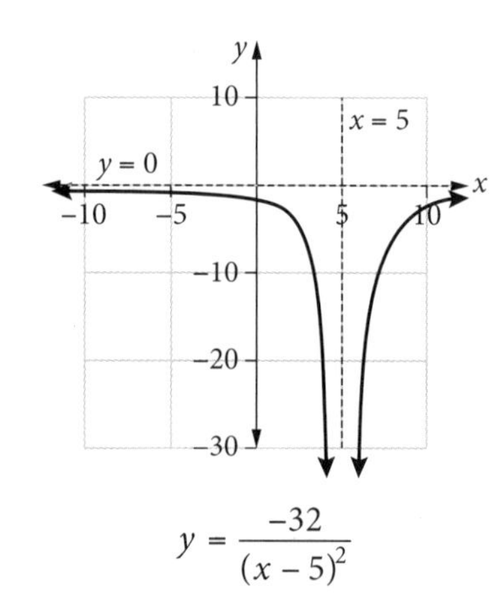


$y = \frac{-32}{(x - 5)^2}$

## 1.8 Inverse functions

This section will cover

- $f$ is one of the functions specified (see page 2), and the inverse transformation.

VCE Mathematics Study Design 2023–2027 p. 99, © VCAA 2022

It was previously shown that exponential and logarithmic functions are inverses of each other.

For an inverse $f^{-1}$ to exist, the original function $f$ must be **one-to-one**.

To create and graph an inverse function $f^{-1}$ from the original function $f$:

- check that $f$ is one-to-one
- interchange $x$ and $y$ values
- solve for $y =$
- name your inverse function as $f^{-1}(x) =$
- ran $f^{-1}$ = dom $f$
- dom $f^{-1}$ = ran $f$
- sketch the graph of $f^{-1}$

- If the function is strictly increasing, i.e. $f(a) < f(b)$ whenever $a < b$, then all points of intersection between $f$ and $f^{-1}$, if they exist, are on the line $y = x$. To find the points of intersection for a strictly increasing function and its inverse, solve $f^{-1} = f$ or $f^{-1} = x$ or $f = x$. However, if $f(x)$ is not strictly increasing, there may be intersections that cannot be found by solving $f(x) = x$ or $f^{-1}(x) = x$. In these cases, solving $f^{-1} = f$ is the correct method.

**Hint**
It is often much simpler algebraically to equate the function or its inverse to the line $y = x$, but this method is only valid when the function is strictly increasing. If in doubt, graph the function and its inverse to check whether there are solutions outside $y = x$.



### Example 19

$f: (0, \infty) \rightarrow R, f(x) = x^2 - 1$

Find the rule for the inverse function and compare the graphs of $f$ and $f^{-1}$.

With a restricted domain $(0, \infty)$, $f$ is a one-to-one function.

### Solution

Interchange $x$ and $y$:

$y = x^2 - 1$ becomes $x = y^2 - 1$

$y^2 = x + 1$

$\therefore y = \pm\sqrt{x+1}$

$f^{-1}(x) = \ldots$ where the upper section of graph is required.

$\therefore f^{-1}(x) = \sqrt{x+1}$

$\text{ran } f^{-1}(x) = \text{dom } f = (0, \infty)$

$\text{dom } f^{-1}(x) = \text{ran } f = (-1, \infty)$

Sketch the graph of $f^{-1}(x) = \sqrt{x+1}$ for domain $(-1, \infty)$.

Comparing the graphs:

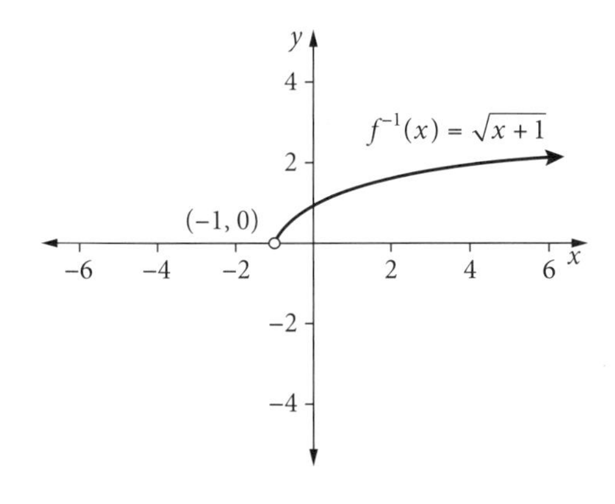


**Hint**
Note the open circle at $x = -1$ because $-1$ is not included in the domain.

Because $f(x)$ is strictly increasing on the restricted domain, point(s) of intersection between $f$ and $f^{-1}$ are on the line $y = x$.

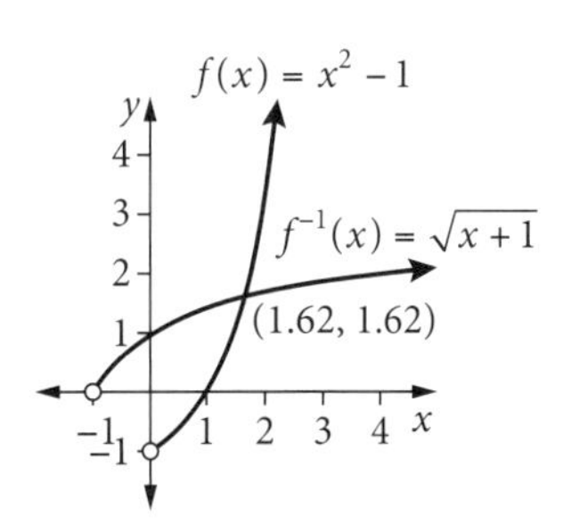


Point of intersection $(1.62, 1.62)$ correct to two decimal places.

# 1.9 Sum, difference and product

This section will cover

- graphs of sum, difference and product.

VCE Mathematics Study Design 2023–2027 p. 99, © VCAA 2022

Graphs of sum, difference and product of functions can be drawn using combinations of the functions previously studied.

- The **sum of two functions** $f(x)$ and $g(x)$ is defined as

  $(f+g)(x) = f(x) + g(x)$
- The **difference of two functions** $f(x)$ and $g(x)$ is defined as

  $(f-g)(x) = f(x) - g(x)$
- The **product of two functions** $f(x)$ and $g(x)$ is defined as

  $(fg)(x) = f(x)g(x)$
- The domain of each is the intersection of the domains of the two functions $f(x)$ and $g(x)$.

To sketch $(f+g)(x)$, use addition of ordinates.

To sketch $(f-g)(x)$, write as $(f+(-g))(x)$ and then use addition of ordinates by first sketching $y = f(x)$ and $y = -g(x)$.

## 1.9.1 Addition of ordinates

The graph of $y = f(x) + g(x)$ can be obtained by sketching the graphs of $y = f(x)$ and $y = g(x)$ on the same set of axes and then adding the $y$ values together for suitable $x$ values.

**Hint**

The most difficult questions are addition of ordinates, when students are not given the equations of $f(x)$ and $g(x)$.

### Example 20

For the graphs shown, sketch $y = f(x) + g(x)$.

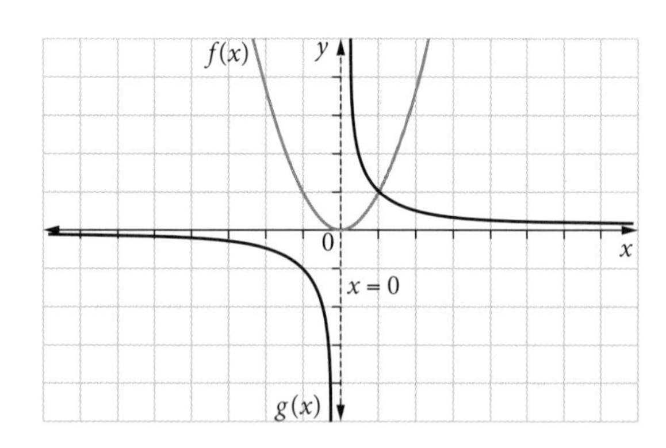


- Select $x$ values and observe separate $y$ values of $f(x)$ and $g(x)$. Estimate the sum of $y$ values.
- A large negative $y = f(x)$ value and a small positive $g(x)$ value will give a smaller negative $f(x) + g(x)$ value.
- dom $(f+g)$ = dom $f \cap$ dom $g$

In this example, dom $f = R$ and dom $g = R \setminus \{0\}$.

### Solution

Graph of $y = f(x) + g(x)$ with asymptote $x = 0$ and $y = f(x)$ also becomes an asymptote.

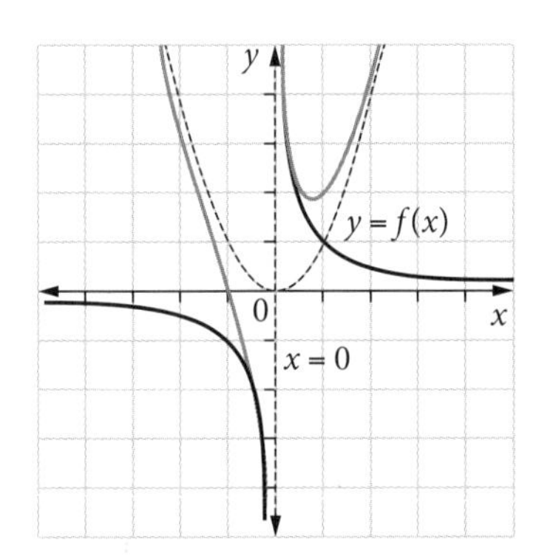


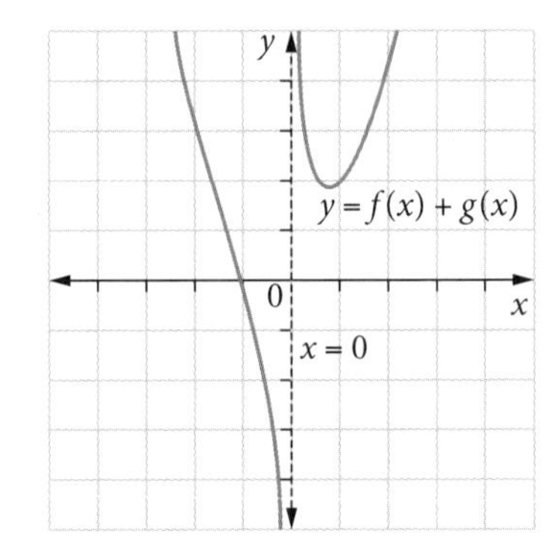


domain $(f + g) = R \setminus \{0\}$



## 1.10 Composite functions

This section will cover

- composite functions involving functions of the types specified (see page 2) (not including composite functions that result in reciprocal or quotient functions).

VCE Mathematics Study Design 2023–2027 p. 99, © VCAA 2022

Composite functions exist if the range of the inner function is contained in, or equal to, the domain of the outer function.

Firstly, sketch graphs and decide the domain and range of each function.

### Example 21

Given $f(x) = x^2$ and $g(x) = 2x + 1$:

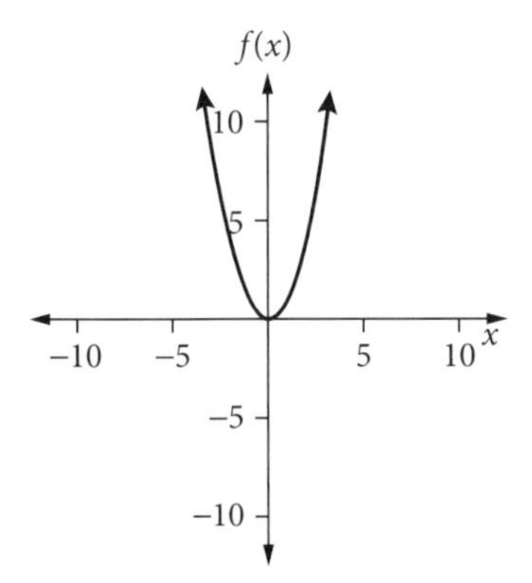


dom $f = R$, ran $f = [0, \infty)$

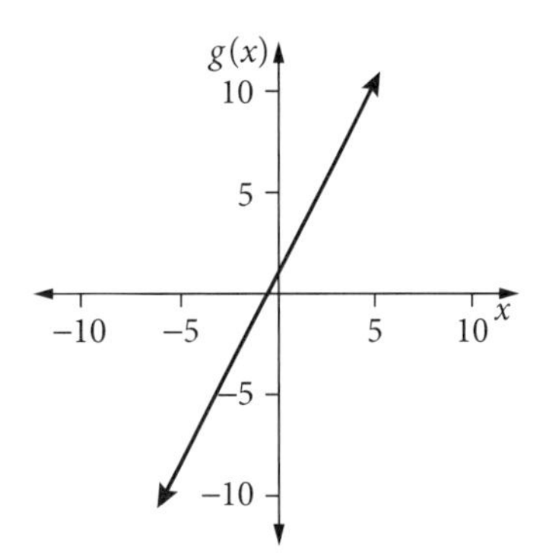


dom $g = R$, ran $g = R$

- $f(g(x))$ is defined if ran $g \subseteq$ dom $f$.
- $g(f(x))$ is defined if ran $f \subseteq$ dom $g$.
- $f(g(x))$ is defined since ran $g \subseteq$ dom $f$: $R \subseteq R$.
- Rule: $f(g(x)) = f(2x + 1) = (2x + 1)^2$
- $g(f(x))$ is defined since ran $f \subseteq$ dom $g$: $[0, \infty) \subseteq R$.
- Rule: $g(f(x)) = g(x^2) = 2x^2 + 1$
- If $f(g(x))$ is defined, then dom $f(g(x))$ = dom $g$.
- If $g(f(x))$ is defined, then dom $g(f(x))$ = dom $f$.

**Hint**
Composites functions are defined if the range of the inner function is contained in or equal to the domain of the outer function. This is a helpful concept but, in the exams, the actual range and domains are required to be stated.

**Hint**
The domain of the composite function is the domain of the inner function, restricted or not as required.

**Example 22**

Investigate the existence of $f(g(x))$ and $g(f(x))$, stating their rule and domain if the composite function exists.

Let $f(x) = x^2 - 2x$ and $g(x) = \frac{1}{x}$.

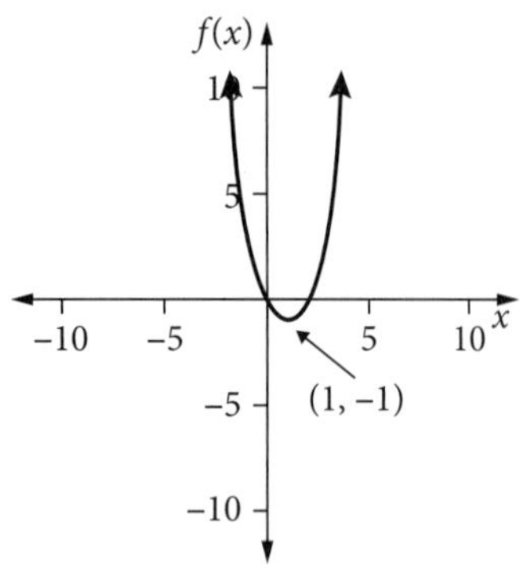


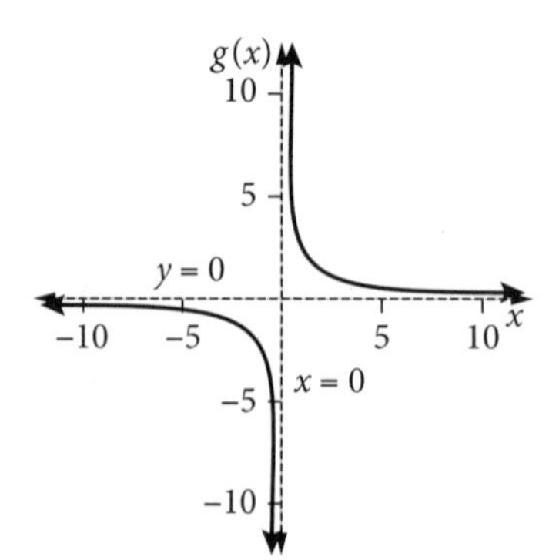


dom $f = R$, ran $f = [-1, \infty)$

dom $g = R \setminus \{0\}$, ran $g = R \setminus \{0\}$

Sketch $f(x)$ and $g(x)$ over their maximal domains.

- $f(g(x))$ is defined since ran $g \subseteq$ dom $f$.
- $R \setminus \{0\} \subseteq R$
- Rule of $f(g(x)) = \frac{1}{x^2} - \frac{2}{x}$
- $g(f(x))$ is not defined since ran $f \not\subseteq$ dom $g$.
- $[-1, \infty) \not\subseteq R \setminus \{0\}$

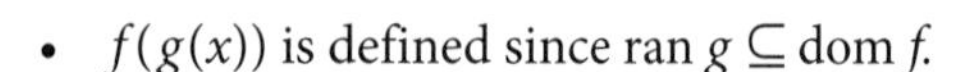

**Hint**
Remember that a function is defined by its rule and domain.

- If $f(g(x))$ is defined, then dom $f(g(x))$ = dom $g$.

  So dom $f(g(x))$ = dom $g = R \setminus \{0\}$.

Graph of $y = f(g(x))$

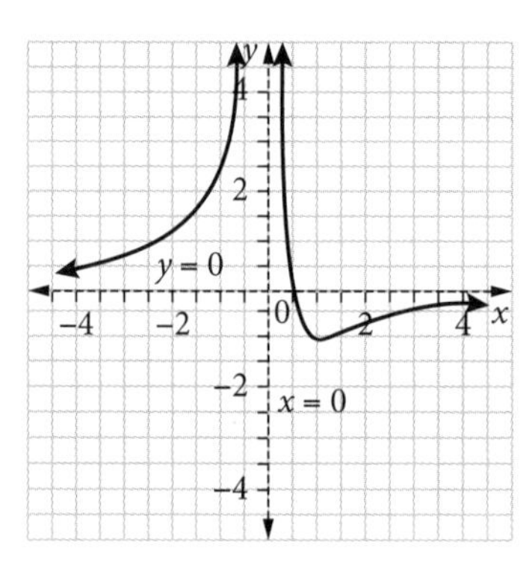


## 1.11 Piecewise (hybrid) functions

**Piecewise (hybrid) functions** are different functions sketched together on the one graph.

- To sketch piecewise functions, sketch each section separately, noting the domain given, and calculating the endpoints. Note that the endpoints can't double up, using ≤ and ≥ at the one point, as this would create a one-to-many relation.

**Example 23**

Sketch $f(x) = \begin{cases} x^2 + 4x, & x \le -2 \\ 2x, & x > -2 \end{cases}$.

**Solution**

Find the endpoints:

$y = x^2 + 4x$ at $x = -2$ gives $y = -4$

$y = 2x$ at $x = -2$ gives $y = -4$

The same $y$ values at the joining endpoint of each function means the graph is continuous.

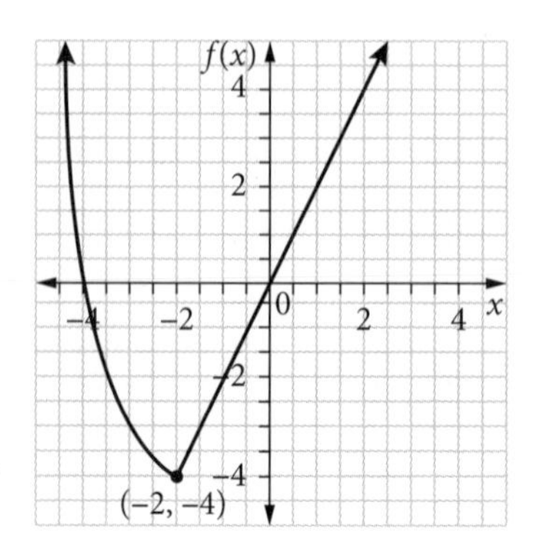

Piecewise functions can be sketched on CAS.

- Because the piecewise function is a many-to-one or one-to-one function, the calculation of the $y$ value will only be evaluated in the section of the graph where it exists. So $x = -3$ exists only on the parabola section and $x = -1$ only on the line section.
- $f(-3) = -3$ and $f(-1) = -3$

**Hint**

When calculating values in piecewise functions, every $x$ value will only have one unique matching $y$ value, as it does in any function. Look for where each section of the graph is defined.

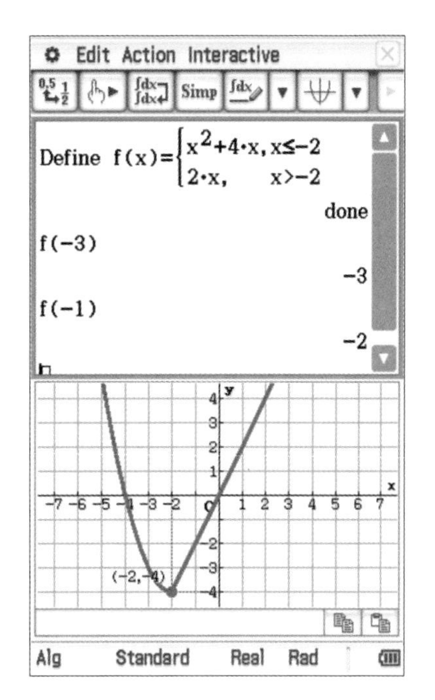


## 1.11.1 Modelling using piecewise (hybrid) functions

### Example 24

Consider the function $h(x) = \begin{cases} e^{2x}, & x \le 0 \\ \dfrac{1}{x^2}, & 0 < x < 2 \\ -x + 3, & x \ge 2 \end{cases}$

The graph of $h(x)$ describes a stylised height, in metres, of water in a river before and after a lock is placed at $x = 0$ and at $x = 2$.

**a** Find the height of water at $x = 0$, $x = 1$, $x = 2$, $x = 3$.

**b** Find the $x$ value(s) when the height of water is 2 metres.

Solve

**i** $e^{2x} = 2 \Leftrightarrow x = \frac{1}{2}\log_e(2)$

**ii** $\frac{1}{x^2} = 2 \Leftrightarrow x = \frac{1}{\sqrt{2}}$ (positive for the given domain)

**iii** $-x + 3 = 2 \Leftrightarrow x = 1$

### Solution

**a** $h(0) = 1$
$h(1) = 1$
$h(2) = 1$
$h(3) = 0$

**b** The line $y = 2$ crosses only one of the graph sections so the only answer is $\frac{1}{x^2} = 2 \Leftrightarrow x = \frac{1}{\sqrt{2}} \approx 0.707$.

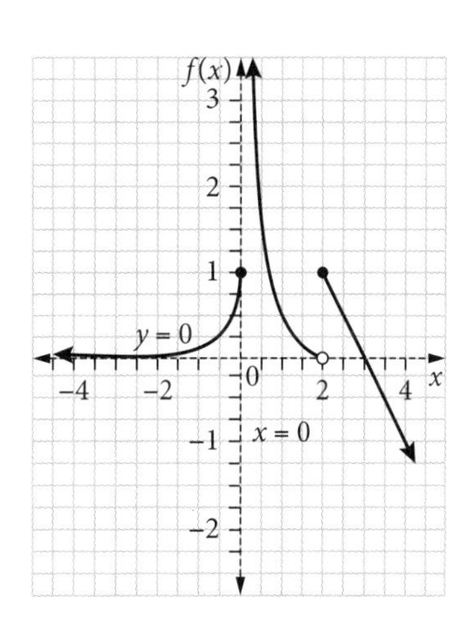

# 1.12 Modelling functions

This section will cover

- modelling of practical situations using polynomial, power, circular, exponential and logarithmic functions, simple transformation and combinations of these functions, including simple piecewise (hybrid) functions.



We can use polynomial, power, circular, exponential and logarithmic functions, simple transformation and combinations of these functions, including piecewise (hybrid) functions, to model practical situations.

### Example 25

A decreasing exponential function of the form $y = ae^{-t} + b$ describes the concentration of a drug (mg) in an animal, for $t$ (hours) after $t \geq 0$ in a particular experiment. This graph has a horizontal asymptote at $y = 2$ and passes through the point $(2, 5e^{-2} + 2)$.

**a** Write down the equation of the function.

**b** Sketch the graph of the function for $t \geq 0$.

**c** State the level of drug in the animal after 3 hours.

### Solution

**a** $y = ae^{-t} + b$

Horizontal asymptote at $y = 2$ gives $b = 2$.

Thus, $y = ae^{-t} + 2$.

Substitute the point $(2, 5e^{-2} + 2)$ to get $a = 5$.

The equation is $y = 5e^{-t} + 2$.

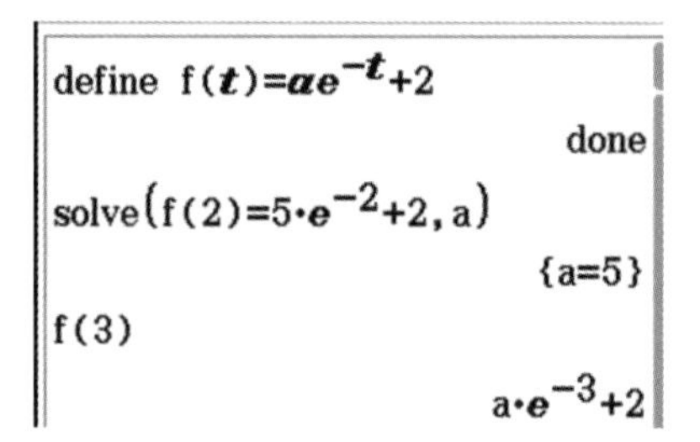


**b** Graph of $y = 5e^{-t} + 2$ for $t \geq 0$.

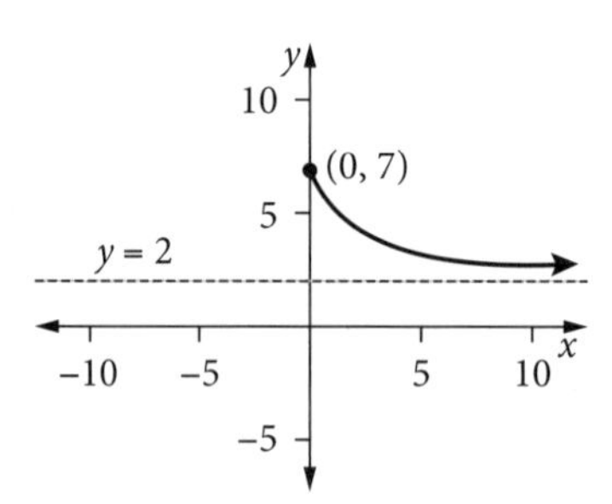


**c** $y = 5e^{-t} + 2$ when $t = 3$

The level of drug in the animal after 3 hours is $5e^{-3} + 2$ mg.

## 1.12.1 Modelling using combinations of functions

Many possible practical modelling situations can now be represented by combinations of different functions since CAS can sketch them easily.

- For example, $y = ax^n e^{-kx} + b$ could be given as a possible model for the amount of medication remaining in the bloodstream after a dose.

If given $a = 2$, $n = \frac{1}{2}$, $k = 0.1$, and $b = 4$, the equation is $y = 2x^{0.5}e^{-0.1x} + 4$ and the graph is shown. There will be a horizontal asymptote at $y = 4$.

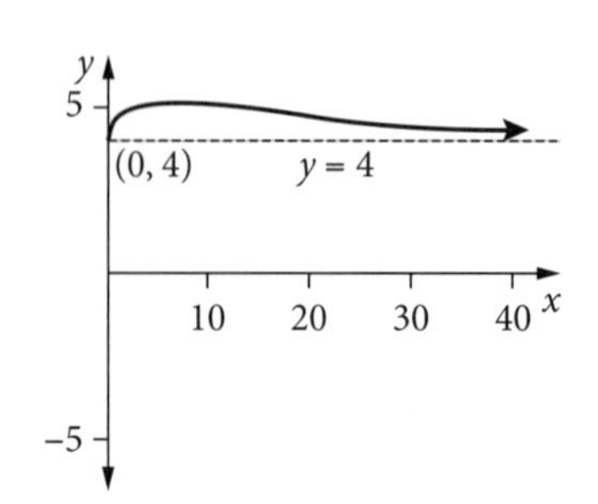

# Glossary

**amplitude** The vertical distance from the centre of a circular function to its maximum and minimum points. The amplitude of $y = \sin(x)$ and $y = \cos(x)$ is 1.

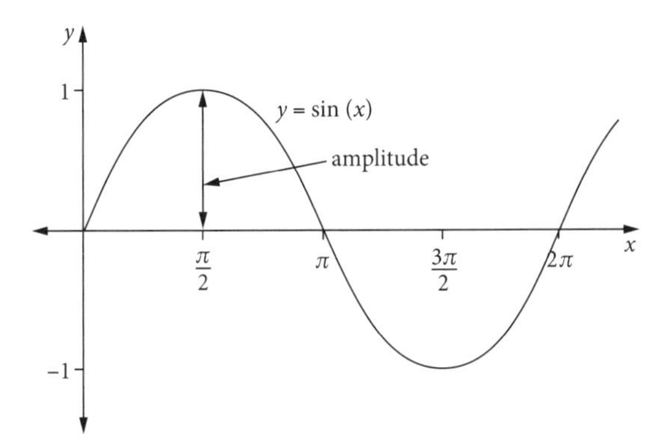


**asymptote** A line, or sometimes a curve, towards which a graph gets infinitesimally close for large $x$ or $y$ values. In the graph below, the lines $x = 2$ and $y = 4$ are asymptotes.

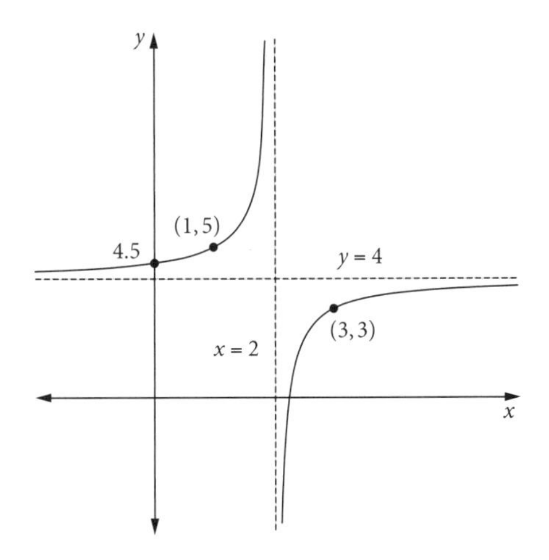


**axis of symmetry** A line about which a graph is reflected equally on both sides.

**base** The number being raised to a power; for example, in $2^5$, the base is 2.

**circular function** A trigonometric function such as $y = \sin(x)$, $y = \cos(x)$ or $y = \tan(x)$.

**co-domain** A set of values that the range of a function or relation can fall in. For example, the function $f: [0, 7] \rightarrow R, g(x) = x^2 + 4$ has a domain $[0, 7]$, co-domain $R$ and range $[4, 53]$.
*See also* **domain**, **range**.

**dependent variable** A variable that is calculated from another variable according to a rule or equation. The variable on the vertical scale of a graph. For $y = f(x)$, $y$ is the dependent variable.

**A+ DIGITAL FLASHCARDS**
Revise this topic's key terms and concepts by scanning the QR code or typing the URL into your browser.

https://get.ga/aplus-vce-maths-methods

**dilation** Sketching and squashing (compressing) of a graph, either from the $x$-axis (vertically) or $y$-axis (horizontally).

**discriminant, Δ** The part $\Delta = b^2 - 4ac$ of the quadratic formula $x = \dfrac{-b \pm \sqrt{b^2 - 4ac}}{2a}$ that gives information about the number and types of roots of the quadratic equation $ax^2 + bx + c = 0$.

**domain** The set of values of the independent variable $x$ in a relation or function.

**exponent** Another name for power. For example, in $2^5$, the exponent is 5.

**exponential decay** A decrease that is happening quickly at first, then gradually more quickly, according to the function $y = ba^x$, where $y$ is the quantity, $b$ is the initial value, $a$ is the growth factor ($a < 1$) and $x$ is the time. Natural exponential decay follows the function $y = be^{kt}$, where $t$ is the time and $k$ is a negative constant.

**exponential function** A function of the form $y = ba^x$, where the variable is in the exponent (power).

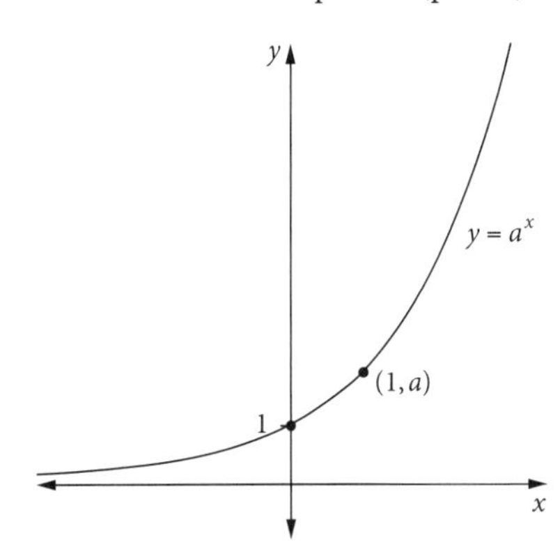


**exponential growth** An increase that is happening slowly at first, then gradually more quickly, according to the function $y = ba^x$, where $y$ is the quantity, $b$ is the initial value, $a$ is the growth factor ($a > 1$) and $x$ is the time. Natural exponential growth follows the function $y = be^{kt}$, where $t$ is the time and $k$ is a positive constant.

**function** A relation such that each $x$ value within the domain has one unique matching $y$ value.

**general form of a linear equation** $ax + by + c = 0$, where $a$, $b$ and $c$ are whole numbers and $a$ is positive.

**general form of a quadratic equation**
$y = ax^2 + bx + c$, where $a$, $b$ and $c$ are rational numbers.

**gradient-intercept form** (of a linear equation)
$y = mx + c$, where $m$ is the gradient and $c$ is the $y$-intercept.

**hyperbola** A discontinuous graph with equation of the form $y = \dfrac{a}{x-b} + c$.

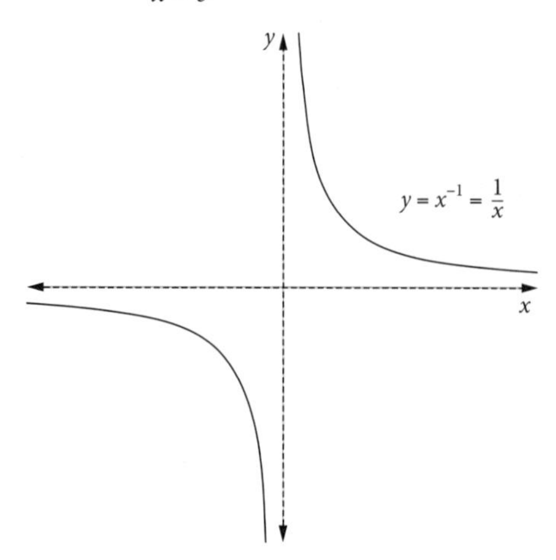


**image** A function after it has been transformed or the $y$ value given the $x$ value and vice versa.

**independent variable** A variable that is used to calculate the value of another variable. The variable on the horizontal scale of a graph. For $y = f(x)$, $x$ is the independent variable.

**interval notation**

| Set notation | Interval notation |
|---|---|
| $x \geq a$ | $[a, \infty)$ |
| $x > a$ | $(a, \infty)$ |
| $x \leq a$ | $(-\infty, a]$ |
| $x < a$ | $(-\infty, a)$ |
| $a < x < b$ | $(a, b)$ |
| $a \leq x \leq b$ | $[a, b]$ |
| $a < x \leq b$ | $(a, b]$ |
| $a \leq x < b$ | $[a, b)$ |

**linear function** A function in the form $y = mx + c$, where the power of $x$ is 1. The graph of a linear function is a straight line.

**linear polynomial** A polynomial of degree 1.

**logarithm** The exponent (power) to which a fixed number, the base, must be raised to give a number or variable. For example, $2^5 = 32$ so $\log_2(32) = 5$.

**logarithm laws** The rules involving the logarithms of $xy$, $\dfrac{x}{y}$, $x^n$ and change of base, related to the index laws.
*See also* **change of base theorem** (Chapter 2).

**logarithmic function** A function of the form $y = \log_a(x)$.

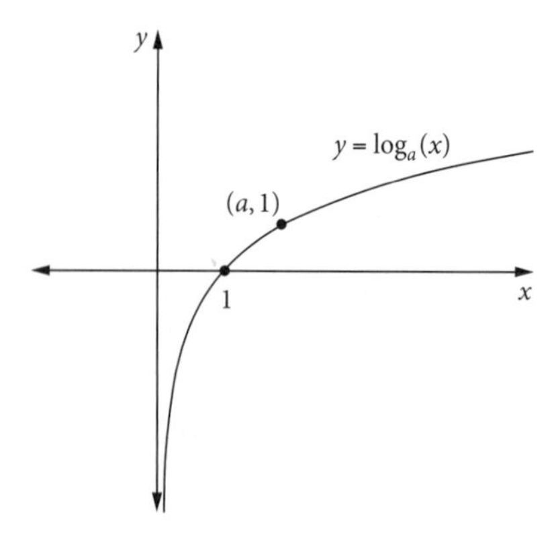


**maximal domain** The largest set of real numbers for which a function rule is defined. Also called natural or implied domain.

**natural exponential function** A function of the form $y = be^x$, where $e$ is the exponential constant approximately equal to 2.718 and the variable $x$ is in the exponent (power).

**natural logarithmic function** A function of the form $y = \log_e(x) = \ln(x)$ such that $e^y = x$, where $e$ is the exponential constant approximately equal to 2.718.

**one-to-one function** A function such that each value in the range has only one ordered pair $(x, y)$; each $y$ value matches with only one $x$ value.

**parabola** The graph of a quadratic function $y = ax^2 + bx + c$.

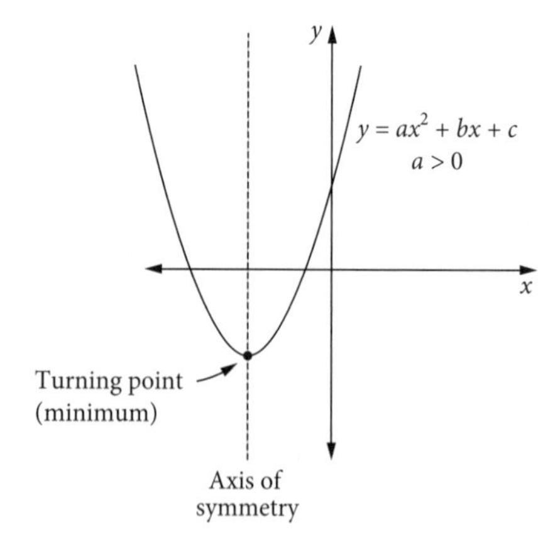

**period** The horizontal distance a circular function moves before repeating itself. The period of $y = \sin(x)$ and $y = \cos(x)$ is $2\pi$.

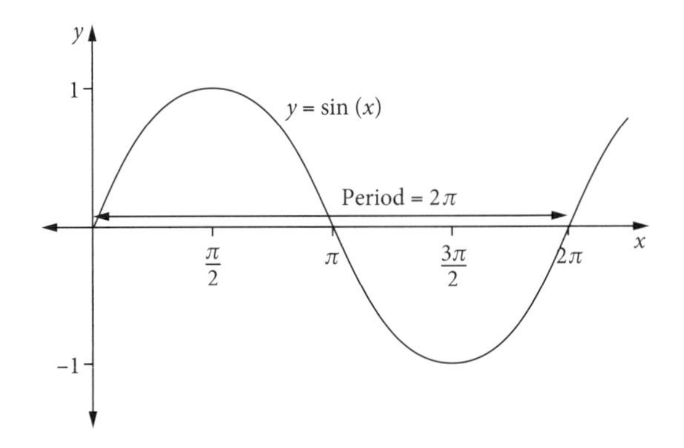


**piecewise (hybrid) function** A function that is a combination of two or more functions, for different values of $x$.

**point-gradient form** (of a linear equation) $y - y_1 = m(x - x_1)$, where $m$ is the gradient and the line passes through the point $(x_1, y_1)$.

**power function** A function in the form $x^n$, where $n$ is a real number.

**quadratic equation** An equation in the form $ax^2 + bx + c = 0$, involving the variable $x^2$.

**quadratic formula** The formula $x = \dfrac{-b \pm \sqrt{b^2 - 4ac}}{2a}$ that gives the solutions of the quadratic equation $ax^2 + bx + c = 0$.

**quadratic function** A function of the form $y = ax^2 + bx + c$, where the highest power of $x$ is 2. The graph of a quadratic function is a parabola.

**range** The set of values of the dependent variable $y$ in a relation or function $y = f(x)$.

**reflection** Mirror-image or 'flipping' of a graph so that it is back-to-front or bottom-to-top.

**relation** A set of ordered pairs $(x, y)$ that describe a relationship between the variables $x$ and $y$.

**root** A value of an equation which, when substituted into the equation, satisfies the equation. It is a solution to the equation.

**transformation** A change of a function and its graph through dilation, reflection or translation.

**translation** Shifting a graph parallel to the $x$-axis (horizontally) or $y$-axis (vertically).

**trigonometric function** *See* **circular function**.

**truncus** A discontinuous graph of an equation of the form $y = \dfrac{a}{(x-b)^2} + c$.

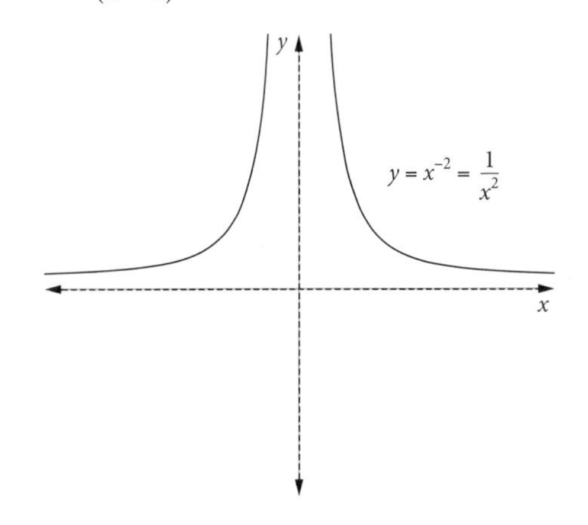


**zeros** The $x$ values of a function $f(x)$ where $f(x) = 0$.

# Exam practice

## Short-answer questions

### Technology free: 20 questions

Solutions to this section start on page 205.

**Question 1** (2 marks)

Sketch the graph of $y = \begin{cases} x^4, & x \geq 1 \\ -1, & x < 1 \end{cases}$.

**Question 2** (2 marks)

For the quadratic function shown, find the equation that defines the function.

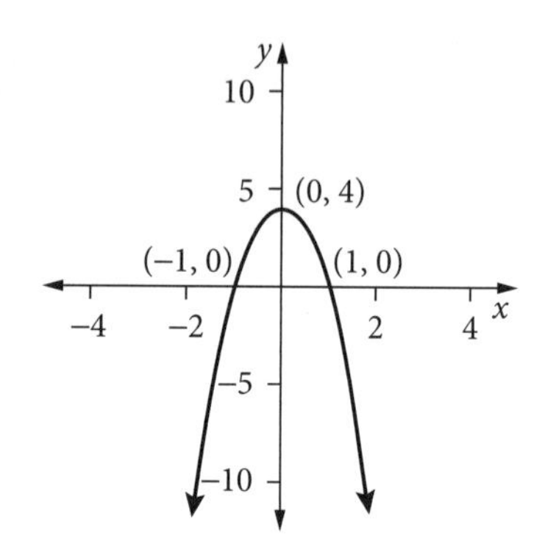


**Question 3** (3 marks)

For the square root function shown, determine the equation and domain that defines the function.

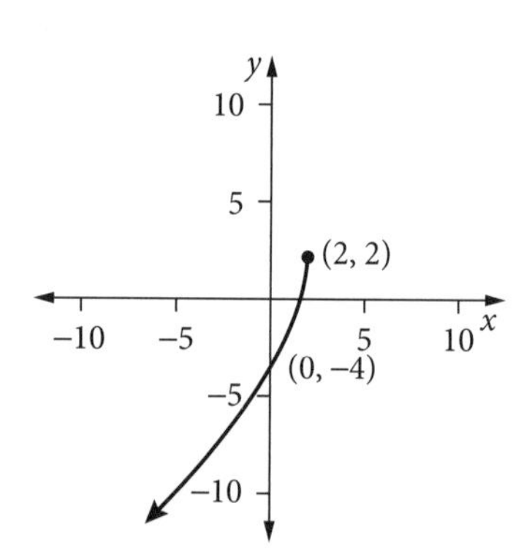


**Question 4** (4 marks)

**a** Sketch the graph of $f(x) = 1 - 5x$, labelling all axial intercepts. 2 marks

**b** On the same set of axes, sketch the graph of the inverse function $f^{-1}$, labelling all axial intercepts clearly. 2 marks

**Question 5** (3 marks)

State the amplitude, period and the range of the graph of $y = 3\sin\left(4x - \frac{\pi}{3}\right)$.

**Question 6** (4 marks)

If the graph of $y = \sqrt{x}$ is reflected in the $x$-axis, dilated by a factor of 2 from the $y$-axis, and then translated 1 unit in the negative direction of the $y$-axis,

**a** give the equation of the image graph 1 mark

**b** sketch both $y = \sqrt{x}$ and the image graph on the same set of axes. 3 marks

**Question 7** (2 marks)

A quartic function is shown.

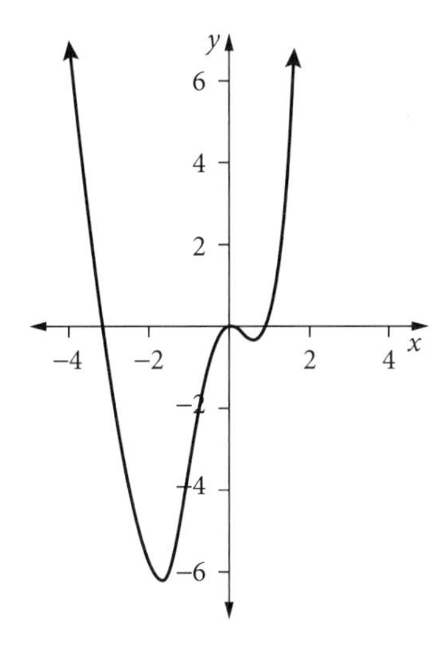


Sketch the inverse relation on the same set of axes.

**Question 8** (2 marks)

Sketch the graph of $y = 3e^{x+1} + 1$, labelling intercepts with the axes and giving the equations of any asymptotes.

**Question 9** (2 marks)

Sketch the graph of $y = 3\log_e(2 - x)$ labelling intercepts with the axes and giving the equations of any asymptotes.

**Question 10** (4 marks)

Sketch $y = \sin(x)$ and $y = \sin(2x)$ on the same set of axes for $0 \le x \le 2\pi$. Hence, using addition of ordinates, sketch the graph of $y = \sin(x) + \sin(2x)$.

**Question 11** (4 marks)

If $f(x) = x^2$ and $g(x) = e^{x-2}$, state, with reason, whether or not the functions $f(g(x))$ and $g(f(x))$ exist.

**Question 12** (4 marks)

**a** Sketch the graph of $y = 5\log_2(2 - x) + 1$, giving the equations of any asymptotes and labelling any axial intercepts. 2 marks

**b** State the domain and range of the graph. 2 marks

**Question 13** (4 marks)

**a** Sketch the graph of $y = -2\tan(\pi x)$ for $-1 \le x \le 1$, giving the equations of any asymptotes and labelling any axial intercepts. 3 marks

**b** State the domain and range of the graph. 1 mark

**Question 14** (4 marks)

**a** Sketch the graph of $y = 3^{-x}$ on a set of axes.

**b** On the same set of axes, sketch the graph of $y = -\log_3(x)$.

**Question 15** (3 marks)

A decreasing exponential function of the form $y = ae^{-x} + b$ has a horizontal asymptote at $y = -2$ and passes through the origin. Write down the equation of the function.

**Question 16** (4 marks)

**a** A cubic function of the form $y = a(x - b)(x - c)(x - d)$ has $x$-intercepts at $x = 3$ and $x = 2$, and passes through the points $(0, 5)$ and $(1, 1)$. Find the equation of the function. 2 marks

**b** Hence sketch the function, labelling axial intercepts. 2 marks

**Question 17** (3 marks)

If $u(x) = 3x - 1$ and $g(u(x)) = [u(x)]^2$,

**a** find the rule for $h(x) = g(u(x))$ 2 marks

**b** find $h(2)$. 1 mark

**Question 18** (6 marks)

The graph of $f(x) = \frac{-2}{(x-3)^2} + 1$ is shown below.

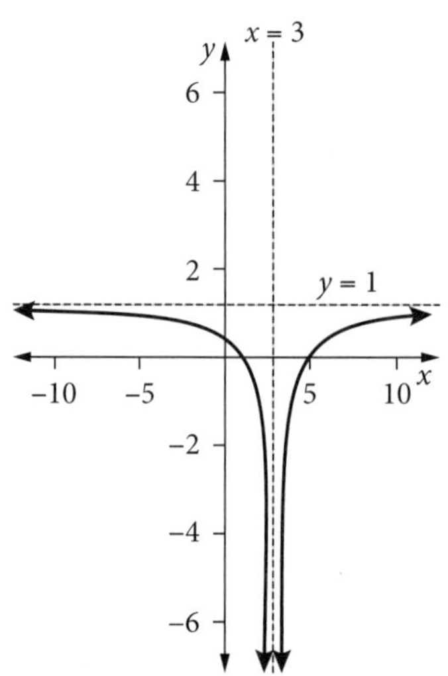


**a** Sketch $y = f(-x)$. 2 marks

**b** Sketch $y = f(1 - x)$. 2 marks

**c** Sketch $y = -f(-x)$. 2 marks

**Question 19** (3 marks)

The coordinates of $A$ and $B$ are $(-6, 7)$ and $(2, -9)$, respectively. A line that is perpendicular to the line $y = -4x + 1$ passes through the midpoint of $A$ and $B$.

Find the equation of the line.

**Question 20** (5 marks)

**a** Sketch the graph of

$$y = \begin{cases} -4, & x \le -1 \\ x + 1, & -1 < x < 3 \\ -\frac{2}{3}x + 6, & x \ge 3 \end{cases}$$

labelling the coordinates of any axial intercepts and endpoints. 3 marks

**b** Evaluate $f(-2)$, $f(0)$ and $f(6)$. 1 mark

**c** Find the $x$ value(s) for which $f(x) = 1$. 1 mark



## Multiple-choice questions

### Technology active: 50 questions

Solutions to this section start on page 208.

**Question 1**

The linear function shown has a gradient of

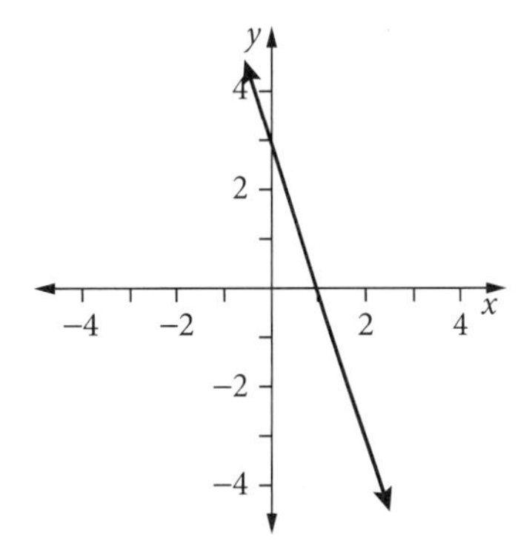

**A** −3

**B** −1

**C** 1

**D** 2

**E** 3

**Question 2**

The quadratic function shown could have a stationary point at

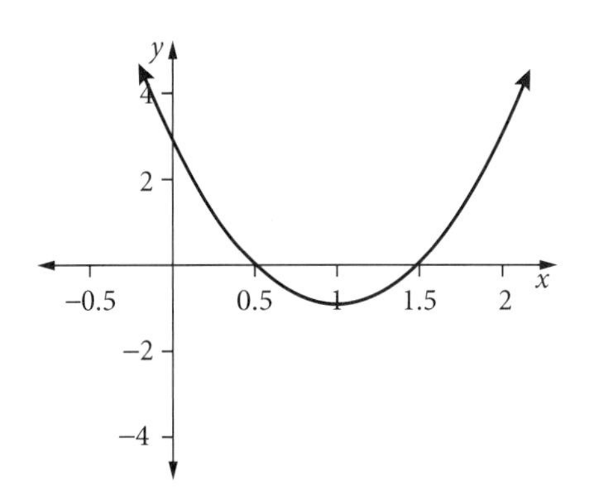


**A** $(0, 3)$

**B** $(0.5, 0)$

**C** $(0.5, 1.5)$

**D** $(1, -1)$

**E** $(1.25, -1)$

**Question 3**

Which of the following is **not** a function?

**A** $\{(x, y): y = x^2 - 3x\}$

**B** $\{(x, y): y = -\sqrt{3x - 1}\}$

**C** $\{(x, y): y = 3x - 1\}$

**D** $\{(x, y): (0, 1), (1, 2), (2, 5), (4, 17)\}$

**E** $\{(x, y): y = \pm\sqrt{x + 3}\}$

**Question 4**

A possible equation for the graph shown could be

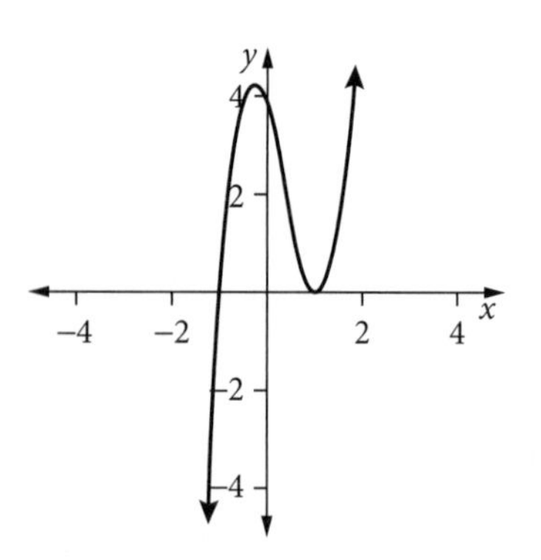


**A** $y = -(x + 1)(x - 1)(x + 1)$

**B** $y = (x + 1)(x - 1)^2$

**C** $y = 4(x + 1)(x - 1)^2$

**D** $y = -(x + 1)^2(x - 1)(x - 2)$

**E** $y = -2(x + 1)(x - 1)^2$

**Question 5**

The inverse of $\{(-1, 4), (0, 3), (1, 1), (2, -1)\}$ is

**A** $\{-1, 0, 1, 2\}$

**B** $\{4, 3, 1, -1\}$

**C** $\{(4, -1), (3, 0), (1, 1), (-1, 2)\}$

**D** $\{(2, -1), (1, 1), (0, 3), (-1, 4)\}$

**E** $\{(-1, 4), (0, 3), (1, 1), (2, -1)\}$

**Question 6**

If $f(x) = 0.5x - 3$, then the rule of the inverse function $f^{-1}$ is

**A** $f^{-1}(x) = \frac{x}{2} + 3$

**B** $f^{-1}(x) = \frac{x}{2} - 3$

**C** $f^{-1}(x) = \frac{1}{0.5x - 3}$

**D** $f^{-1}(x) = 2x + 6$

**E** $f^{-1}(x) = \frac{x + 6}{2}$

**Question 7**

If $f: (0, 3) \to R$, where $f(x) = 0.5x - 3$, then the inverse function $f^{-1}$ is defined by

**A** $f^{-1}(x) = \frac{x}{2} + 3, x \in (0, 3)$

**B** $f^{-1}(x) = \frac{x}{2} - 3, x \in (0, 3)$

**C** $f^{-1}(x) = \frac{1}{0.5x - 3}, x \in (1.5, 3)$

**D** $f^{-1}(x) = 2x + 6, x \in (0, 3)$

**E** $f^{-1}(x) = 2x + 6, x \in (-3, -1.5)$

**Question 8**

The range of $\left\{(x, y): y = -\sqrt{9 - x^2}\right\}$ is given by

**A** $[-3, 0]$

**B** $[0, 3]$

**C** $(-3, 0)$

**D** $[-9, 0]$

**E** $(-\infty, 3)$

**Question 9**

The graph of $y = \frac{1}{(2x - 5)^2} + 2$ has asymptotes at

**A** $x = 2, y = \frac{5}{2}$

**B** $x = -\frac{2}{5}, y = -2$

**C** $x = -\frac{5}{2}, y = 2$

**D** $x = \frac{5}{2}, y = 2$

**E** $x = 5, y = 2$

**Question 10**

If $f(x) = \frac{1}{x - 1} - 1$, then the maximal domain is

**A** $\{x: x \geq 1\}$

**B** $\{x: x < 1\}$

**C** $R \setminus \{0\}$

**D** $R \setminus \{1\}$

**E** $R \setminus \{-1\}$

**Question 11**

The graph of $y = -e^{\frac{x}{2}} + 1$ cuts the $x$-axis at

**A** 1 **B** 0 **C** $\log_e\left(\frac{1}{2}\right)$ **D** $\log_e(2)$ **E** $\log_e(1)$

**Question 12**

The coordinates of the point of intersection of the graphs of $y = 3e^{-x} + 2$ and $y = 5$ are

**A** $(0, 1)$ **B** $(0, 3)$ **C** $(0, 5)$ **D** $(0, 7)$ **E** $(5, 0)$

**Question 13**

If $f(x) = x + 2$ and $g(x) = x^2$, then $g(f(x))$ equals

**A** $x^2 + 2$ **B** $x^4$ **C** $(x + 2)^4$

**D** $x + 4$ **E** $x^2 + 4x + 4$

**Question 14**

If $f(x) = 2x + 1$ and $g(x) = 3\log_e(x)$, when testing the existence of $f(g(x))$ it is true that

**A** $R\setminus\{0\} \subseteq R$ **B** $(-\infty, 1) \not\subset R\setminus\{0\}$ **C** $R \subseteq R$

**D** $(0, \infty) \not\subset R\setminus\{0\}$ **E** $R \subseteq R\setminus\{0\}$

**Question 15**

The period of $y = 2\sin(\pi x)$ is

**A** 1 **B** 2 **C** $\frac{\pi}{2}$ **D** $\pi$ **E** $2\pi$

**Question 16**

The period of $y = -3\tan(\pi x)$ is

**A** 1 **B** 2 **C** $\frac{\pi}{2}$ **D** $\pi$ **E** $2\pi$

**Question 17**

The graph shown could be modelled by which of the following functions, where $a \in R$?

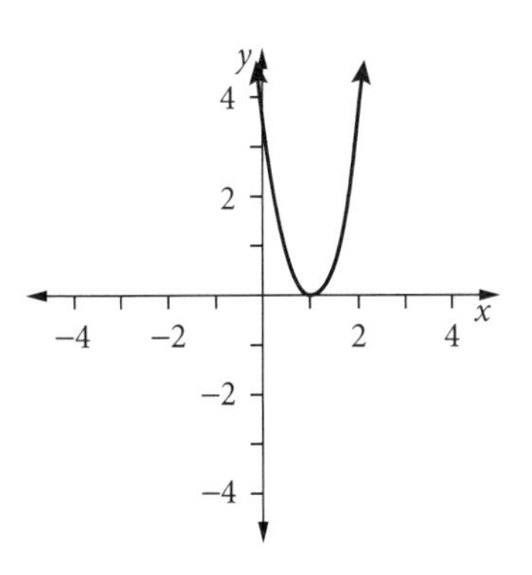


**A** $y = a(x - 2)^2 + 3$

**B** $y = ax(x - 1)^2$

**C** $y = a(x - 1)^2 + 2$

**D** $y = a(x + 1)^2 + 3$

**E** $y = 2a(x - 1)^2$

**Question 18**

The transformations required to change $y = \sqrt{x}$ to $y = 2 + 3\sqrt{x - 1}$ are

**A** dilation by a factor of 2 from the $x$-axis, translation of 1 unit to the right, and 3 units down

**B** dilation by a factor of 3 from the $y$-axis, translation of 1 unit to the right, and 2 units up

**C** dilation by a factor of 3 from the $y$-axis, translation of 1 unit to the left, and 2 units up

**D** dilation by a factor of 3 from the $x$-axis, translation of 1 unit to the right, and 2 units up

**E** dilation by a factor of $\frac{1}{3}$ from the $y$-axis, translation of 1 unit to the right, and 2 units up

**Question 19**

If $y = g(x)$ for $x \in [-1, 3]$ and $y = h(x)$ for $x \in [0, 4]$, then $y = 2g(x) - 4h(x)$ is defined for the domain

**A** $x \in (0, 6)$ **B** $x \in [0, 3]$ **C** $x \in [-1, 4]$

**D** $x \in [-2, 0]$ **E** $x \in (-1, 4)$

**Question 20**

A piecewise function is given by

$$f(x) = \begin{cases} x + 2, & x \leq 0 \\ \dfrac{x^2}{4}, & 0 < x \leq 1 \\ -3, & x > 1 \end{cases}$$

The value of $f(3)$ is

**A** $-3$ **B** $\frac{9}{4}$ **C** 3 **D** 5 **E** 9

**Question 21**

The graph of a function is shown. Which of the following is most likely to be the graph of the inverse function?

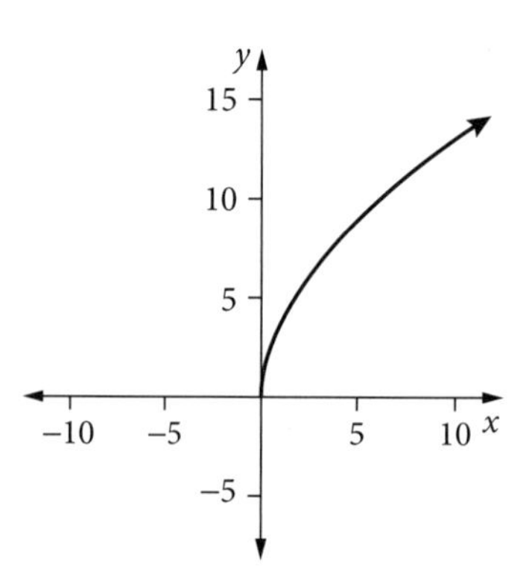


**A**

**B**

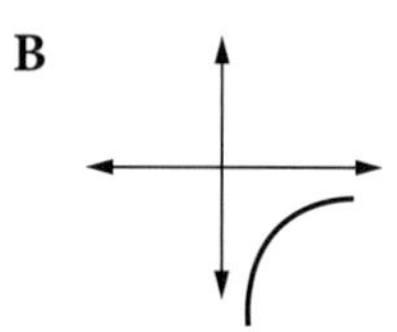

**C**

**D**

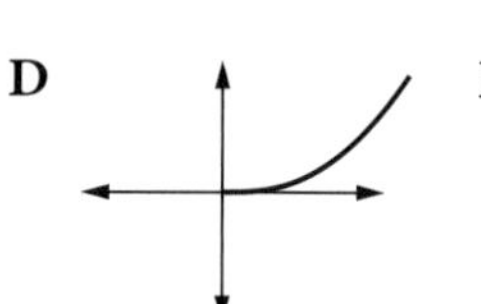

**E**

**Question 22**

If $f(x) = -\sqrt{2(x + 2)} - 1$, then the range of the inverse function $f^{-1}$ is

**A** $(-\infty, 1)$ **B** $(-\infty, -1)$ **C** $(-1, \infty)$ **D** $[-4, \infty)$ **E** $[-2, \infty)$

**Question 23**

The graph of $y = 2(2^{x-1}) - 8$ has $x$ and $y$ axes intercepts, respectively, as

**A** $x = 3, y = -7$ **B** $x = \frac{1}{2}, y = -4$ **C** $x = 1, y = -8$

**D** $x = \frac{1}{2}, y = 4$ **E** $x = 7, y = -3$

### Question 24

Let $f(x) = x^2 - 2x$ and $g(x) = \frac{1}{x^2}$ for their maximal domains.

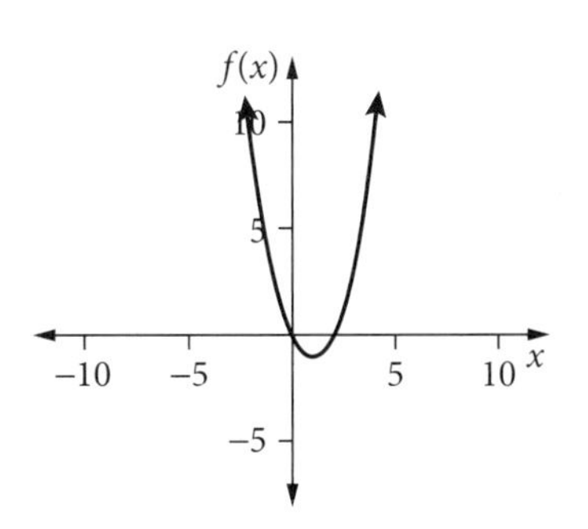


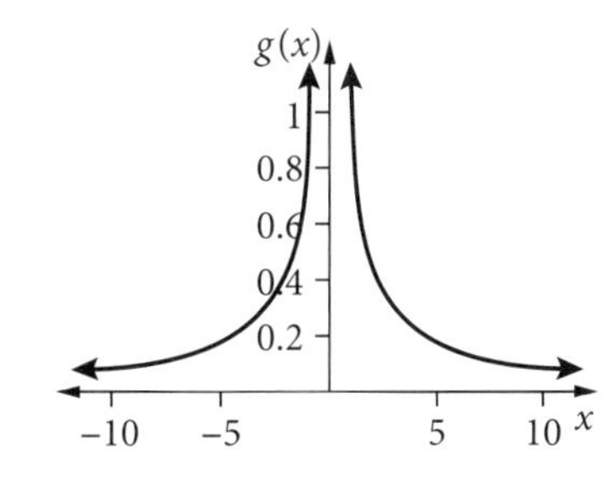


To find $f(g(x))$, the incorrect statement is

**A** $\text{dom } f(g(x)) = \text{dom } g = R \setminus \{0\}$

**B** $\text{dom } f(g(x)) = \text{dom } f = R$

**C** $\text{dom } f = R$, $\text{ran } f = [-1, \infty)$

**D** $\text{dom } g = R \setminus \{0\}$, $\text{ran } g = (0, \infty)$

**E** $f(g(x))$ is defined since $(0, \infty) \subseteq R$

### Question 25

The line $2y - 6 = x$ passes through the point $\left(\frac{1}{2}, b\right)$. The value of $b$ is

**A** $-\frac{11}{4}$ **B** $-\frac{5}{4}$ **C** $\frac{11}{4}$ **D** $\frac{13}{4}$ **E** 7

### Question 26

The range of the piecewise function $f(x) = \begin{cases} x^2 + 2x, & x \le -1 \\ 2x^3, & x > -1 \end{cases}$ is

**A** $(-\infty, 1)$ **B** $(-2, \infty)$ **C** $(-1, \infty)$ **D** $[-4, \infty)$ **E** $[-2, \infty)$

### Question 27

The rule of the image of $y = x^{\frac{2}{3}}$ when there is a dilation by a factor of 3 from the $y$-axis is

**A** $y = x$ **B** $y = x^{\frac{2}{3}}$ **C** $y = 3x^{\frac{2}{3}}$ **D** $y = \left(\frac{x}{3}\right)^{\frac{2}{3}}$ **E** $y = (3x)^{\frac{2}{3}}$

### Question 28

The period and amplitude of the graph of $y = -3\sin(2x)$ are

| | Period | Amplitude |
|---|---|---|
| **A** | $\pi$ | −3 |
| **B** | 2 | 3 |
| **C** | $2\pi$ | −3 |
| **D** | $4\pi$ | 3 |
| **E** | $\pi$ | 3 |

### Question 29

The rule of the function $y = -\frac{2}{\sqrt{x}}$ when it is reflected in the $x$-axis is

**A** $y = \frac{2}{\sqrt{x}}$ **B** $y = \frac{2}{\sqrt{-x}}$ **C** $y = -\frac{2}{\sqrt{-x}}$ **D** $y = \frac{1}{\sqrt{-x}}$ **E** $y = -\frac{1}{\sqrt{x}}$

### Question 30

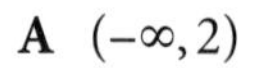

The graph of $y = \sin(x) - \cos(2x)$ is shown.

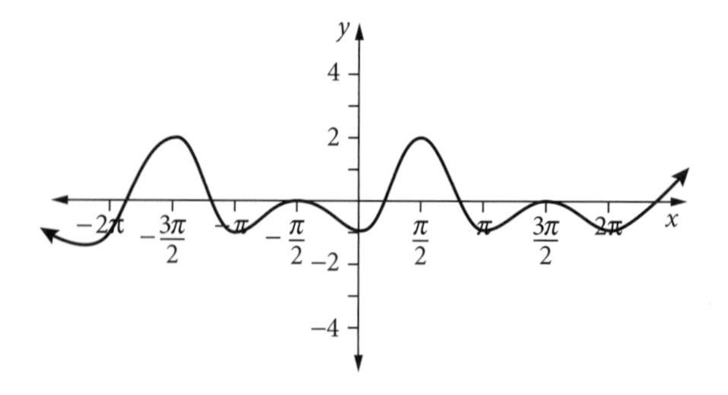


The range of the graph is

**A** $(-\infty, 2)$ **B** $[-1.125, 2]$

**C** $(-1, \infty)$ **D** $[-1.125, \infty)$

**E** $[-1, 2)$

### Question 31

The graph of $y = 3\tan(3x)$ for domain $x \in [0, \pi]$ has asymptotes at

**A** $x = -\frac{\pi}{2}, \frac{\pi}{2}$ **B** $x = \frac{\pi}{6}, \frac{\pi}{2}, \frac{5\pi}{6}$ **C** $x = \frac{\pi}{2}, \frac{3\pi}{2}$

**D** $x = \frac{\pi}{4}, \frac{3\pi}{4}$ **E** $x = \frac{\pi}{3}, \frac{2\pi}{3}$

### Question 32

The image of the point $(-2, 3)$ under the transformations:

reflection in $y$-axis, then dilation by a factor of 5 from the $x$-axis, is

**A** $(2, -3)$ **B** $(-2, 3)$ **C** $(-10, 3)$ **D** $(2, 15)$ **E** $(-2, 30)$

### Question 33

The equation of the graph shown, where $a$ is positive, could be given by

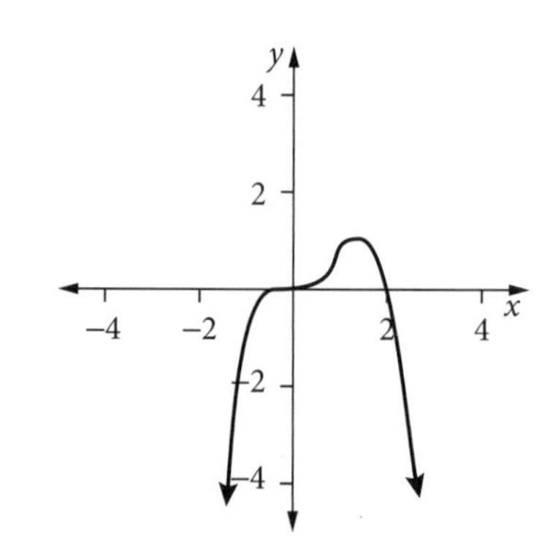


**A** $y = -ax^3(x - 2)$

**B** $y = -x^3(x - a)$

**C** $y = ax(x - 2)^3$

**D** $y = x^3(x - a)$

**E** $y = ax^2(x - 2)^2$

### Question 34

The graph shown is of the form $y = a\sin(n(x - b))$.

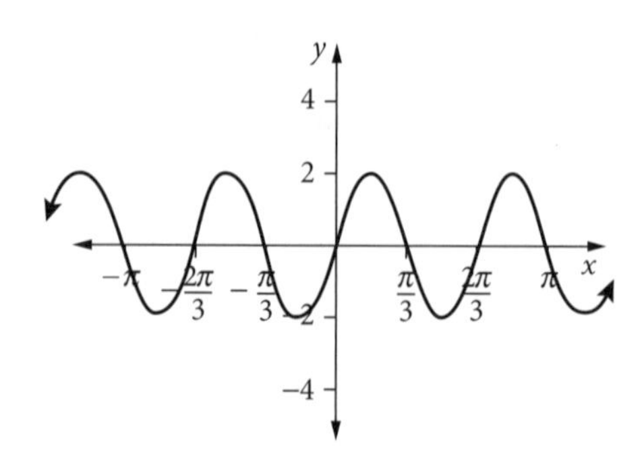


The graph has the equation

**A** $y = 2\sin(3(x - \pi))$

**B** $y = -2\sin(3(x - \pi))$

**C** $y = -2\sin(3x)$

**D** $y = -2\sin(2(x - \pi))$

**E** $y = \sin(3(x - \pi))$

### Question 35

If $f(x) = 3\log_e(x-2)$, then which of the following statements is **not** true?

**A** $3f(x+1)$ has a vertical asymptote at $x = 1$

**B** $f(x-1)$ has a vertical asymptote at $x = 3$

**C** $f(x) + 1$ has a vertical asymptote at $x = 2$

**D** $f(x)$ and $2f(x)$ have the same vertical asymptote at $x = 2$

**E** $f(x)$ and $f(3x)$ have the same vertical asymptote at $x = 2$

### Question 36

The graph shown has the equation

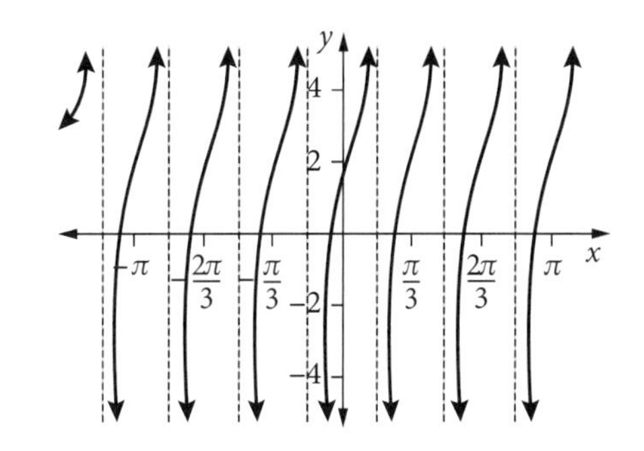


**A** $y = \tan\left(\frac{x}{3}\right)$

**B** $y = 3\tan(3x)$

**C** $y = 3\tan(2x)$

**D** $y = \tan(3x) + 2$

**E** $y = \tan\left(\frac{x}{3}\right) + 2$

### Question 37

If $f(x) = 10x - 2$ and $g(x) = f(f(x))$, then the rule for the inverse function $g^{-1}$ is given by

**A** $g^{-1}(x) = \frac{x}{100} + \frac{11}{50}$

**B** $g^{-1}(x) = 10x - 2$

**C** $g^{-1}(x) = \frac{1}{10x-2}$

**D** $g^{-1}(x) = 100x - 22$

**E** $g^{-1}(x) = \frac{1}{100x-22}$

### Question 38

A possible equation for the graph below, where $m, n > 0$, is

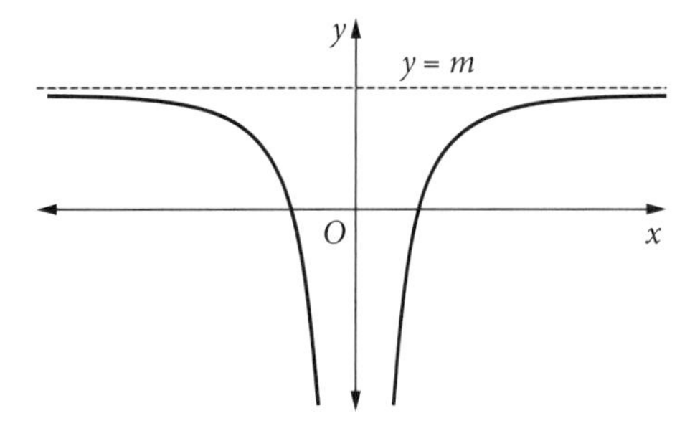


**A** $y = m + \frac{n}{x^2}$

**B** $y = m - \frac{n}{x}$

**C** $y = m - \frac{n}{x^2}$

**D** $y = \frac{n}{x^2} - m$

**E** $y = -\frac{n}{x^2} - m$

### Question 39

The graph of $y = -\cos\left(4x - \frac{\pi}{2}\right)$ has been transformed from the graph of $y = \cos(x)$ by

**A** a reflection in the $x$-axis, dilation by a factor of $\frac{1}{4}$ from the $y$-axis and a horizontal translation of $\frac{\pi}{8}$ units to the right

**B** a reflection in the $x$-axis, dilation by a factor of $\frac{1}{4}$ from the $y$-axis and a horizontal translation of $\frac{\pi}{2}$ units to the right

**C** a reflection in the $x$-axis, dilation by a factor of 4 from the $y$-axis and a horizontal translation of $\frac{\pi}{8}$ units to the right

**D** a reflection in the $x$-axis, dilation by a factor of 4 from the $y$-axis and a horizontal translation of $\frac{\pi}{2}$ units to the right

**E** a reflection in the $y$-axis, dilation by a factor of $\frac{1}{4}$ from the $x$-axis and a horizontal translation of $\frac{\pi}{8}$ units to the right

**Question 40**

The function $f(x) = 3x^2$ is dilated by a factor of 2 from the $x$-axis and reflected in the $y$-axis and then translated 1 unit down. The image equation that results is

**A** $y = 6x^2$  **B** $y = 3x^2$  **C** $y = 3x^2 - 1$

**D** $y = 6x^2 - 1$  **E** $y = -6x^2 - 1$

**Question 41**

The graph of $y = \log_e(x)$ has been transformed to the graph of $y = \log_e(5 - 2x)$. The transformations required were

**A** reflection in the $y$-axis, dilation by a factor of 2 from the $x$-axis, translation of $\frac{5}{2}$ units to the left

**B** reflection in the $y$-axis, dilation by a factor of 2 from the $x$-axis, translation of $\frac{5}{2}$ units to the right

**C** reflection in the $x$-axis, dilation by a factor of 2 from the $x$-axis, translation of $\frac{5}{2}$ units to the right

**D** reflection in the $x$-axis, dilation by a factor of $\frac{1}{2}$ from the $x$-axis, translation of $\frac{5}{2}$ units to the left

**E** reflection in the $y$-axis, dilation by a factor of $\frac{1}{2}$ from the $y$-axis, translation of $\frac{5}{2}$ units to the right

**Question 42**

The transformation which maps the curve with equation $y = \sin(x)$ to the curve with equation $y = 2\sin(x - 3) + 1$ follows the mapping notation

**A** $(x, y) \to (x + 3, 2y + 1)$  **B** $(x, y) \to (x - 3, 2y + 1)$  **C** $(x, y) \to (x + 3, 2y - 1)$

**D** $(x, y) \to \left(x + 3, \frac{1}{2}y - 1\right)$  **E** $(x, y) \to \left(x - 3, \frac{1}{2}y - 1\right)$

**Question 43**

The range of $y = 11 - 4\cos(\pi(3 - x))$ is

**A** $[-11, 11]$  **B** $[7, 11]$  **C** $[7, 15]$  **D** $[15, 24]$  **E** $[3, \pi]$

**Question 44**

A model which fits the graph shown is of the form $f(x) = A\sin(nx) + B$.

The values of $A$, $n$ and $B$, respectively are

**A** $-2, 1, 4$

**B** $-2, 1, 6$

**C** $-2, 2\pi, 6$

**D** $2, 2\pi, 6$

**E** $2, 1, 8$

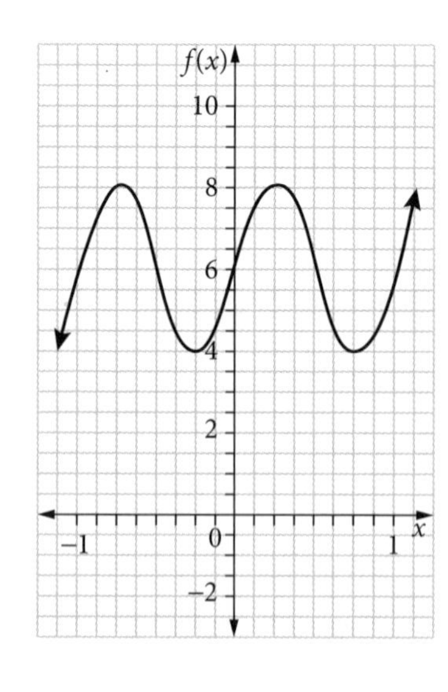

### Question 45

For the transformations in the order

- dilate by 2 units from the $y$-axis
- reflect over the $y$-axis
- translate by 1 unit in the negative direction of the $x$-axis,

the image of the curve with equation $y = x^4$ is

**A** $y = \frac{(x+1)^4}{16}$ **B** $y = -\frac{(x+1)^4}{16}$ **C** $y = 2(2x+1)^4$

**D** $y = \frac{(x+1)^4}{4}$ **E** $y = -\frac{(x+1)^4}{4}$

### Question 46

$y = ax^n e^{-kx} + b$ could be given as a possible model for the amount of medication remaining in the bloodstream after a dose of medication.

For the values $a = 2$ and $n = \frac{1}{2}$, the graph goes through the points $(0, 1)$ and $(2, 2)$. The values of $k$ and $b$ are

**A** $k = \frac{3}{4}, b = 1$ **B** $k = \frac{3}{4}\log_e(2), b = 1$ **C** $k = \frac{3}{4}\log_e(2), b = \log_e(2)$

**D** $k = \frac{3}{4}\ b = 2$ **E** no solution

### Question 47

The graphs of $f^{-1}(x) = \sqrt{2x-1}$ and $f(x)$ are given.

The point(s) of intersection between $f$ and $f^{-1}$ is/are

**A** $(-\sqrt{2}, \sqrt{2})$

**B** $(2+\sqrt{2}, \sqrt{2})$

**C** $(2, 2)$

**D** $(1, 1)$ and $\left(\frac{1}{2}, \frac{1}{2}\right)$

**E** $(1, 1)$

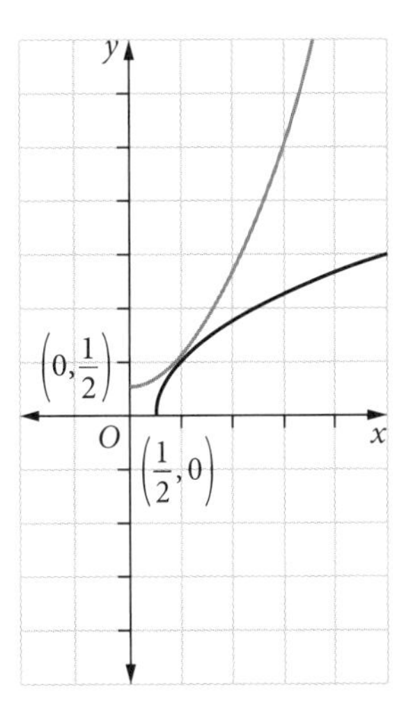


### Question 48

The equation of the graph shown could be

**A** $y = 3\log_e(x-2)$

**B** $y = -3\log_e(x) + 2$

**C** $y = -3e^{-x} + 2$

**D** $y = -3e^{x} + 2$

**E** $y = 3e^{-x} + 2$

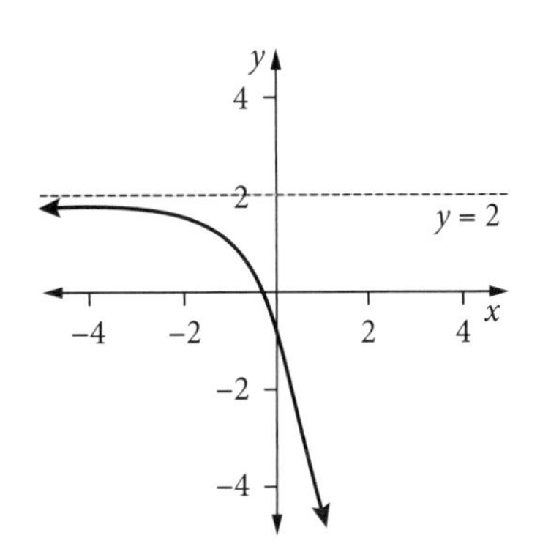

**Question 49**

The equation of the parabola that joins the points $(0, 1)$, $(2, 3)$, $(5, 7)$ is

**A** $y = \dfrac{x^2 + 13x + 15}{15}$ **B** $y = \dfrac{x^2 + 13x + 1}{15}$ **C** $y = \dfrac{3x^2}{8} - \dfrac{x}{2} + \dfrac{1}{8}$

**D** $y = -\dfrac{x^2}{24} + \dfrac{7x}{6} - \dfrac{9}{8}$ **E** $y = -\dfrac{5x^2}{12} + \dfrac{17x}{4} - \dfrac{23}{6}$

**Question 50**

The equation of the graph shown could be

**A** $y = x^2$ **B** $y = x^{\frac{2}{3}}$ **C** $y = x^{\frac{3}{2}}$

**D** $y = 2x^2$ **E** $y = x^3$

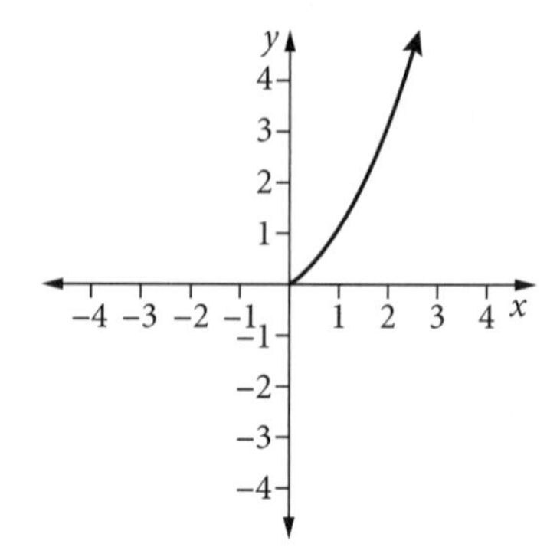


## Extended-answer questions

### Technology active: 5 questions

Solutions to this section start on page 211.

**Question 1** (15 marks)

A polynomial is defined as $P(x) = ax^3 + bx^2 + 2x + 10$.

**a** It is known that $x - 2$ is a factor of $P(x)$ and that if $P(x)$ is divided by $x + 1$, the remainder is $-3$. Show that $a = \dfrac{5}{2}$ and $b = -\dfrac{17}{2}$. 3 marks

**b** Hence find the solutions for the equation $y = P(x) = 0$. 3 marks

**c** Sketch the graph of $y = P(x)$, labelling the coordinates of any axial intercepts. 4 marks

A section of the graph drawn in part **c** represents a proposed ski slope, where $y$ metres is the height above ground level and $x$ metres is the horizontal distance from the end of the ski slope. The ski slope ends at the point $(0, 10)$.

Consider $f: [0, 5] \rightarrow R, f(x) = \dfrac{5}{2}x^3 - \dfrac{17}{2}x^2 + 2x + 10$

**d** Find the maximum height of the ski slope. 2 marks

**e** How high will the engineers have to build the platform for the end of the slope? 1 mark

**f** The engineers do not permit the ski slope to be opened if the path goes underground. Will the ski slope go ahead? 2 marks

**Question 2** (10 marks)

A jet of water from a hose is pointed over the fence by children into their neighbour's garden. The water stream from the hose starts at a distance from the ground of 2.88 metres. Amy, who is holding the hose, is standing at a horizontal distance of 14 metres from the fence. The water goes over the fence and lands on the other side in the neighbour's garden at a horizontal distance of 4 metres from the fence.

**a** Use the function $h(x) = a + b(x + 6)^2$ to model the path of the water in the air, where $a$ and $b$ are constants and the variables, $h$ and $x$, are in metres. Find the values of $a$ and $b$. 3 marks

**b** Sketch the function $h(x) = a + b(x + 6)^2$ that models the path of the water in the air, with your values of $a$ and $b$. Use the domain $x \in [-14, 4]$. 2 marks

**c** The neighbours are sick of the children next door and have built a fence that is 5 metres high. Will the water go over the fence? Why/why not? 2 marks

**d** There are power lines above the children's garden that are 10 metres off the ground. Will the water hit the power lines? 1 mark

**e** The children's dog likes to jump up and lap at the water in the air. The dog, Scruffy, can jump 2 metres in the air to reach the water. Where does Scruffy manage to drink the water? 2 marks

**Question 3** (12 marks)

**a** If $3x^3 + ax^2 + 5x = 3(x + b)^3 + c$, find all possible values of $a$, $b$ and $c$. 4 marks

**b** Letting $f_1(x) = 3(x + b)^3 + c$, and using values from part **a**, where $b < 0$ and $c > 0$, find the $x$ coordinates of the point(s) where $f(x) = 0$. 1 mark

**c** Letting $f_2(x) = 3(x + b)^3 + c$, and using values from part **a**, where $b > 0$ and $c < 0$, find the $x$ coordinates of the point(s) where $f(x) = 0$. 1 mark

**d** Hence, find where $f_1(x) = f_2(x)$. 1 mark

**e** Sketch the graphs of $f_1(x)$ and $f_2(x)$ on the same axes, labelling point(s) of intersection. 3 marks

**f** Sketch the inverse of graph $f_1(x)$ on the same axes, labelling point(s) of intersection. 2 marks

CHAPTER 1 – EXAM PRACTICE

**Question 4** (9 marks)

The temperature of a cup of tea cools according to the rule $T = T_0 \times 2^{-kt}$, where $T$ is the temperature in degrees Celsius and $t$ is time in hours. The original temperature of the cup of tea is 90°C.

**a** What is $T_0$? 1 mark

It takes 30 minutes for the temperature to halve.

**b** What fraction of the original temperature is the temperature of the cup of tea after 60 minutes? 3 marks

**c** What is the temperature of the cup of tea after 60 minutes? 1 mark

**d** A different drink's temperature follows the formula $T = T_0 \times 2^{-kt} + 20$, where $T$ is the temperature in degrees Celsius and $t$ is time in hours. It takes 20 minutes for the temperature to halve and the original temperature of the cup of tea is 90°C.

Find the value of $k$ in this case, giving your answer correct to two decimal places. 2 marks

The formula is changed to suit another drink. The graph of $T = T_0 \times 2^{-kt}$ has a sequence of transformations applied to it, in the order given.

- The graph is translated 15 units in the negative direction of the $t$-axis.
- It is then translated 7 units in the negative direction of the $T$-axis.
- It is then dilated by a factor of 5 from the $t$-axis.
- It is then dilated by a factor of $\frac{1}{2}$ from the $T$-axis.
- The graph is then reflected in the $T$-axis.

**e** After all these transformations are applied, what is the new rule for $T(t)$? 2 marks

**Question 5** (14 marks)

Consider the function $f(x) = ae^{x-2} + c$ where $a, c \in R$.

**a** State the type of function and the domain and range, in terms of $c$, of $f$. 3 marks

**b** Find the values of $a$ and $c$ if $f(0) = 2$ and $f(-2) = 0$. 2 marks

**c** Define the function $f$ in the form $f$: dom $\to R, f(x) = \ldots$ 1 mark

**d** Sketch the graph of $f$ labelling axial intercepts and the equation of the asymptote. 3 marks

**e** Find the rule and domain of the inverse $f^{-1}$ of the function $f$. 2 marks

**f** Hence find the points of intersection, if any, between the graphs of $f^{-1}$ and $f$. 1 mark

**g** Show your answer to part **f** on a graph. 2 marks

# Chapter 2
# Area of Study 2: Algebra, number and structure

## Content summary notes

**Algebra, number and structure**

**Algebra of functions**

- Composite functions
- Inverse functions
- Recognising functionalities
- Exponentials and logarithms
- Circular functions
- Incorporated in other areas of study

**Algebra of polynomials**

- Equating coefficients
- Factor theorem
- Remainder theorem
- Rational root theorem

**Solutions of equations**

- Appropriate solution process
- Factorisation
- Inverse operations
- Graphical approach
- Numerical approach
- Algebraic approach
- Exact and non-exact solutions
- Newton's method
- Literal equations

**Systems of equations**

- Simultaneous equations
- Simultaneous linear equations
- Graphical approach
- Numerical approach
- Algebraic approach

In this area of study students cover the algebra of functions, including composition of functions, inverse functions and the solution of equations. They also study the identification of appropriate solution processes for solving equations, and systems of simultaneous equations, presented in various forms. Students also cover recognition of equations and systems of equations that are solvable using inverse operations or factorisation, and the use of graphical and numerical approaches for problems involving equations where exact value solutions are not required, or which are not solvable by other methods. This content is to be incorporated as applicable to the other areas of study.

This area of study includes:

- solution of polynomial equations with real coefficients of degree $n$ having up to $n$ real solutions, including numerical solutions
- functions and their inverses, including conditions for the existence of an inverse function, and use of inverse functions to solve equations involving exponential, logarithmic, circular and power functions
- composition of functions, where $f$ composite $g$, $f \circ g$, is defined by $(f \circ g)(x) = f(g(x))$ given $r_g \subseteq d_f$
- solution of equations of the form $f(x) = g(x)$ over a specified interval, where $f$ and $g$ are functions of the type specified in the 'Functions, relations and graphs' area of study, by graphical, numerical and algebraic methods, as applicable

- solution of literal equations and general solution of equations involving a single parameter
- solution of simple systems of simultaneous linear equations, including consideration of cases where no solution or an infinite number of possible solutions exist (geometric interpretation only required for two equations in two variables).

## 2.1 Algebra of polynomials

This section will cover

- solution of polynomial equations with real coefficients of degree $n$ having up to $n$ real solutions, including numerical solutions.

VCE Mathematics Study Design 2023–2027 p. 99, © VCAA 2022

The first section reviews the algebra of polynomials, equating coefficients, and solving polynomial equations with real coefficients up to degree $n$, having up to $n$ real solutions. Polynomials have only whole number powers.

A polynomial of **degree** $n$ can be written as $P(x) = a_0 + a_1x + a_2x^2 + \ldots + a_{n-1}x^{n-1} + a_nx^n$, where $a_0, a_1, a_2 \ldots a_n$ are real constant coefficients and $n$ is a positive integer or zero.

- A **monic polynomial** has its leading term with a coefficient of 1. For example, the degree 3 monic polynomial $P(x) = x^3 + 2x^2 + 6x + 3$ has coefficients of 1, 2, 6 and 3.
- A non-monic polynomial has the leading term with a coefficient other than 1. For example, the degree 4 non-monic polynomial $Q(x) = 5x^4 + 2x^2 + x + 6$ has coefficients of 5, 0, 2, 1 and 6.

### Example 1

$f(x) = 2 + 3x$ is a polynomial of degree 1 (linear function).

$g(x) = -3 + x + 5x^2$ is a polynomial of degree 2 (quadratic function).

$h(x) = \sqrt{x} + \frac{1}{x} = x^{\frac{1}{2}} + x^{-1}$ is not a polynomial, because powers other than positive integers or zero are used.

- The factorised form of the polynomial allows an easier manipulation of the algebra and graphing of the function.

### Example 2

The polynomial $f(x) = 3x^3 - 2x^2 - 6x + 4$ can be written in factorised form as $f(x) = (x^2 - 2)(3x - 2)$.

This is also equal to $f(x) = (x - \sqrt{2})(x + \sqrt{2})(3x - 2)$.

$f(x)$ is a polynomial of degree 3 with its graph having $x$-axis intercepts at $x = \pm\sqrt{2}$ and $x = \frac{2}{3}$.

CAS can factorise, using **rfactor**. If surds are involved, solve for roots and graph the polynomial. The graph of the polynomial below is of a positive cubic function.

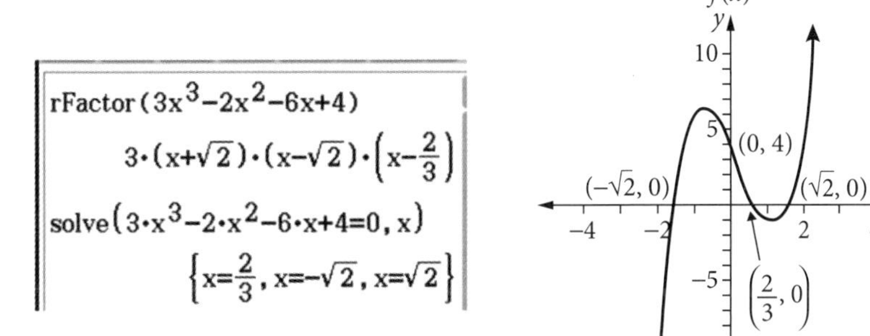
rFactor$(3x^3-2x^2-6x+4)$

$3\cdot(x+\sqrt{2})\cdot(x-\sqrt{2})\cdot\left(x-\frac{2}{3}\right)$

solve$(3\cdot x^3-2\cdot x^2-6\cdot x+4=0, x)$

$\left\{x=\frac{2}{3}, x=-\sqrt{2}, x=\sqrt{2}\right\}$

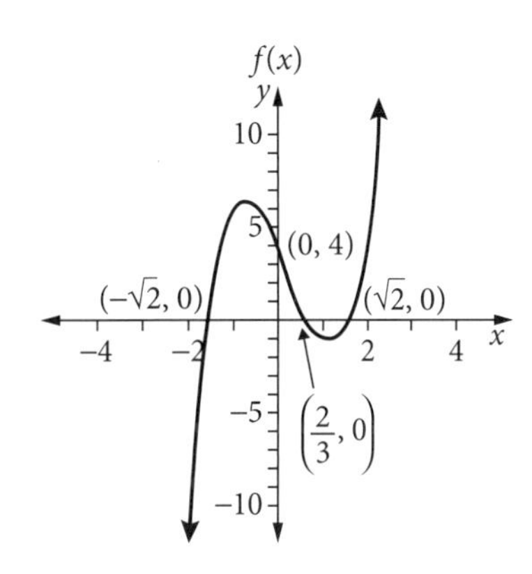


We use the letters $P$, $Q$ or $R$ to name polynomials.

**Example 3**

Let $P(x) = x^3 - 4x^2 - 11x + 30$.

Factorised, $P(x) = (x - 5)(x - 2)(x + 3)$.

To sketch this polynomial, solve $P(x) = 0$ to get $x = 5$, $x = 2$ or $x = -3$, to find the $x$-intercepts of the graph.

Note: The constant term, +30, which is the $x^0$ term, is the $y$-intercept.

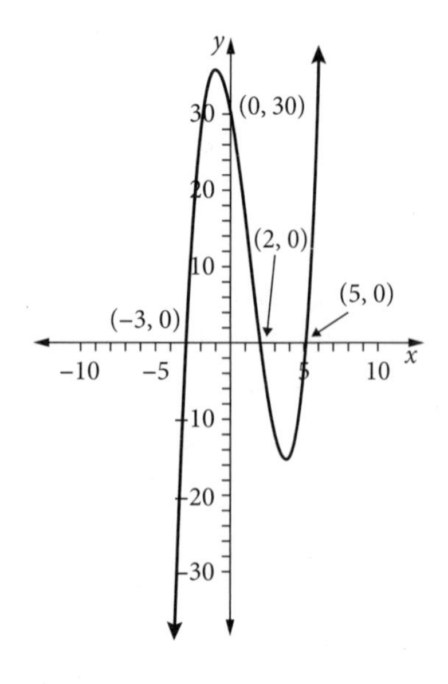


## 2.1.1 Equating coefficients

If two polynomials are equal, then the degrees of the polynomials are equal with the corresponding coefficients of the terms also equal.

- If $P(x) = Q(x)$ where $P(x) = a_0 + a_1x + a_2x^2 + \ldots + a_nx^n$ and $Q(x) = b_0 + b_1 x + b_2 x^2 + \ldots + b_n x^n$, then $a_0 = b_0$, $a_1 = b_1$, …, $a_n = b_n$.

**Example 4**

If $3x^2 + 6x + 5 = a(x + b)^2 + c$, find the values of $a$, $b$ and $c$.

**Solution**

$$3x^2 + 6x + 5 = a(x^2 + 2bx + b^2) + c$$

giving $\quad 3x^2 + 6x + 5 = ax^2 + 2abx + ab^2 + c$.

Equating coefficients: $3 = a$ (coefficients of $x^2$)

$6 = 2ab$ (coefficients of $x^1$)

$5 = ab^2 + c$ (coefficients of $x^0$)

Hence $a = 3$, $b = 1$, $c = 2$.

## 2.1.2 The factor and remainder theorem

When a polynomial $P(x)$ is divided by $(ax+b)$, the remainder is found by evaluating $P\left(-\frac{b}{a}\right)$.

### 2.1.2.1 The factor theorem: the special case where the remainder is zero

$(ax+b)$ is a factor of $P(x)$ if and only if $P\left(-\frac{b}{a}\right)=0$.



---

**Example 5**

Find the remainder when $P(x)=x^3+3x^2-5x+7$ is divided by $x+1$.

**Solution**

Solving $x+1=0$ gives $x=-1$. This is called the 'zero' of the bracket.

Substitute $x=-1$ into $P(x)$:

$P(-1)=(-1)^3+3(-1)^2-5(-1)+7=14$

Hence the remainder is 14.

**Hint**
This is easier than doing long division to find the remainder.

CAS uses the **propFrac** command to perform long division when factorising and finding the remainder.

The division $\frac{P(x)}{x+1}=\frac{x^3+3x^2-5x+7}{x+1}=x^2+2x-7+\frac{14}{x+1}$.

propFrac$\left(\frac{x^3+3x^2-5x+7}{x+1}\right)$

$x^2+2\cdot x+\frac{14}{x+1}-7$

Writing the answer in the form $Q(x)+\frac{R(x)}{D(x)}$, the quotient is $x^2+2x-7$ with the remainder of 14 and divisor $x+1$.

---

**Example 6**

Show that when $P(x)=x^3-4x^2-11x+30$ is divided by $x-5$, the remainder is zero.

**Solution**

$P(5)=(5)^3-4(5)^2-11(5)+30=0$

This means that $(x-5)$ is a factor of $P(x)$.

Using various forms of **long division**, the other factors can be found.

- Long division

$$
\begin{array}{r|l}
 & x^2 + x - 6 \\
x-5 & x^3-4x^2-11x+30 \\
 & \underline{x^3-5x^2} \\
 & x^2-11x \\
 & \underline{x^2-5x} \\
 & -6x+30 \\
 & \underline{-6x+30} \\
 & 0
\end{array}
$$

as expected the remainder = 0

- **Synthetic division**

$$\begin{array}{c|cccc} & 1 & -4 & -11 & 30 \\ 5 & & 5 & 5 & -30 \\ \hline & 1 & 1 & -6 & 0 \end{array}$$

as expected the remainder = 0

- Equating terms/box method

  We expect factors of the original cubic to be (linear) × (quadratic)

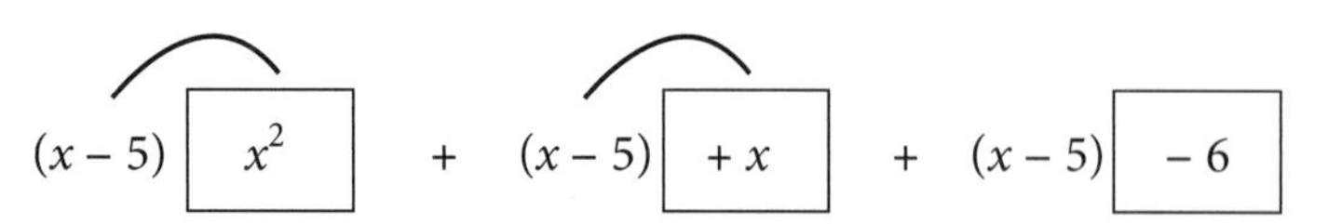


> **Hint**
> The box method requires multiplying step-by-step, adjusting the numbers to produce the given polynomial.

$x^3 - 5x^2 + x^2 - 5x - 6x + 30$

giving $(x - 5)(x^2 + x - 6) = (x - 5)(x + 3)(x - 2)$.

- Following one of the above methods, we get

$$\begin{aligned} x^3 - 4x^2 - 11x + 30 &= (x - 5)(x^2 + x - 6) \\ &= (x - 5)(x + 3)(x - 2) \end{aligned}$$

CAS will find all factors:

```
factor(x^3-4x^2-11x+30)
                    (x+3)·(x-2)·(x-5)
```

- The **factor theorem** then assists in finding the solution of equations.

Solve: $x^3 - 4x^2 - 11x + 30 = 0$

Using your answer from above: $x^3 - 4x^2 - 11x + 30 = (x - 5)(x - 2)(x + 3)$

Using the **Null Factor Law**: $\therefore x = 5$ or $x = 2$ or $x = -3$

Using CAS:

```
solve(x^3-4x^2-11x+30=0, x)
                    {x=-3, x=2, x=5}
```

Using the above information, we can graph $y = x^3 - 4x^2 - 11x + 30$ with $x$-intercepts $x = -3$, $x = 2$ and $x = 5$, and $y$-intercept $y = 30$.

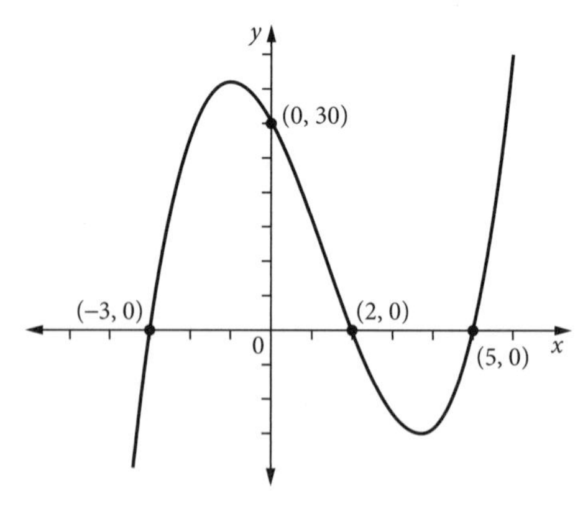

### Example 7

Using the factor theorem, factorise the polynomial $Q(x) = 5x^4 - 12x^3 - 31x^2 - 6x + 8$ and hence, find the roots of the graph $y = 5x^4 - 12x^3 - 31x^2 - 6x + 8$.

### Solution

Consider the factors ±1, ±2, ±4, ±8 of the constant term 8.

Try $Q(1) = 5 - 12 - 31 - 6 + 8 = -36 \neq 0$

Try $Q(-1) = 5 + 12 - 31 + 6 + 8 = 0$

$\therefore (x + 1)$ is a factor of $Q(x)$.

Using synthetic division to find the cubic factor:

| | 5 | −12 | −31 | −6 | 8 |
|---|---|---|---|---|---|
| −1 | | −5 | 17 | 14 | −8 |
| | 5 | −17 | −14 | 8 | 0 |

the remainder = 0

giving $Q(x) = (x + 1)(5x^3 - 17x^2 - 14x + 8)$.

Factorise again:

Let $P(x)= 5x^3 - 17x^2 - 14x + 8$.

Again, consider the factors ±1, ± 2, ± 4, ± 8 of the constant term 8.

Try $P(1) = 5 - 17 - 14 + 8 = -18 \neq 0$

Try $P(-1) = -5 - 17 + 14 + 8 = 0$

**Hint**
We already know from above that $P(1)$ does not equal zero.

$\therefore (x + 1)$ is a factor of $P(x)$.

Using synthetic division again to find the quadratic factor:

| | 5 | −17 | −14 | 8 |
|---|---|---|---|---|
| −1 | | −5 | 22 | −8 |
| | 5 | −22 | 8 | 0 |

the remainder = 0

giving $Q(x) = (x + 1)(x + 1)(5x^2 - 22x + 8)$.

Factorise the quadratic factor:

$Q(x) = (x + 1)^2(5x - 2)(x - 4)$

Roots of the quartic graph $y = (x + 1)^2(5x - 2)(x - 4)$ are $x = -1$, $x = \frac{2}{5}$, $x = 4$.

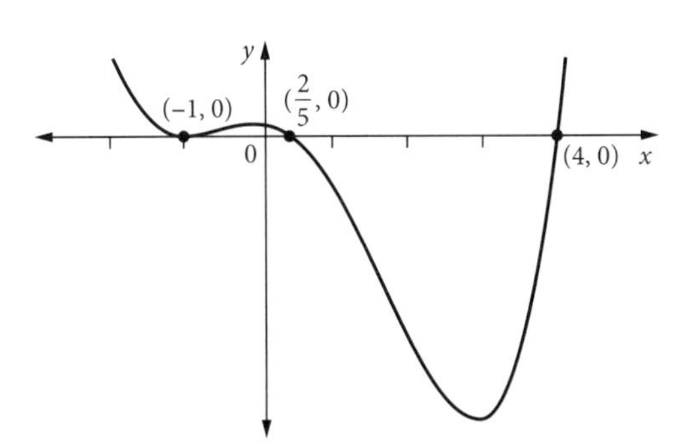


**Hint**
Note that the repeated factor $(x + 1)^2$ makes the graph 'touch' the $x$-axis at the point (−1, 0).

### 2.1.2.2 The rational root theorem

For the non-monic polynomial $P(x) = qx^2 + ax + p$, where $q$, $a$, $p$ are real constants, we can predict rational roots (if they exist) in the form of $\dfrac{\text{factors of } p}{\text{factors of } q}$.

---

**Example 8**

Consider $P(x) = 15x^2 + 4x - 4$ where $p = -4$ and $q = 15$.

Possible roots are $\dfrac{\pm1, \pm2, \pm4}{\pm1, \pm3, \pm5}$.

Try $P\left(\dfrac{2}{5}\right) = 15\left(\dfrac{2}{5}\right)^2 + 4\left(\dfrac{2}{5}\right) - 4 = 0$

$\therefore \left(x - \dfrac{2}{5}\right)$ or $(5x - 2)$ is a factor.

Try $P\left(-\dfrac{2}{3}\right) = 15\left(-\dfrac{2}{3}\right)^2 + 4\left(-\dfrac{2}{3}\right) - 4 = 0$

$\therefore \left(x + \dfrac{2}{3}\right)$ or $(3x + 2)$ is a factor.

$P(x) = (5x - 2)(3x + 2)$ giving the roots $x = \dfrac{2}{5}, x = -\dfrac{2}{3}$.

### 2.1.2.3 Newton's method of solving equations

A method for finding a numerical solution to a polynomial equation is **Newton's method**.

$$x_{n+1} = x_n - \frac{f(x_n)}{f'(x_n)}$$

- the first approximation $x_1$ is given by $x_1 = x_0 - \dfrac{f(x_0)}{f'(x_0)}$, where $x_0$ is the initial estimate.
- the second approximation $x_2$ is given by $x_2 = x_1 - \dfrac{f(x_1)}{f'(x_1)}$.
- and so on.

---

**Example 9**

Use Newton's method to approximate a zero for the cubic function $y = x^3 - 4x + 1$.

If $x_0 = 2$, find the value of $x_1$.

**Solution**

Differentiate $y = x^3 - 4x + 1$.

$f'(x) = 3x^2 - 4$

Calculate $f(x_0)$ and $f'(x_0)$, where $x_0 = 2$.

$f(2) = 2^3 - 4(2) + 1 = 1$

$f'(2) = 3(2)^2 - 4 = 8$

Use Newton's method to find $x_1$.

$$\begin{aligned} x_{n+1} &= x_n - \frac{f(x_n)}{f'(x_n)} \\ x_1 &= x_0 - \frac{f(x_0)}{f'(x_0)} \\ &= 2 - \frac{f(2)}{f'(2)} \\ &= 2 - \frac{1}{8} \\ &= \frac{15}{8} \\ &= 1.875 \end{aligned}$$

**Note**

Note that

$f(1.875) = 1.8753 - 4 \times 1.875 + 1$

$\approx 0.0917\ldots$

$\approx 0$

## 2.2 Functionality

This section uses simple functional relations to characterise properties of functions, including symmetry and algebraic equivalence, specifically including the exponential and logarithm laws.

You can be asked to verify or show some properties of functions using function notation.

**Exponential example**

If $f(x) = a^x$ and $a \in N$, verify that $f(x + y) = f(x)f(y)$.

LHS $= f(x + y) = a^{x+y}$

RHS $= f(x)f(y) = a^x \times a^y = a^{x+y} =$ LHS

```
define f(x)=a^x
                              done
judge(f(x+y)=f(x)f(y)
                              TRUE
```

## 2.3 Exponentials and logarithms

### 2.3.1 Exponential laws and logarithm laws

#### 2.3.1.1 Exponential laws

1. $a^x \times a^y = a^{x+y}$
2. $a^x \div a^y = a^{x-y}$
3. $(a^x)^y = a^{xy}$
4. $(ab)^x = a^x b^x$
5. $a^0 = 1$
6. $a^{\frac{p}{q}} = \sqrt[q]{a^p}$

#### 2.3.1.2 Logarithm laws

1. $\log_a(mn) = \log_a(m) + \log_a(n)$
2. $\log\left(\frac{m}{n}\right) = \log_a(m) - \log_a(n)$
3. $\log_a\left(\frac{1}{n}\right) = \log_a(n^{-1}) = -\log_a(n)$
4. $\log_a(m^p) = p\log_a(m)$
5. $\log_a(a) = 1$
6. $\log_a(1) = 0$

These laws can be used to simplify exponential and logarithmic expressions and solve equations.

**Example 10**

Simplify the expression

$3\log_2(x) - \log_2(x + 1) + \log_2(y)$.

**Solution**

$3\log_2(x) - \log_2(x + 1) + \log_2(y)$
$= \log_2(x^3) - \log_2(x + 1) + \log_2(y)$
$= \log_2(x^3y) - \log_2(x + 1)$
$= \log_2\left(\frac{x^3y}{x+1}\right)$

**Hint**

Mistakes can happen when asked to simplify expressions such as $\frac{\log_2(x)}{\log_2(\sqrt{x})}$. This is a log divided by a log, not an example of the second logarithm law.

$$\frac{\log_2(x)}{\log_2(\sqrt{x})} = \frac{\log_2(x)}{\frac{1}{2}\log_2(x)} = \frac{1}{\frac{1}{2}} = 2$$



CHAPTER 2

**Example 11**

Simplify the expression $\log_3\left(\frac{x}{\sqrt{x}}\right)$ using logarithm laws.

**Solution**

$$\log_3\left(\frac{x}{\sqrt{x}}\right) = \log_3(x) - \frac{1}{2}\log_3(x)$$
$$= \frac{1}{2}\log_3(x)$$

**Example 12**

Simplify the expression $\left(\frac{x}{m^2}\right)(xm)^{-1}\left(\frac{m^3}{x^2}\right)$ where $x$, $m$ does not equal zero.

**Solution**

$$\left(\frac{x}{m^2}\right)(xm)^{-1}\left(\frac{m^3}{x^2}\right) = \left(\frac{x}{m^2}\right)\frac{1}{xm}\left(\frac{m^3}{x^2}\right)$$
$$= \frac{xm^3}{x^3m^3}$$
$$= \frac{1}{x^2}$$

## 2.3.2 Change of base rule

If $y = \log_a(b)$, then $y = \dfrac{\log_e(b)}{\log_e(a)} = \dfrac{\log_{10}(b)}{\log_{10}(a)}$.

This is helpful because CAS often has expressions that automatically revert to base $e$.

**Example 13**

Solve the equation $2^x = 5$.

**Solution**

$2^x = 5 \Leftrightarrow x = \log_2(5)$

CAS gives the answer as

$$x = \frac{\log_e(5)}{\log_e(2)}$$

```
solve(2^x=5, x)
                         {x = ln(5)/ln(2)}
log_2(5)
                               ln(5)/ln(2)
```

Both forms of the answer are correct.

**Example 14**

The expression $\dfrac{\log_m(n)}{\log_n(m)}$ where $m, n \in Z^+$ can be written as

**A** $\dfrac{2\log_e(n)}{2\log_e(m)}$ **B** $\dfrac{(\log_e(n))^2}{(\log_e(m))^2}$ **C** $\dfrac{(\log_e(m))^2}{(\log_e(n))^2}$ **D** $\dfrac{\log_{10}(n)}{\log_{10}(m)}$ **E** $\dfrac{\log_e(n^2)}{\log_e(m^2)}$

#### Solution

Using change of base rule:

If $y = \log_a(b)$, then $y = \dfrac{\log_e(b)}{\log_e(a)} = \dfrac{\log_{10}(b)}{\log_{10}(a)}$.

$$\text{We get}= \frac{\log_m(n)}{\log_n(m)} = \frac{\frac{\log_e(n)}{\log_e(m)}}{\frac{\log_e(m)}{\log_e(n)}}$$

$$= \frac{\log_e(n)}{\log_e(m)} \div \frac{\log_e(m)}{\log_e(n)}$$

$$= \frac{\log_e(n)}{\log_e(m)} \times \frac{\log_e(n)}{\log_e(m)}$$

$$= \frac{(\log_e(n))^2}{(\log_e(m))^2}$$

This can be confused with $\dfrac{2\log_e(n)}{2\log_e(m)}$ (option **A**), which is not the answer.

The correct answer is **B**.

## 2.4 Functions and their inverses

This section will cover

- functions and their inverses, including conditions for the existence of an inverse function, and use of inverse functions to solve equations using exponential, logarithmic, circular and power functions.

VCE Mathematics Study Design 2023–2027 p. 99, © VCAA 2022

A reminder of inverse functions is below.

#### Example 15

Consider $f: (-\infty, -2] \rightarrow R, f(x) = x^2 + 4x$.

**a** Sketch the graphs of $f$ and $f^{-1}$ on the same set of axes.

**b** Find the rule $f^{-1}$ and state the domain and range.

#### Solution

Sketch $y = f(x)$ and reflect it along the line $y = x$.

Swap coordinates, for example, $(-4, 0)$ on $f$ becomes $(0, -4)$ on $f^{-1}$.

Also $(-2, -4)$ on $f$ becomes $(-4, -2)$ on $f^{-1}$.

Interchange $x$ and $y$:

$x = y^2 + 4y$

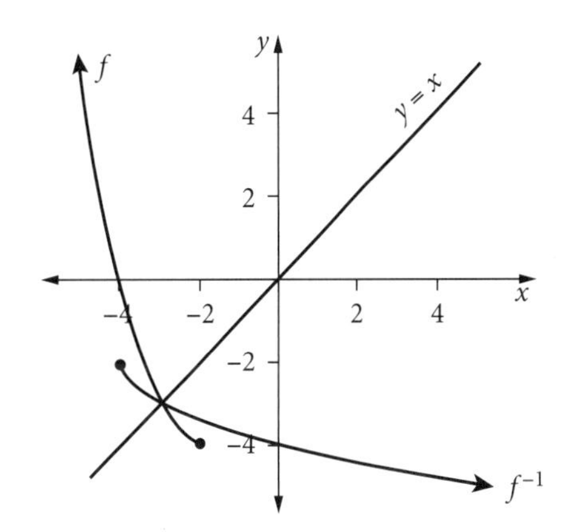

Make $y$ the subject (complete the square):

$$x = (y^2 + 4y + 4) - 4$$
$$x = (y + 2)^2 - 4$$
$$x + 4 = (y + 2)^2$$
$$y + 2 = \pm\sqrt{x + 4}$$
$$y = \pm\sqrt{x + 4} - 2$$

**Hint**
Remember to consider the + and – before you decide which branch it is.

$\text{dom}\, f = (-\infty, -2]$

$\text{ran}\, f = [-4, \infty)$

Consider the original graph to see, by the shape of the inverse, which branch to select.

By inspection, $f^{-1}(x) = -\sqrt{x+4} - 2$.

Hence,

$f^{-1}: [-4, \infty] \to R, f^{-1}(x) = -\sqrt{x+4} - 2$ $\quad$ $\text{ran}\, f^{-1} = (-\infty, -2]$

**note** $\text{dom}\, f^{-1} = \text{ran}\, f$ $\quad$ **note** $\text{ran}\, f^{-1} = \text{dom}\, f$

When asked for $f^{-1}$, give the rule and the domain.

When asked for the rule of $f^{-1}$, you only need to give the rule, not the domain.

## 2.4.1 Conditions for the existence of an inverse function

The inverse function $f^{-1}$ will only exist if the original function $f$ is one-to-one.

This is because the inverse of a many-to-one function becomes one-to-many, which is not a function.

---

### Example 16

The parabola $f(x) = 2(x - 1)^2 + 3$ will only have an inverse function if the domain is restricted so that $f$ is one-to-one.

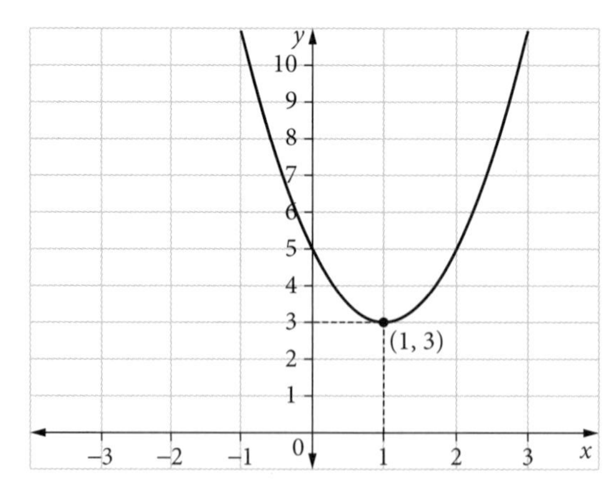

When asked for the largest domain for the existence of $f^{-1}$, either of the domains $[1, \infty)$ or $(-\infty, 1]$ are suitable.

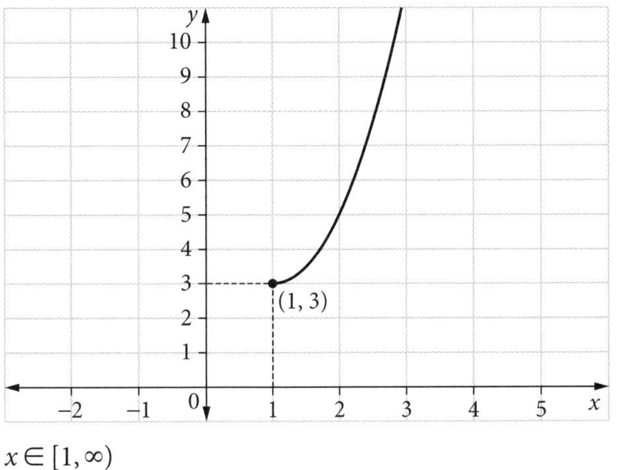


$x \in [1, \infty)$

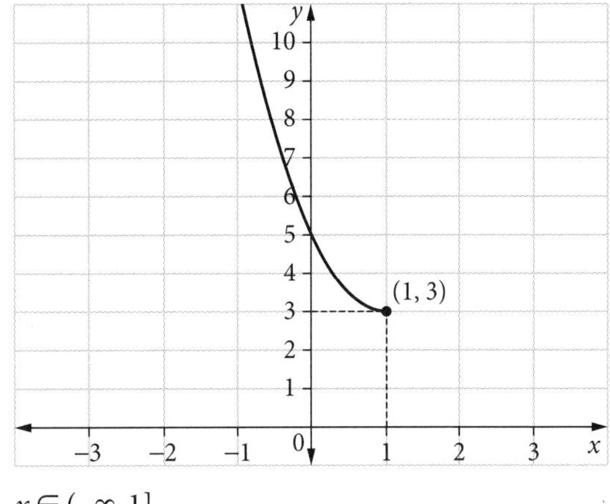


$x \in (-\infty, 1]$

## 2.5 Circular functions

### 2.5.1 Radians

When calculating sin, cos or tan of an angle, always check the mode of your calculator to see if it is in degree or radian mode.

Radian mode is always used in any calculus question and increasingly used in many other questions.

- To convert degrees to radians, multiply by $\frac{\pi}{180}$.
- To convert radians to degrees, multiply by $\frac{180}{\pi}$.

CAS can convert between degrees and radians in one step.

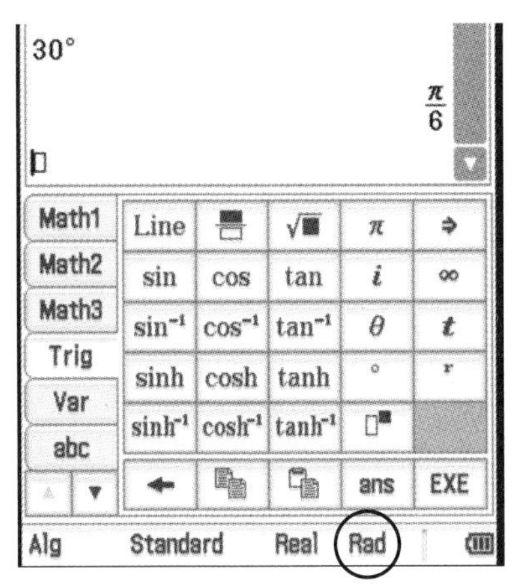


Converting from degrees to radians.

Converting from radians to degrees.

## 2.5.2 The unit circle

Most of the trigonometry information can be demonstrated using the unit circle.

- $\cos(\theta)$ is the length cut off on the $x$-axis
- $\sin(\theta)$ is the length cut off on the $y$-axis
- $\tan(\theta)$ is the length cut off on the tangent

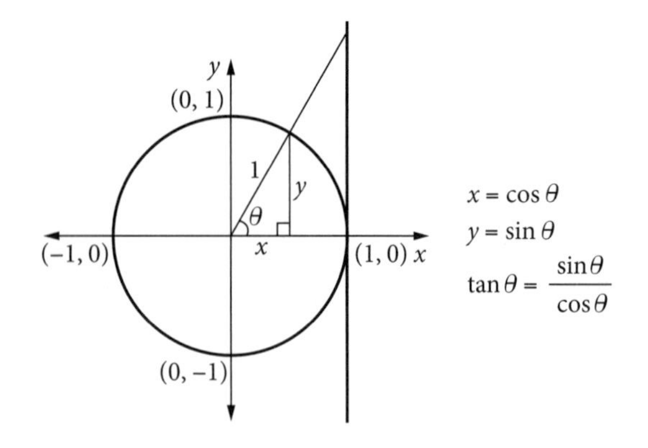


### 2.5.2.1 Symmetry of the unit circle

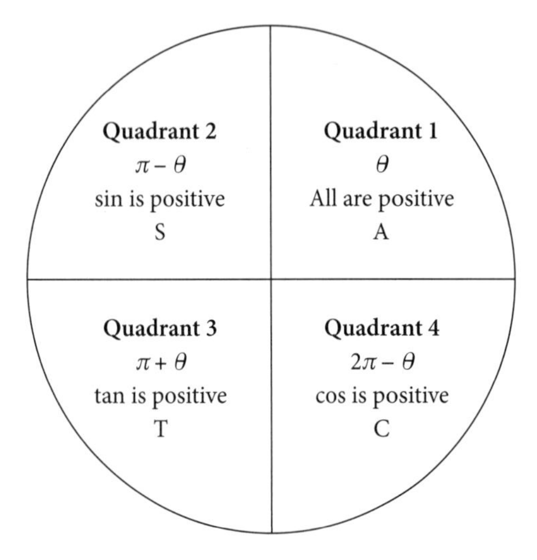


**Note:** $\theta$ is measured from the horizontal axis.

 anticlockwise is the positive direction.

Another way to think about the symmetry of the unit circle is the 'butterfly diagram'.

Unit circle showing the 'wings of a butterfly'.

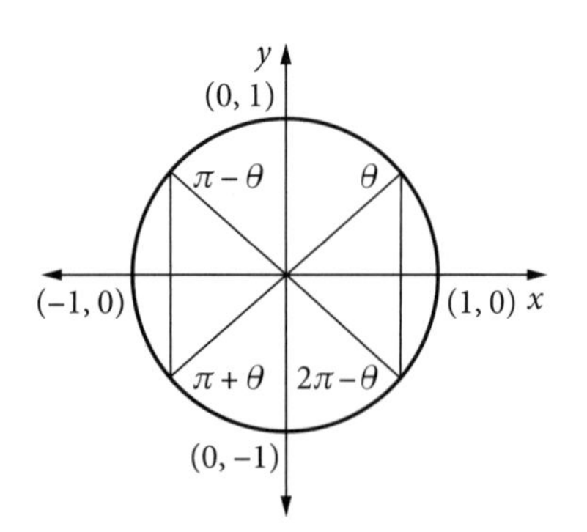


---

**Example 17**

Simplify $\sin(\pi - \theta)$.

**Solution**

$\pi - \theta$ takes us to the 2nd quadrant.

sin is positive in the 2nd quadrant, but is the same length as sin in the 1st quadrant.

$\sin(\pi - \theta) = \sin(\theta)$

**Example 18**

Simplify $\cos(\pi + \theta)$.

**Solution**

$\pi + \theta$ takes us to the 3rd quadrant.

cos is negative in the 3rd quadrant, but is the same length as cos in the 1st quadrant.

$\cos(\pi + \theta) = -\cos(\theta)$

## 2.5.3 Exact values on the axis

Exact values on the axes for $0, \frac{\pi}{2}, \pi, \frac{3\pi}{2}, 2\pi$ … can be calculated using the axis lengths on the unit circle.

**Example 19**

Evaluate $\sin\left(\frac{3\pi}{2}\right)$.

**Solution**

$\frac{3\pi}{2}$ position has coordinates $(0, -1)$.

$y = -1$ is the $y$ value after turning through $\frac{3\pi}{2}$ in an anti-clockwise direction.

$\sin\left(\frac{3\pi}{2}\right) = -1$

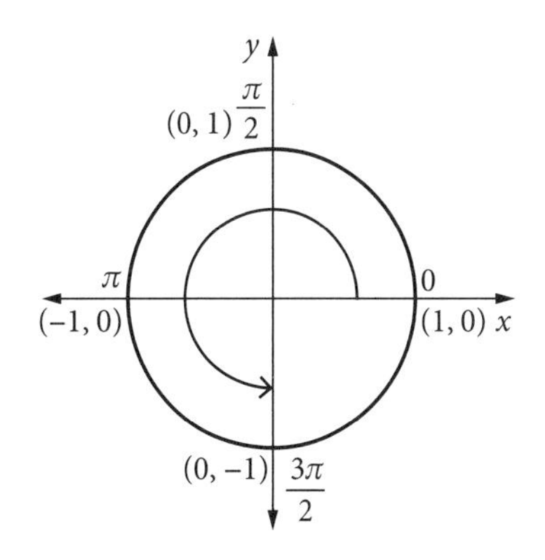


**Example 20**

Evaluate $\cos(-\pi)$.

**Solution**

$-\pi$ position has coordinates $(-1, 0)$.

$x = -1$ is the $x$ value after turning through $\pi$ in a clockwise direction.

This is the same as $\cos(\pi) = -1$, turning anti-clockwise through $\pi$.

$\cos(-\pi) = -1$

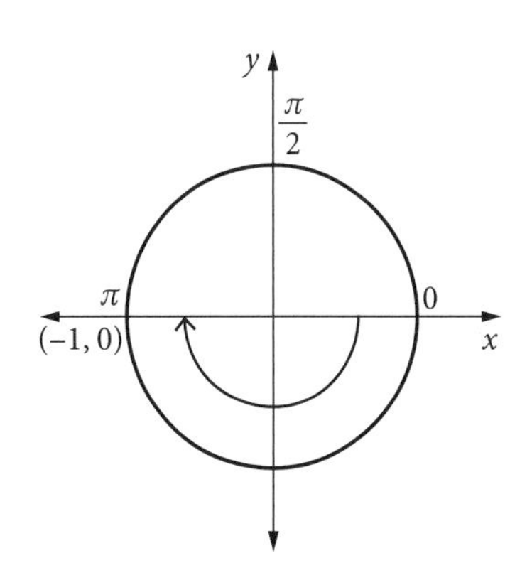

## 2.5.4 Exact values in the quadrants

Exact values in a quadrant for $\frac{\pi}{6}, \frac{\pi}{4}, \frac{\pi}{3}$ are given in the following table.

| $\theta$ | $\frac{\pi}{6}$ (30°) | $\frac{\pi}{4}$ (45°) | $\frac{\pi}{3}$ (60°) |
|---|---|---|---|
| **sin** | $\frac{1}{2}$ | $\frac{\sqrt{2}}{2}$ | $\frac{\sqrt{3}}{2}$ |
| **cos** | $\frac{\sqrt{3}}{2}$ | $\frac{\sqrt{2}}{2}$ | $\frac{1}{2}$ |
| **tan** | $\frac{\sqrt{3}}{3}$ | 1 | $\sqrt{3}$ |

A way to visualise the exact values and their symmetry using $\frac{\pi}{6}$, $\frac{\pi}{4}$, $\frac{\pi}{3}$ over the 4 quadrants in the unit circle is below. The labelled coordinates are the $(x, y)$ values.

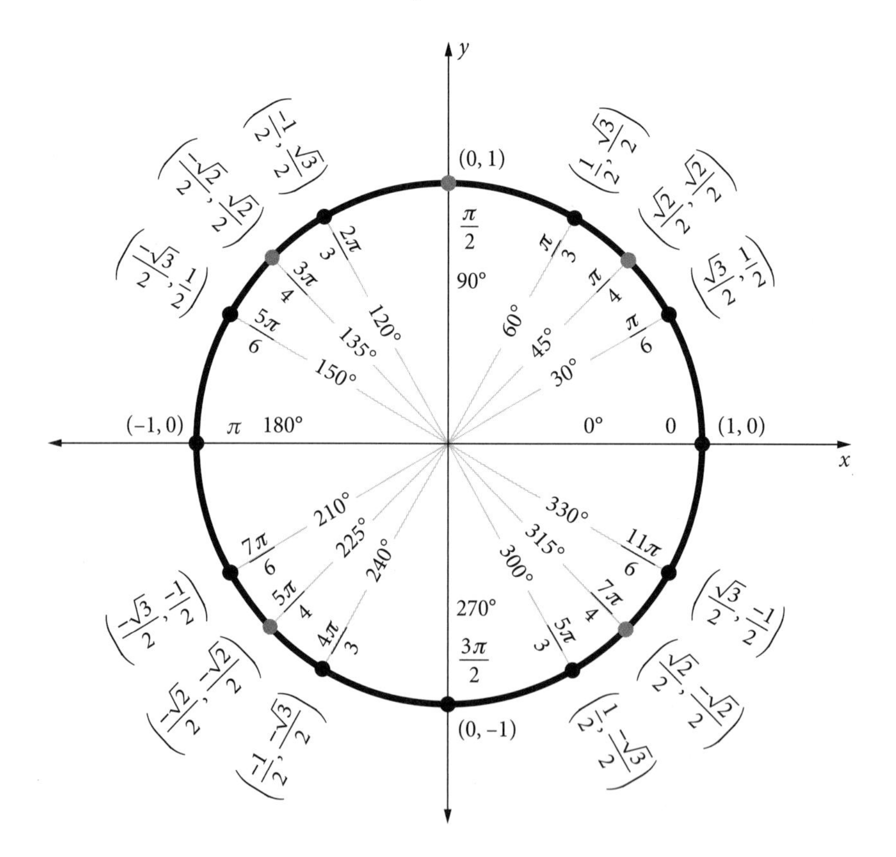

**Example 21**

Evaluate $\tan\left(\frac{11\pi}{6}\right)$.

**Solution**

Firstly, consider in which quadrant $\frac{11\pi}{6}$ lies.

An angle in the 4th quadrant.

The angles in the 4th quadrant are expressed as $2\pi - \theta$.

$$\tan\left(\frac{11\pi}{6}\right) = \tan\left(2\pi - \frac{\pi}{6}\right)$$

$$\tan\left(2\pi - \frac{\pi}{6}\right) = -\tan\left(\frac{\pi}{6}\right) \quad \text{(tan is negative in 4th quadrant)}$$

$$\tan\left(\frac{11\pi}{6}\right) = -\tan\left(\frac{\pi}{6}\right) = -\frac{1}{\sqrt{3}}$$

**Hint**

$\frac{1}{\sqrt{3}} = \frac{\sqrt{3}}{3}$. Both are equally correct.

## 2.5.5 Complementary circular functions

When we simplify or evaluate expressions that do not follow the symmetry of the unit circle, consider complementary pairs:

- sine ↔ cosine
- secant ↔ cosecant
- tangent ↔ cotangent

**Hint**

Note the prefix 'co' for 'complementary'. In Maths Methods we consider only sine, cosine, tangent and secant.

Consider the angle expressions $\frac{\pi}{2} \pm \theta, \frac{3\pi}{2} \pm \theta$.

$\sin\left(\frac{\pi}{2} + \theta\right)$ does not follow the 'butterfly' symmetry diagram.

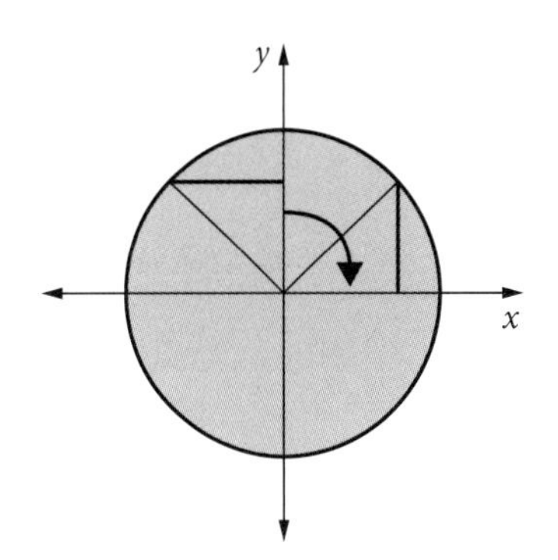


We can see that the cos length becomes the sin length and vice versa as we rotate the triangle to the 1st quadrant.

This is because for $\frac{\pi}{2} + \theta$, complementary pairs come into play.

The original sin length was positive, giving us the equivalence $\sin\left(\frac{\pi}{2} + \theta\right) = \cos(\theta)$.

**Example 22**

Simplify $\cos\left(\frac{\pi}{2} + \theta\right)$.

**Solution**

$\frac{\pi}{2} + \theta$ takes us to the 2nd quadrant where cos is negative.

cos becomes its complement, sin.

$$\cos\left(\frac{\pi}{2} + \theta\right) = -\sin(\theta)$$

## 2.5.6 Trigonometric identities

We can use properties of the unit circle to develop trigonometric identities.

Pythagoras' theorem $x^2 + y^2 = 1$ gives:

- $\cos^2(x) + \sin^2(x) = 1$
- $\cos^2(x) = 1 - \sin^2(x)$
- $\sin^2(x) = 1 - \cos^2(x)$
- $\tan(x) = \dfrac{\sin(x)}{\cos(x)}$

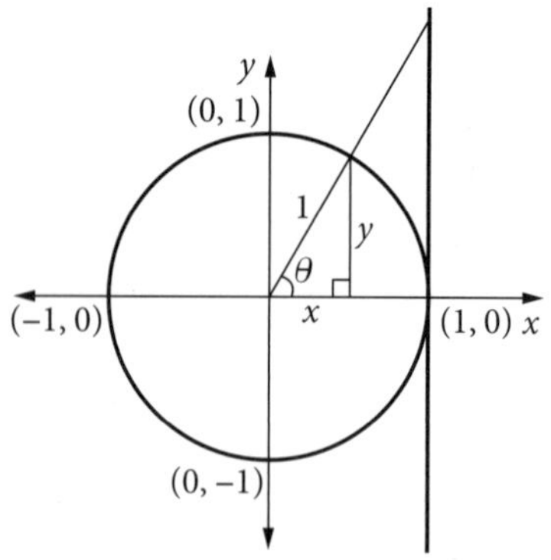


$x = \cos\theta$

$y = \sin\theta$

$\tan\theta = \dfrac{\sin\theta}{\cos\theta}$

---

### Example 23

Given $\cos(\theta) = -\frac{1}{3}$ where $\theta \in \left[\frac{\pi}{2}, \pi\right]$, find

**a** $\sin(\theta)$

**b** $\tan(\theta)$

### Solution

#### Method 1: Using trig. identities

**a**

$$\cos^2(\theta) + \sin^2(\theta) = 1$$

$$\left(-\frac{1}{3}\right)^2 + \sin^2(\theta) = 1$$

$$\therefore \sin^2(\theta) = 1 - \frac{1}{9}$$

$$\therefore \sin(\theta) = \pm\sqrt{\frac{8}{9}} = \pm\frac{2\sqrt{2}}{3}$$

where $\theta \in \left[\frac{\pi}{2}, \pi\right]$ (2nd quadrant)

$$\sin(\theta) = \frac{2\sqrt{2}}{3}$$

**b**

$$\tan(\theta) = \frac{\sin(\theta)}{\cos(\theta)}$$

$$\therefore \tan(\theta) = \frac{\frac{2\sqrt{2}}{3}}{-\frac{1}{3}}$$

where $\theta \in \left[\frac{\pi}{2}, \pi\right]$ (2nd quadrant)

$$\tan(\theta) = -2\sqrt{2}$$

#### Method 2: Using right-angled triangles

Given $\cos(\theta) = -\frac{1}{3}$ where $\theta \in \left[\frac{\pi}{2}, \pi\right]$:

**a** In a triangle, use $\cos(\theta) = \frac{1}{3}$

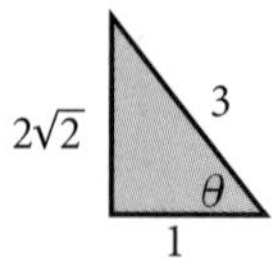


$$\therefore \sin(\theta) = \frac{O}{H} = \frac{2\sqrt{2}}{3} \text{ where } \theta \in \left[\frac{\pi}{2}, \pi\right].$$

**b**

$$\tan(\theta) = \frac{O}{A} = \frac{2\sqrt{2}}{1}$$

$$\therefore \tan(\theta) = -2\sqrt{2} \text{ where } \theta \in \left[\frac{\pi}{2}, \pi\right].$$

# 2.6 Solving equations using inverse functions

We can solve equations using inverses involving exponential, logarithmic, circular and power functions.

It was previously shown that exponential and logarithmic functions are inverses of each other and that it is possible to create and graph an inverse function $f^{-1}$ from the original function $f$.

## 2.6.1 Exponential functions

Solve for $x$ in the equation $e^{x+1} = 3$ by considering its inverse function, the logarithmic function.

$e^{x+1} = 3 \Leftrightarrow x + 1 = \log_e(3)$

$x = \log_e(3) - 1$ is the solution to the equation $e^{x+1} = 3$.

## 2.6.2 Logarithmic functions

Solve for $x$ in the equation $\log_{10}(x - 1) = 4$ by considering its inverse function, the exponential function.

$\log_{10}(x - 1) = 4 \Leftrightarrow x - 1 = 10^4$

$x = 10^4 + 1 = 10\,001$ is the solution to the equation $\log_{10}(x - 1) = 4$.

---

**Example 24**

Solve for $x$, $\log_x(25) = 2$.

**Solution**

$$\log_x(25) = 2 \Leftrightarrow x^2 = 25$$
$$x = \pm\sqrt{25} = \pm 5$$

Answer: $x = 5$ (base cannot be negative)

**Hint**
In log equations, always test the solutions in the original log equation to determine which solutions exist.

---

**Example 25**

Solve $0.3^x = 4$.

**Hint**
You cannot make 0.3 and 4 the same base, so take $\log_{10}()$ or $\log_e()$ of both sides.

**Solution**

$$0.3^x = 4$$
$$\log_e(0.3^x) = \log_e(4)$$
$$\therefore x\log_e(0.3) = \log_e(4)$$
$$x = \frac{\log_e(4)}{\log_e(0.3)}$$
$$= -1.151 \quad \text{(to three decimal places)}$$

```
simplify(solve(0.3^x=4,x))
                    {x=-2·ln(2)/ln(10/3)}
solve(0.3^x=4,x)
                    {x=-1.151433285}
```

**Hint**
CAS answers can vary widely when solving and simplifying logarithm equations and expressions.

### 2.6.2.1 Solving exponential equations

Consider these options:

1 Make sure both sides of the equation are the same base.

2 Simplify using index laws.

3 If bases are the same, equate indices.

**Example 26**

Solve for $x$.

$27^{x-2} \times 3^x = 9$

**Solution**

$$\begin{aligned} 27^{x-2} \times 3^x &= 9 \\ (3^3)^{(x-2)} \times 3^x &= 3^2 && \text{(make all bases the same)} \\ 3^{3x-6} \times 3^x &= 3^2 && \text{(use index laws)} \\ 3^{4x-6} &= 3^2 \\ 4x - 6 &= 2 && \text{(equate indices)} \\ \therefore x &= 2 \end{aligned}$$

### 2.6.2.2 Solving logarithmic equations

Consider these options:

1 Simplify using log laws.

2 log = log — If you have only log terms on both sides, equate the expressions.

3 log = number — If you have all log terms on one side and a number on the other, write an equivalent exponential statement.

**Example 27**

**Method: log = log**

Solve for $x$.

$2\log_3(x) - \log_3(x-1) = \log_3(x+2)$

**Solution**

Simplify. $\log_3(x^2) - \log_3(x-1) = \log_3(x+2)$

Write as log = log. $\log_3\left(\frac{x^2}{x-1}\right) = \log_3(x+2)$

Equate expressions. $\frac{x^2}{x-1} = x+2$

$$\begin{aligned} x^2 &= (x+2)(x-1) \\ x^2 &= x^2 + x - 2 \\ 0 &= x - 2 \end{aligned}$$

Answer: $x = 2$

**Hint**

Check your answer(s) in the original equation, making sure it doesn't create log (–ve) or log (0).

**Example 28**

**Method: log = number**

Solve for $x$.

$2\log_3(x) - \log_3(x-1) = 2$

**Solution**

Simplify. $\log_3(x^2) - \log_3(x-1) = 2$

Write as log = number. $\log_3\left(\dfrac{x^2}{x-1}\right) = 2$

Write as equivalent exponential statement. $\Leftrightarrow 3^2 = \dfrac{x^2}{x-1}$

$$x^2 = 9x - 9$$

$$\therefore x^2 - 9x + 9 = 0$$

$$x = \frac{9 \pm 3\sqrt{5}}{2} \quad \text{(both answers exist)}$$

**Hint**

Check your answer(s) in the original equation.

## 2.6.3 Solving equations of the quadratic type

Solve for $x$, $e^{2x} - 4e^x + 3 = 0$.

These questions require substitution and recognising that you have to factorise a quadratic trinomial to get the answer.

Let a 'dummy variable' $A = e^x$.

$e^{2x} - 4e^x + 3 = 0$ becomes $A^2 - 4A + 3 = 0$.

Factorising gives $(A-3)(A-1) = 0$.

$A = 3$ or $A = 1$

Substituting back, $A = e^x$:

$e^x = 3$ or $e^x = 1$

$e^x = 3 \Leftrightarrow x = \log_e(3)$

$e^x = 1 \Leftrightarrow x = \log_e(1) = 0$

$x = \log_e(3)$ or $x = 0$

**Hint**

In this case, $x = 0$ can be considered a 'trivial' solution.

Learn to recognise these types of equations, by noticing the patterns in the powers.

Note that $3e^{-x} + e^x - 4 = 0$ is the same question as above since $\dfrac{3}{e^x} + e^x - 4 = 0$.

$\therefore 3 + e^{2x} - 4e^x = 0$

Rearranging, $e^{2x} - 4e^x + 3 = 0$ is the same equation as before.

## 2.6.4 Application questions involving exponential functions

The general form of the equation is $N = N_0e^{kt}$, where $N$ is the amount of a quantity at time $t$.

Both $N_0$ and $k$ are real constants. $N_0$ stands for the initial amount when $t = 0$ and $k$ is the growth or decay constant. The usual method for these sorts of questions is to use the information provided to find the constants $N_0$ and $k$ first, often using simultaneous equations.

**Example 29**

The mass of a particle of bacteria decays following the equation $M = M_0e^{-kt}$, where $M$ mgs and $t$ weeks.

The initial mass is 20 mg and after 10 weeks the mass is 5 mg.

Find the values of $M_0$ and $k$.

**Solution**

Substitute $M_0 = 20$: $M = 20e^{-kt}$

Substitute $t = 10, M = 5 \Leftrightarrow 5 = 20e^{-10k}$

Solve for $k$. $5 = 20e^{-10k}$

$$\Rightarrow k = \frac{1}{5}\log_e(2)$$

$M_0 = 20,\ k = \frac{1}{5}\log_e(2) \approx 0.139$

```
solve(5=20·e^(-10·k), k)
                         {k=ln(2)/5}
solve(5=20·e^(-10·k), k)
                         {k=0.1386294361}
```

## 2.6.5 Solving circular functions equations

To solve trigonometric equations, use your knowledge of circular functions.

Consider the function $f(x) = 2\cos(2x) - \sqrt{3}$.

Solve for $x$ in the equation $2\cos(2x) - \sqrt{3} = 0$, where $x \in [0, 2\pi]$.

$$2\cos(2x) - \sqrt{3} = 0$$

$$\cos(2x) = \frac{\sqrt{3}}{2}$$

$$2x = \cos^{-1}\left(\frac{\sqrt{3}}{2}\right)$$

$$x = \frac{1}{2}\cos^{-1}\left(\frac{\sqrt{3}}{2}\right)$$

$$= \frac{1}{2} \times \frac{\pi}{6}$$

**Hint**
There will be more solutions in the domain $[0, 2\pi]$

Basic or **reference angle** is $\frac{\pi}{6}$.

In a **unit circle**, tick the quadrants where cos is positive.

The answer can be found using the symmetry of the unit circle.

When writing down possible solutions, go around the circle twice (2 times $2\pi$) to allow for dividing by 2 in the last step.

Or consider $x \in [0, 2\pi]$ becomes $2x \in [0, 4\pi]$.

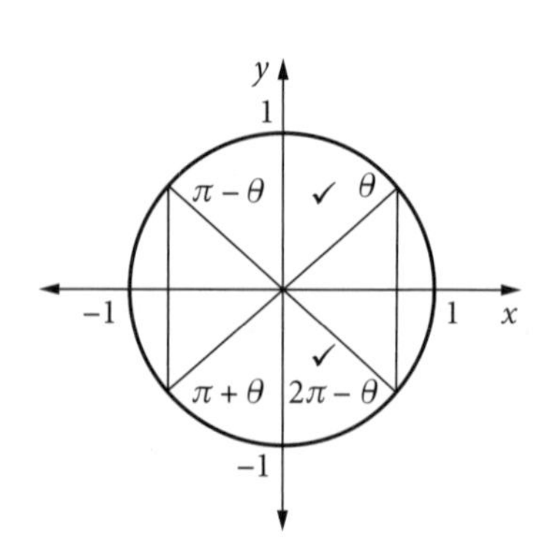


$$2x = \frac{\pi}{6}, 2\pi - \frac{\pi}{6}, 2\pi + \frac{\pi}{6}, 4\pi - \frac{\pi}{6}$$

$$2x = \frac{\pi}{6}, \frac{11\pi}{6}, \frac{13\pi}{6}, \frac{23\pi}{6}$$

$$x = \frac{\pi}{12}, \frac{11\pi}{12}, \frac{13\pi}{12}, \frac{23\pi}{12}$$

### Example 30

Solve for $x$ in the equation $2\sin(x) = 1$, where $x \in [-\pi, \pi]$.

### Solution

$$\sin(x) = \frac{1}{2}$$

$$\therefore x = \sin^{-1}\left(\frac{1}{2}\right)$$

$$= \frac{\pi}{6}$$

Basic or reference angle $= \frac{\pi}{6}$.

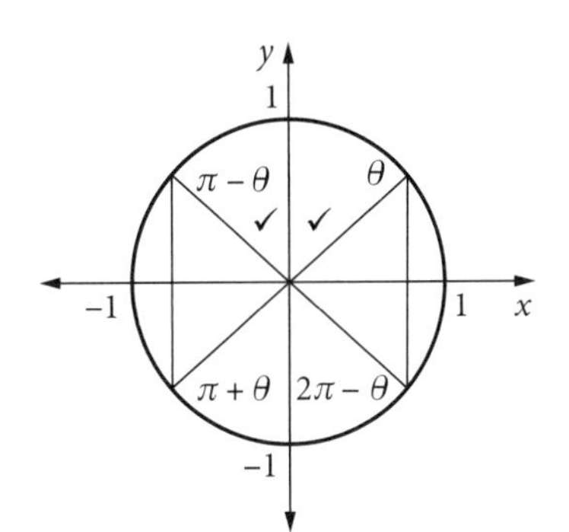


Tick quadrants where sin is positive.

Two solutions found for 0 to $\pi$ and 0 to $-\pi$.

$$x = \frac{\pi}{6}, \pi - \frac{\pi}{6}$$

Answer: $x = \frac{\pi}{6}, \frac{5\pi}{6}$

### Example 31

Solve for $x$ in the equation $\sqrt{2}\cos(2x) + 1 = 0, x \in [0, 2\pi]$.

### Solution

$$\cos(2x) = -\frac{1}{\sqrt{2}}$$

$$\therefore 2x = \cos^{-1}\left(\frac{1}{\sqrt{2}}\right)$$

$$x = \frac{1}{2} \times \frac{\pi}{4}$$

Basic or reference angle $= \frac{\pi}{4}$.

Tick the quadrants where cos is negative.

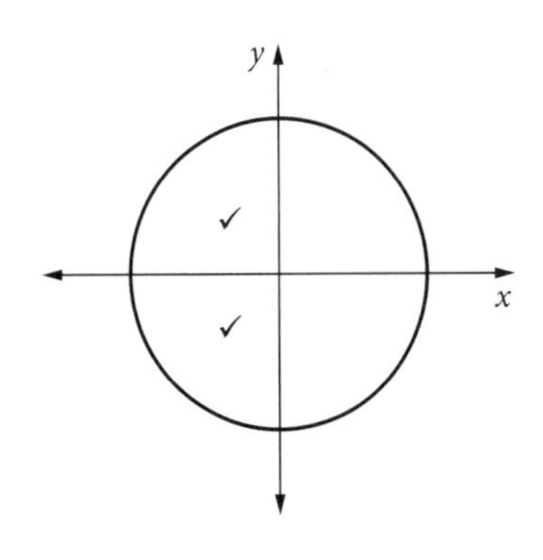


$$2x = \pi - \frac{\pi}{4}, \pi + \frac{\pi}{4}, 2\pi + \left(\pi - \frac{\pi}{4}\right), 2\pi + \left(\pi + \frac{\pi}{4}\right)$$

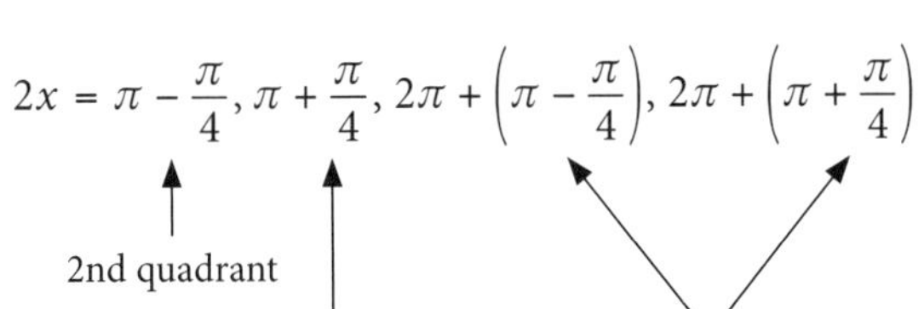


$$2x = \frac{3\pi}{4}, \frac{5\pi}{4}, \frac{11\pi}{4}, \frac{13\pi}{4}$$

$$x = \frac{3\pi}{8}, \frac{5\pi}{8}, \frac{11\pi}{8}, \frac{13\pi}{8}$$

**Hint**

Check the domain provided to ensure that all of your solutions are in the required domain.

CAS can find answers in one step. For particular solutions, we need to define the domain.

```
solve(√2·cos(2·x)+1=0 | 0≤x≤2·π, x)
```
$$\left\{x=\frac{3\cdot\pi}{8}, x=\frac{5\cdot\pi}{8}, x=\frac{11\cdot\pi}{8}, x=\frac{13\cdot\pi}{8}\right\}$$

If you do not define the domain, the answer will have constants in it, shown in different ways on different CAS.

```
solve(√2·cos(2·x)+1=0, x)
```
$$\left\{x=\pi\cdot\text{constn}(1)-\frac{3\cdot\pi}{8}, x=\pi\cdot\text{constn}(2)+\frac{3\cdot\pi}{8}\right\}$$

This is the general solution of a trig. equation and is written as

$$x = \pm\frac{3\pi}{8} + \pi k,\ k \in Z$$

Particular answers can be found by substituting integer values of $k$.

- $k = 0, x = \pm\frac{3\pi}{8}$
- $k = 1, x = \pm\frac{3\pi}{8} + \pi = \frac{11\pi}{8}, \frac{5\pi}{8}$
- $k = 2, x = \pm\frac{3\pi}{8} + 2\pi = \frac{19\pi}{8}, \frac{13\pi}{8}$

**Hint**

In general for sin and cos, the general solution includes $+2\pi k, k \in Z$.

For tan, the general solution includes $+\pi k, k \in Z$.

Here, general solution includes $+\frac{2\pi k}{2} = +\pi k$

## 2.6.6 Solving power function equations

Consider the function $f(x) = 3(x + 1)^4$.

Solve for $x$ in the equation $3(x + 1)^4 = 16$.

Rearrange to get $(x + 1)^4 = \frac{16}{3}$

$$\therefore x + 1 = \pm\left(\frac{16}{3}\right)^{\frac{1}{4}}$$

Thus, $x + 1 = \pm\frac{2}{3^{\frac{1}{4}}}$

Answer: $x = \pm\frac{2}{3^{\frac{1}{4}}} - 1$

```
solve(3·(x+1)^4=16, x)
```
$$\left\{x=\frac{-2}{3^{\frac{1}{4}}}-1, x=\frac{2}{3^{\frac{1}{4}}}-1\right\}$$

## 2.7 Composite functions

This section will cover

- composition of functions, where $f$ composite $g$, $f \circ g$ is defined by $(f \circ g)(x) = f(g(x))$ given $r_g \subseteq d_f$.

VCE Mathematics Study Design 2023–2027 p. 99, © VCAA 2022

The composite function exists only if the range of the inner function is contained in, or equal to, the domain of the outer function. $f(g(x))$ can also be written as $f \circ g$ and, in turn, $g(f(x))$ can be written as $g \circ f$.

### Example 32

Consider $f(x) = x^4$ and $g(x) = e^{x+1}$.

State whether the function $f(g(x))$ exists and, if it exists, find $f(g(x))$.

### Solution

Test: Is the range of the inner contained in, or equal to, the domain of the outer?

Test $r_g \subseteq d_f$.

$f(x) = x^4$

dom $f = R$, ran $f = [0, \infty)$

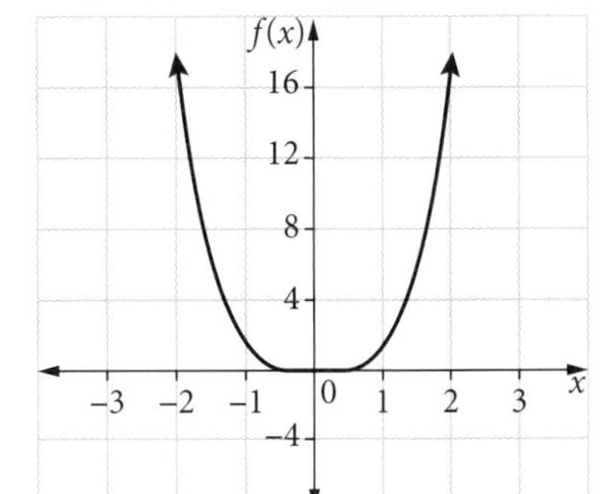


$g(x) = e^{x+1}$

dom $g = R$, ran $g = (0, \infty)$

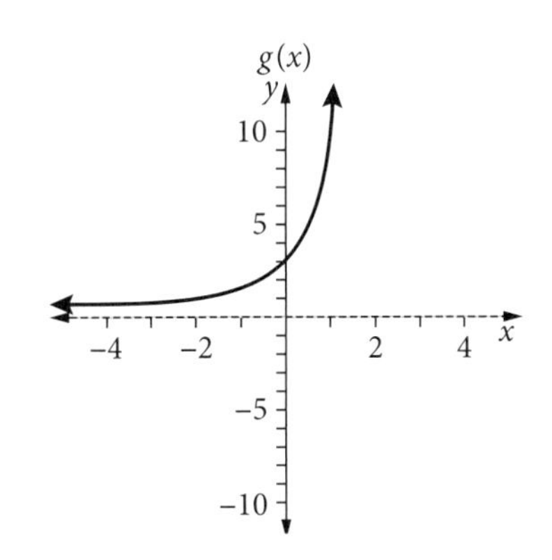


Test if $r_g \subseteq d_f$:

$(0, \infty) \subseteq R$, so $f(g(x))$ exists.

domain of the composite function = the domain of the inner function

dom $f(g(x))$ = dom $g = R$

Rule of $f(g(x)) = (e^{x+1})^4 = e^{4x+4}$

The function is defined by

$f(g(x))\colon R \to R$, where $f(g(x)) = e^{4x+4}$

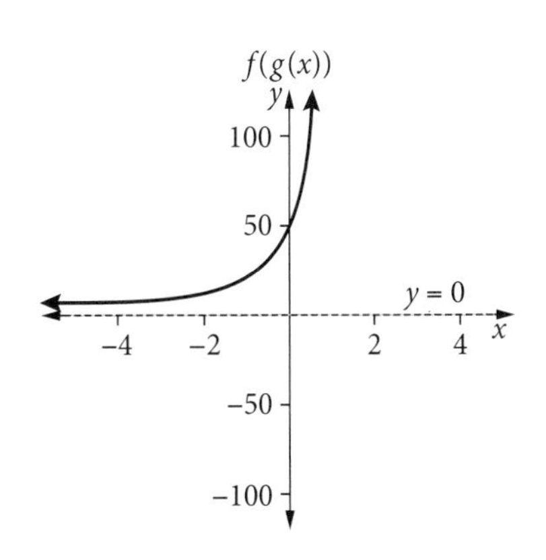


9780170465366

CHAPTER 2

### Example 33

Consider $f(x) = x^2$ and $g(x) = \log_e(x + 1)$.

State whether the function $g(f(x))$ exists and, if it exists, find $g(f(x))$.

### Solution

Test: Is the range of the inner function contained in, or equal to, the domain of the outer function?

Test $r_f \subseteq d_g$:

$f(x) = x^2$

dom $f = R$, ran $f = [0, \infty)$

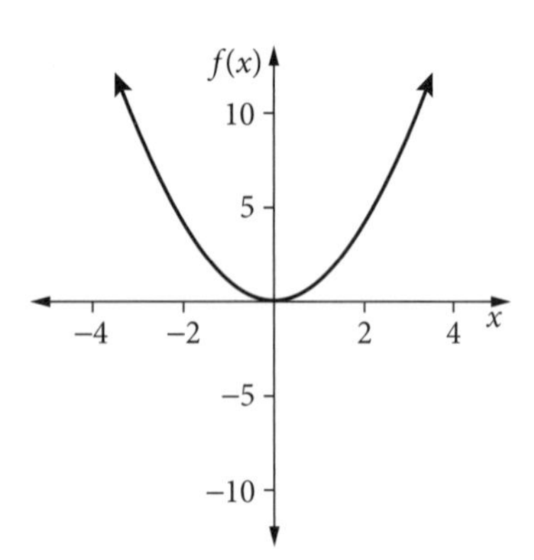


$g(x) = \log_e(x + 1)$

dom $g = (-1, \infty)$, ran $g = R$

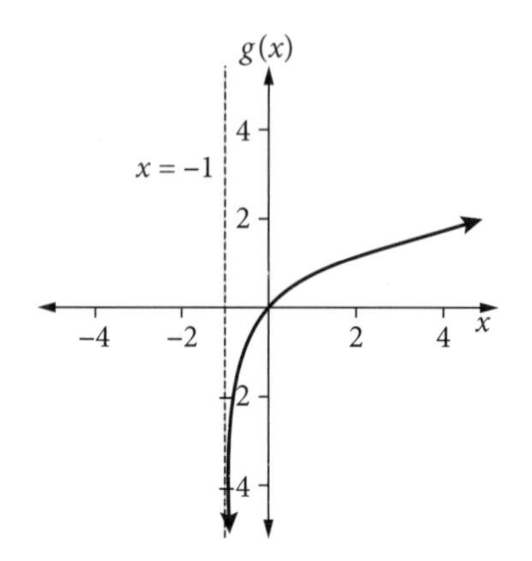


Test if $r_f \subseteq d_g$:

$[0, \infty) \subseteq (-1, \infty)$ so $g(f(x))$ exists.

domain of the composite function = the domain of the inner function

dom $g(f(x))$ = dom $f = R$

Rule of $g(f(x)) = \log_e(x^2 + 1)$

The function is defined by

$g(f(x)): R \to R$, where $g(f(x)) = \log_e(x^2 + 1)$

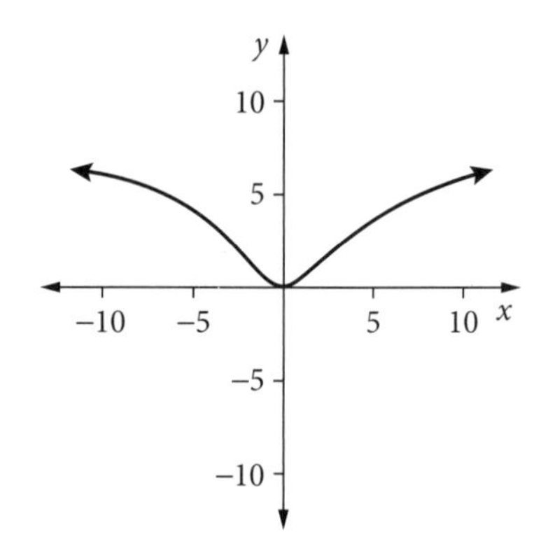

### 2.7.1.1 Restricting a domain for a composite function to exist

Consider the case when the composite function does not exist.

---

**Example 34**

Consider $f(x) = \frac{1}{x}$ and $g(x) = \sqrt{x}$. State whether the function $f(g(x))$ exists.

If $f(g(x))$ doesn't exist, restrict $r_g$ so that it does exist and define the restricted $f(g(x))$.

**Solution**

Test: Is the range of the inner contained in, or equal to, the domain of the outer?

Test $r_g \subseteq d_f$:

$f(x) = \frac{1}{x}$

dom $f = R \setminus \{0\}$, ran $f = R \setminus \{0\}$

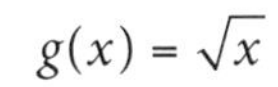

$g(x) = \sqrt{x}$

dom $g = [0, \infty)$, ran $g = [0, \infty)$

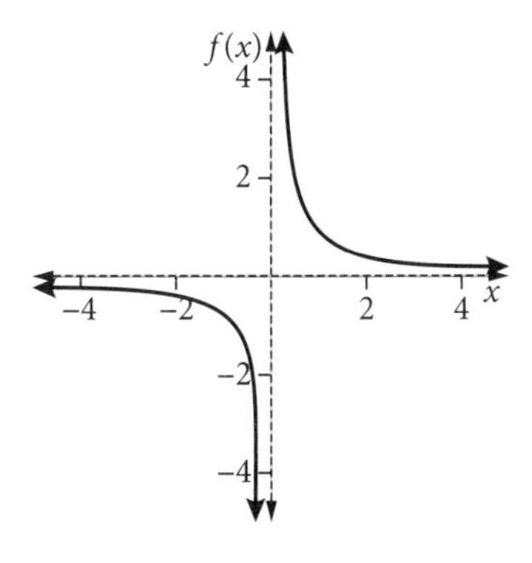


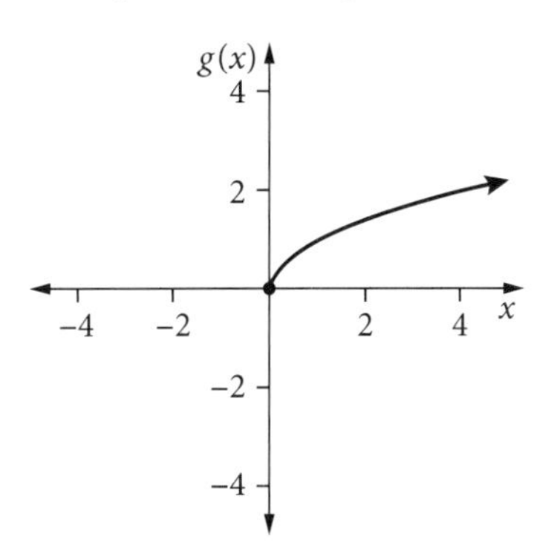


Test if $r_g \subseteq d_f$:

$[0, \infty) \nsubseteq R \setminus \{0\}$, so $f(g(x))$ does not exist.

Let's restrict $r_g$ so that it does exist.

Change $[0, \infty)$ to $(0, \infty)$ = ran $g$ gives dom $g = (0, \infty)$.

This gives $(0, \infty) \subseteq R \setminus \{0\}$, so $f(g(x))$ does exist.

domain of the composite function = the domain of the inner function

In this case: dom $f(g(x))$ = restricted dom $g = (0, \infty)$

Rule of $f(g(x)) = \frac{1}{\sqrt{x}}$

After the restriction, the function is defined by $f(g(x))$: $(0, \infty) \rightarrow R$, where $f(g(x)) = \frac{1}{\sqrt{x}}$.

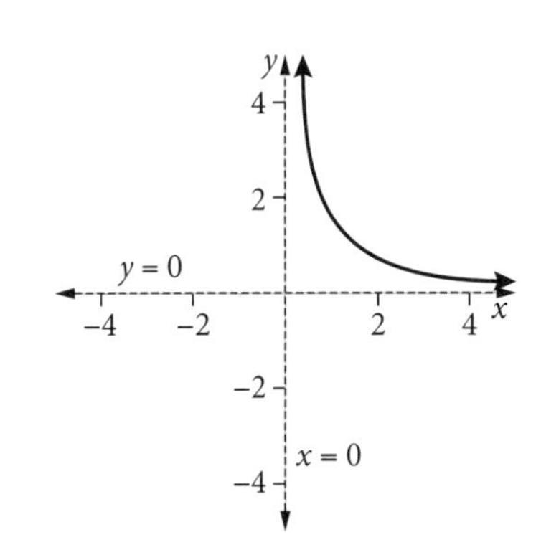

## 2.8 Graphical, numerical and algebraic methods

This section will cover

- solution of equations of the form $f(x) = g(x)$ over a specified interval, where $f$ and $g$ are functions of the type specified in the 'Functions, relations and graphs' area of study, by graphical, numerical and algebraic methods, as applicable
- solution of literal equations and general solution of equations involving a single parameter.

VCE Mathematics Study Design 2023–2027 p. 99, © VCAA 2022

### 2.8.1.1 Graphical methods

Solving $2(x-1)^2 - 5 = 0$ finds the $x$-intercepts of graph $y = 2(x-1)^2 - 5$.

$x = -0.58, 2.58$ to two decimal places.

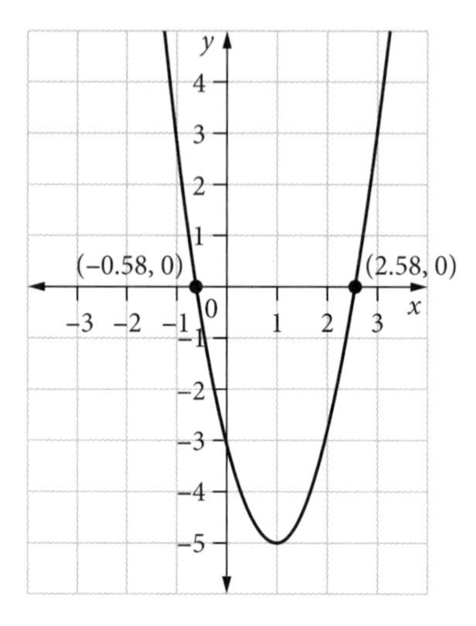


Solving $3(x+2)^4 = 48$ finds the intersections between $y = 48$ and $y = 3(x+2)^4$.

$(-4, 48)$ and $(0, 48)$

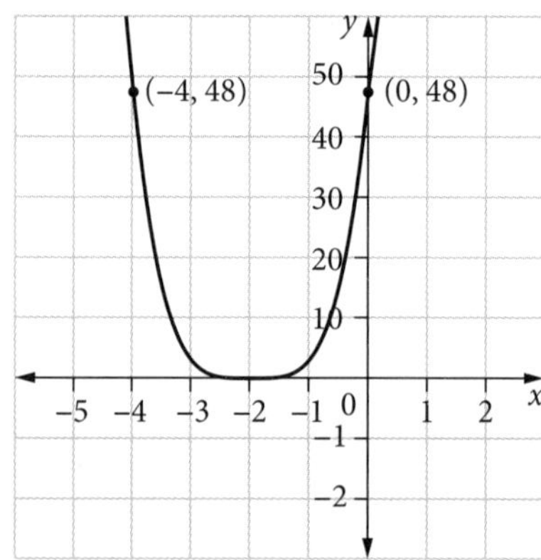


Solving $\log_{10}(x-5) = -2$ finds the intersection between $y = -2$ and $y = \log_{10}(x-5)$.

$(5.01, -2)$

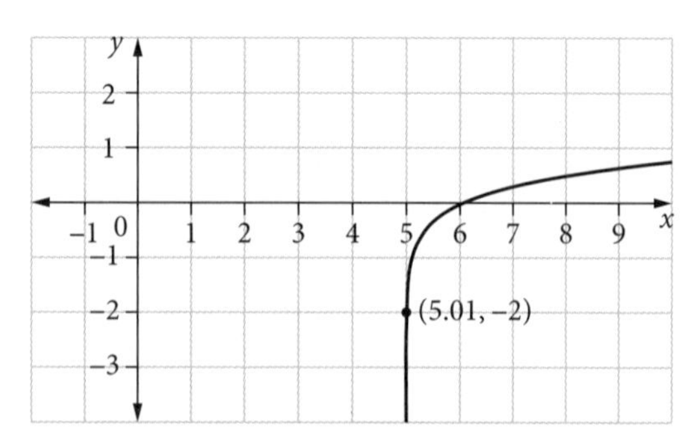


$\sqrt{3}\tan(2x) = -1, x \in [0, \pi]$ finds the intersections between $y = -1$ and $y = \sqrt{3}\tan(2x)$.

$(1.31, -1)$ and $(2.88, -1)$ to two decimal places

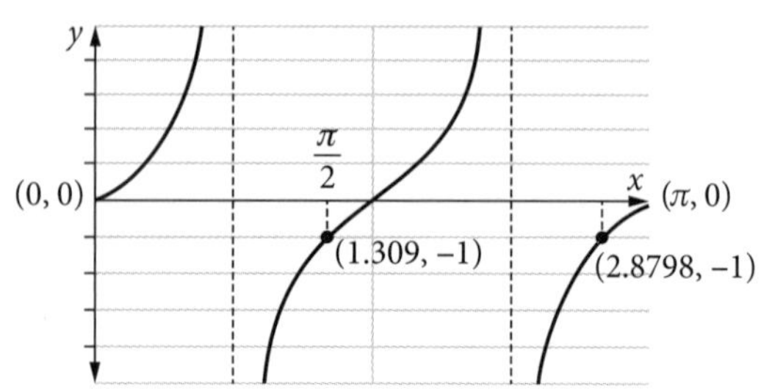

### 2.8.1.2 Numerical methods

When solving equations, assume that the default is an exact answer even if the question does not state this.

However, sometimes exact methods may not apply, or you may be expected to use a graphical or numerical approach when solving the equation to a required number of decimal places.

---

**Example 35**

Find the coordinates of the points of intersection of the graphs $y = 3\sin(2x)$ and $y = e^{-x} + 1$ for $x \in [0, 5]$, correct to four decimal places.

**Solution**

There is no algebraic method for exactly solving $3\sin(2x) = e^{-x} + 1$.

Use the **intersect** facility on CAS using the window $x \in [0, 5]$.

By choosing the window for $x \in [0, 5]$, you can see that there are four solutions.

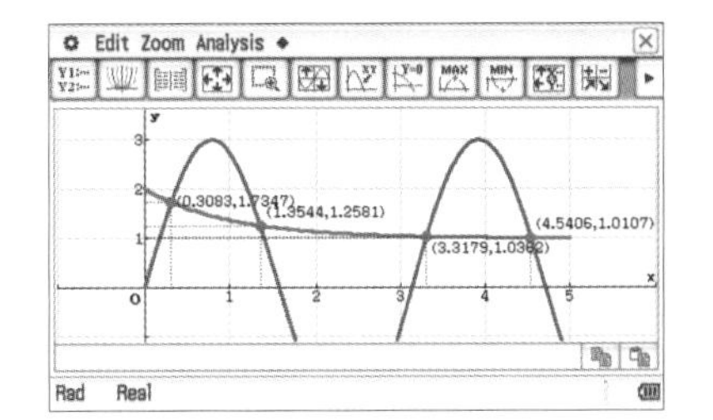


The four solutions are:

$(0.3083, 1.7347), (1.3544, 1.2581), (3.3179, 1.0362), (4.5406, 1.0107)$

```
solve(3·sin(2·x)=e^-x+1|0≤x≤5,x)
{x=0.3082824177,x=1.35442326,x=3.317930073,▶
```

```
solve(3·sin(2·x)=e^-x+1|0≤x≤5,x)
◀7,x=1.35442326,x=3.317930073,x=4.540583553}
```

The $x$ values will be given on CAS. Find the corresponding $y$ values by substituting into either $y = 3\sin(2x)$ or $y = e^{-x} + 1$.

### 2.8.1.3 Algebraic methods

Solve equations such as:

$2(x-1)^2 - 5 = 0$

$3(x+2)^4 = 48$

$\log_{10}(x-5) = -2$

$\sqrt{3}\tan(2x) = -1, x \in [0, \pi]$

Ensure that you can do these equations both by hand and using CAS.

$$2(x-1)^2 - 5 = 0$$
$$(x-1)^2 = \frac{5}{2}$$
$$x - 1 = \pm\sqrt{\frac{5}{2}}$$
$$x = 1 \pm \sqrt{\frac{5}{2}}$$
$$= 1 \pm \frac{\sqrt{10}}{2}$$

```
solve(2·(x-1)^2-5=0,x)
{x=-√10/2+1, x=√10/2+1}
```

$$3(x+2)^4 = 48$$
$$(x+2)^4 = 16$$
$$x + 2 = \pm 2$$
$$x = 0, x = -4$$

```
solve(3(x+2)^4=48,x)
{x=-4,x=0}
```

$\log_{10}(x-5) = -2$

$\Leftrightarrow 10^{-2} = x - 5$

$\frac{1}{100} = x - 5$

$x = 5 + \frac{1}{100}$

$x = \frac{501}{100}$

```
solve(log10(x-5)=-2, x)
                                {x=501/100}
```

$\sqrt{3}\tan(2x) = -1, x \in [0, \pi]$

$\tan(2x) = -\frac{1}{\sqrt{3}}$ (solutions in 2nd and 4th quadrants)

Basic angle $= \tan^{-1}\left(\frac{1}{\sqrt{3}}\right) = \frac{\pi}{6}$

$2x = \pi - \frac{\pi}{6}, 2\pi - \frac{\pi}{6}$

$2x = \frac{5\pi}{6}, \frac{11\pi}{6}$

$\therefore x = \frac{5\pi}{12}, \frac{11\pi}{12}$

```
solve(√3·tan(2·x)=-1 | 0≤x≤π, x)
                                {x=5·π/12, x=11·π/12}
```

## 2.8.2 Literal equations

**Literal equations** are equations involving letters of the alphabet. Equations that include variables, where each variable signifies a meaning/quantity, are called literal equations. Some common examples of literal equations are formulas in geometry like the area of a square: $A = s^2$, or the well-known Mass-Energy equation: $E = mc^2$. There are 3 variables in this literal equation, namely $E$, $m$, $c$, with each variable signifying a physical quantity.

```
solve(E=m·c^2, m)
                                {m=E/c^2}
```

For example:

Solve for a in the equation for the area of a trapezium

$A = \frac{1}{2}h(a+b)$

$2A = h(a+b)$

$\frac{2A}{h} = a + b$

$a = \frac{2A}{h} - b$

# 2.9 Simultaneous linear equations

This section will cover

- solution of simple systems of simultaneous linear equations, including consideration of cases where no solution or an infinite number of possible solutions exist (geometric interpretation only required for two equations in two variables).

The solution to simultaneous equations allows us to analyse how two straight lines intersect.

There are three possibilities:

- One unique solution (different gradients)
- No solutions (parallel lines: same gradient, different $y$-intercepts)
- Infinitely many solutions (the same line: same gradient, same $y$-intercept)

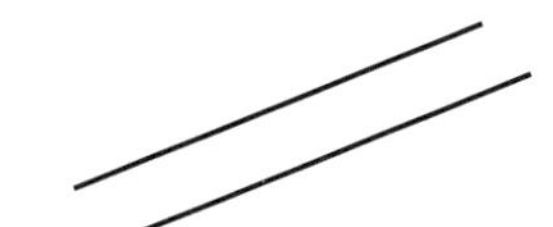

Equating gradients of the lines allow us to decide which of the three types the simultaneous equations represent.

---

**Example 36**

Solve the simultaneous equations:

$2x - 3y = 8 \qquad \ldots[1]$

$x + y = 2 \qquad \ldots[2]$

**Solution**

**a** By elimination: $2x - 3y = 8$

[2] × 3: $3x + 3y = 6$

Adding gives $5x = 14$

$x = \frac{14}{5}$ and by substitution $y = -\frac{4}{5}$.

**b** By substitution: $2x - 3y = 8$

$x + y = 2 \Rightarrow y = -x + 2$ by substitution into $2x - 3y = 8$

$2x - 3(-x + 2) = 8$

gives $2x + 3x = 8 + 6$

so $x = \frac{14}{5}$ and $y = -\frac{4}{5}$

**c** Using CAS:

$\begin{cases} 2x-3y=8 \\ x+y=2 \end{cases} \Big| x, y$

$\left\{x=\frac{14}{5}, y=-\frac{4}{5}\right\}$

**d** Graphically:

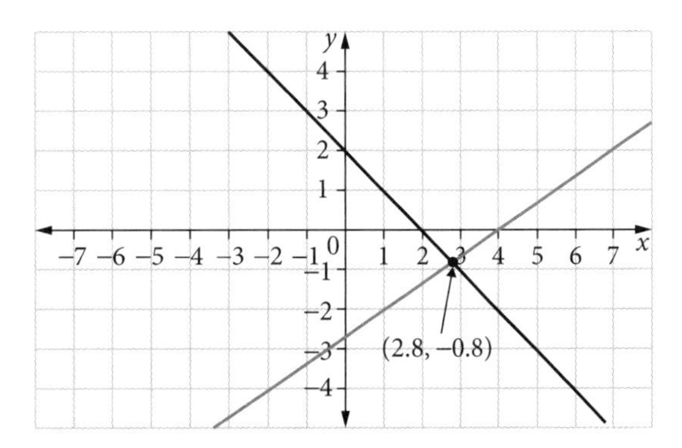


Linear simultaneous equations not only allow us to solve the equations but also to analyse values of constants for which two straight lines intersect or do not intersect.

## 2.9.1 Unique solution, one solution or infinitely many solutions

### 2.9.1.1 Algebraic method

Simultaneous linear equations will have one unique solution except for the cases when the gradients of the lines are equal. In these cases, there can be no unique solution. This means there will be

**i** infinitely many solutions

or

**ii** no solution.

Situation (**i**) arises when the two equations represent the same line.

Situation (**ii**) arises when the two equations represent parallel lines.

---

**Example 37**

**a** Show that the simultaneous equations $x + 2y = 10$ and $y = 3x - 2$ have one unique solution.

**b** Find the value(s) of $m$ such that the simultaneous equations $mx + 2y = 20$ and $(m + 1)x - 2y = 10$ have no solution, where $m$ is a real constant.

**Solution**

**a** $x + 2y = 10$ becomes $y = -\frac{1}{2}x + 5$.

gradient $= -\frac{1}{2}$

For $y = 3x - 2$, gradient $= 3$.

gradients: $-\frac{1}{2} \neq 3$

So must have one unique solution.

Equate lines.

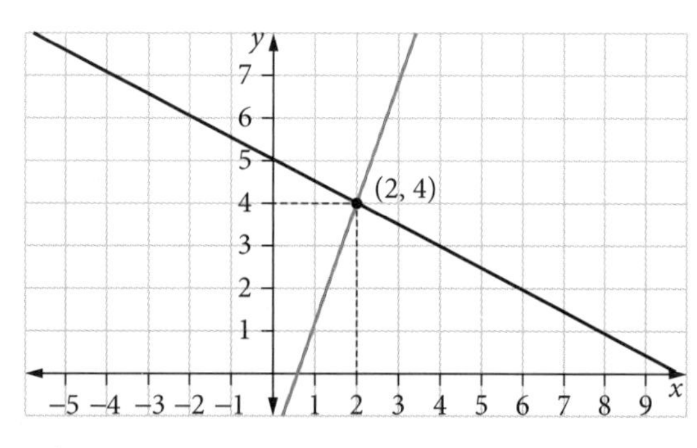


$-\frac{1}{2}x + 5 = 3x - 2$ gives $x = 2$, $y = 4$

one unique solution.

**b** Rearrange equations to the form $y = mx + c$.

$y = -\frac{m}{2}x + 10$ and $y = \frac{m+1}{2}x - 5$

Compare gradients.

gradient $= -\frac{m}{2}$ and gradient $= \frac{m+1}{2}$

Equate gradients for the two possible cases.

$-\frac{m}{2} = \frac{m+1}{2}$ gives $m = -\frac{1}{2}$.

Equating gradients means there is not a *unique* solution to these equations. Now we need to check whether we have no solutions (same gradient, different $y$-intercepts), or infinitely many solutions (same gradient, same $y$-intercept).

| | |
|---|---|
| Test $m = -\frac{1}{2}$ in both equations. | Substitute $m = -\frac{1}{2}$ into $y = -\frac{m}{2}x + 10$ becomes $y = \frac{1}{4}x + 10$. Substitute $m = -\frac{1}{2}$ into $y = \frac{(m+1)}{2}x - 5$ becomes $y = \frac{1}{4}x - 5$. |
| State conclusions for $m = -\frac{1}{2}$. | $m = -\frac{1}{2}$ gives the same gradient with different $y$-intercepts. Equations have no solution, where $m = -\frac{1}{2}$. |

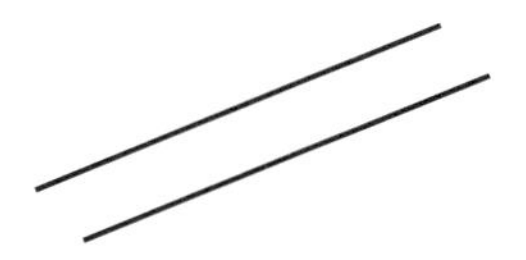

**No unique solution: Two possibilities**

No solutions (parallel lines: same gradient, different $y$-intercepts)

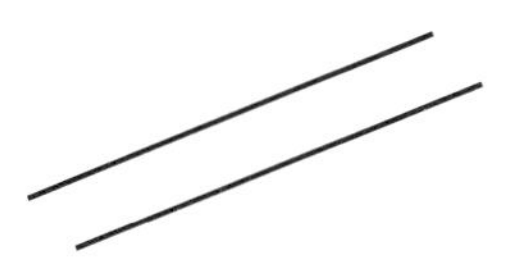

Infinite solutions (the same line: same gradient, same $y$-intercept)

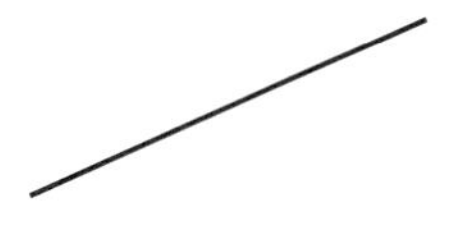

**Not** one unique solution for $m = -\frac{1}{2}$.

solve(m·x+2·y=20, y)

$\left\{y=\frac{-m\cdot x}{2}+10\right\}$

solve((m+1)·x−2·y=10, y)

$\left\{y=\frac{m\cdot x}{2}+\frac{x}{2}-5\right\}$

solve$\left(\frac{-m}{2}=\frac{m+1}{2}, m\right)$

$\left\{m=-\frac{1}{2}\right\}$

### Example 38

**a** Consider the simultaneous equations

$$2x + 6y = 10$$
$$-4x - 12y = -20$$

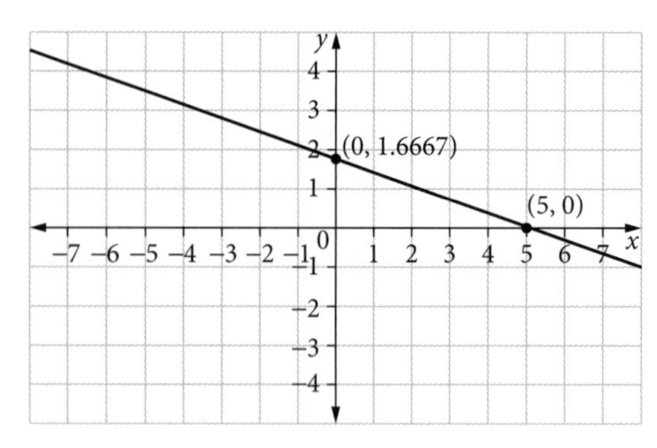


Rearrange each equation in $y = mx + c$ form.

They both look like $y = -\frac{x}{3} + \frac{5}{3}$. This situation gives **infinitely many solutions.**

**b** Consider the simultaneous equations

$$2x + 6y = 1$$
$$-4x - 12y = -20$$

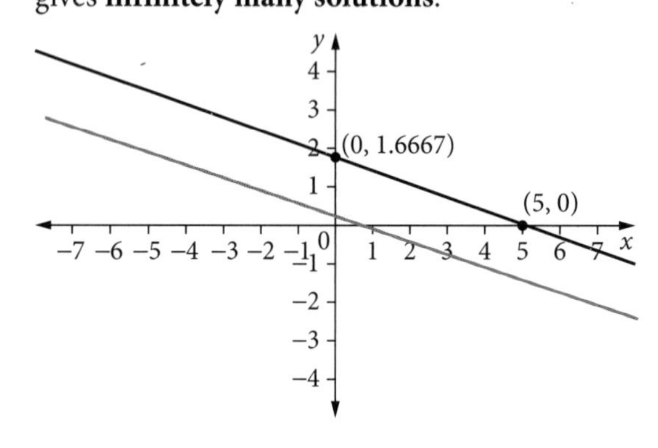


Rearrange each equation in $y = mx + c$ form.

These equations look like

$$y = -\frac{x}{3} + \frac{1}{6} \text{ and } y = -\frac{x}{3} + \frac{5}{3}.$$

The same gradient and different $y$-intercept gives parallel lines. This situation gives **no solutions.**

# Glossary

**change of base theorem** The formula $\log_b(x) = \frac{\log_a(x)}{\log_a(b)}$ that allows us to change the base of a logarithm to any base, for example, $\log_2(10) = \frac{\log_e(10)}{\log_e(2)}$.

**coefficient** The number multiplying a variable. For example, in $x^3 - 3x^2 + 6x$, the coefficient of $x^2$ is −3.

**composite function** A function of a function, when the results of one function are substituted into another function.

**constant** A value that is fixed. A number without a variable. For example, in $4x^2 + 3\sqrt{2}x + 9$, the constant term is 9.

**cubic polynomial** A polynomial of degree 3, for example, $P(x) = -2x^3 + x^2 - 10x - 7$.

**degree of a polynomial** The highest power of the variable in a polynomial. For example, $x^3 - 3x^2 + 6x$ has degree 3.

**difference of two cubes** The expression $a^3 - b^3$ that can be factorised into $(a - b)(a^2 + ab + b^2)$.

**equating coefficients** (of identical polynomials) Applying the rule that if $P(x) = a_nx^n + a_{n-1}x^{n-1} + a_{n-2}x^{n-2} + \ldots + a_1x^1 + a_0$ and $Q(x) = b_nx^n + b_{n-1}x^{n-1} + b_{n-2}x^{n-2} + \ldots + b_1x^1 + b_0$ are equal for all values of $x$, then $a_n = b_n$, $a_{n-1} = b_{n-1}$, $a_{n-2} = b_{n-2}$, and so on up to $a_0 = b_0$.

**exact trigonometric values** The trigonometric ratios of $\left(\frac{\pi}{6}\right)$ 30°, $\left(\frac{\pi}{4}\right)$ 45°, $\left(\frac{\pi}{3}\right)$ 60° and their multiples, which can be expressed as exact rational or surd values such as $\frac{1}{2}$ or $\frac{\sqrt{3}}{2}$.

**exponential equation** An equation involving $a^x$ where the variable is in the exponent (power).

**exponential growth** An increase that is happening slowly at first, then gradually more quickly, according to the function $y = ba^x$, where $y$ is the quantity, $b$ is the initial value, $a$ is the growth factor ($a > 1$) and $x$ is the time. Natural exponential growth follows the function $y = be^{kt}$, where $t$ is the time and $k$ is a positive constant.

**A+ DIGITAL FLASHCARDS**

Revise this topic's key terms and concepts by scanning the QR code or typing the URL into your browser.

https://get.ga/aplus-vce-maths-methods

**exponential laws**

1 $a^x \times a^y = a^{x+y}$
2 $a^x \div a^y = a^{x-y}$
3 $(a^x)^y = a^{xy}$
4 $(ab)^x = a^x b^x$
5 $a^0 = 1$
6 $a^{\frac{p}{q}} = \sqrt[q]{a^p}$

**factor theorem** For a polynomial $P(x)$, if $P(a) = 0$, then $x - a$ is a factor of the polynomial.

**inverse function** The function $y = f^{-1}(x)$ that is the inverse of $y = f(x)$, which links the values of $y = f(x)$ back to $x$. So if $f(3) = -5$, for example, then $f^{-1}(-5) = 3$. An inverse function exists only if the original function is one-to-one.

**leading coefficient** (of a polynomial) The term of a polynomial that contains the highest power. For example, $x^3 - 3x^2 + 6x$ has a leading coefficient of 1 (the coefficient of $x^3$).

**leading term** (of a polynomial) The term of a polynomial that contains the highest power. For example, $x^3 - 3x^2 + 6x$ has a leading term of $x^3$.

**literal equation** An equation that has many variables, such as $ax + b = c$.

**logarithm and index laws** The rules involving the logarithms of $xy, \frac{x}{y}, x^n$ and change of base, related to the index laws.

1 $\log_a(mn) = \log_a(m) + \log_a(n)$
2 $\log\left(\frac{m}{n}\right) = \log_a(m) - \log_a(n)$
3 $\log_a\left(\frac{1}{n}\right) = \log_a(n^{-1}) = -\log_a(n)$
4 $\log_a(m^p) = p\log_a(m)$
5 $\log_a(a) = 1$
6 $\log_a(1) = 0$

*See also* **change of base theorem**.

**logarithmic equation** An equation involving $y = \log_a(x)$.

**long division** A method of dividing one polynomial by another, using a repeated process. *See also* **synthetic division**.

**monic polynomial** A polynomial whose leading coefficient is 1, for example, $x^3 - 3x^2 + 6x$.

**Newton's method** A formula for finding the zeros of a function (or the roots of an equation) that involves making a first guess, then using the formula $x_{n+1} = x_n - \frac{f(x_n)}{f'(x_n)}$ repeatedly to improve on the approximation of the zero.

**null factor law** If $ab = 0$, then either $a = 0$, $b = 0$ or $a = b = 0$.

**polynomial** An expression involving a sum of terms of the form $P(x) = a_n x^n + a_{n-1} x^{n-1} + a_{n-2} x^{n-2} + \ldots + a_1 x^1 + a_0$, where $n$ is a whole number and $a_0, a_1, a_2 \ldots a_{n-1}, a_n$ are real numbers and $a_n \neq 0$.

**quadratic polynomial** A polynomial of degree 2.

**quartic polynomial** A polynomial of degree 4.

**radian** A measure of angle defined as arc lengths around the circumference of a circle of radius 1: $\pi$ radians = 180°.

**reference angle** A first quadrant solution for a trigonometric equation like $\cos(nx) = c$ where the sign of $c$ is ignored.

**remainder theorem** If a polynomial $P(x)$ is divided by $x - a$, the remainder is $P(a)$.

**simultaneous equations** A set of equations that are all satisfied by the same values. Graphically, the solution is the point of intersection of the graphs of the equations.

**sum of two cubes** The expression $a^3 + b^3$ that can be factorised into $(a + b)(a^2 - ab + b^2)$.

**synthetic division** A method of dividing one polynomial by another, using a repeated process on a table of values, based on long division.

**transformation** A change of a function and its graph through dilation, reflection or translation.

**translation** Shifting a graph parallel to the $x$-axis (horizontally) or $y$-axis (vertically).

**unit circle** A circle on the number plane with centre $(0, 0)$ and radius 1 that is used to define the trigonometric ratios for angles of any size, measured from the positive $x$-axis.

# Exam practice

## Short-answer questions

### Technology free: 22 questions

Solutions to this section start on page 215.

**Question 1** (2 marks)

Solve for $x$: $\sqrt{9x-4} = 2$

**Question 2** (3 marks)

For the quadratic function shown, find

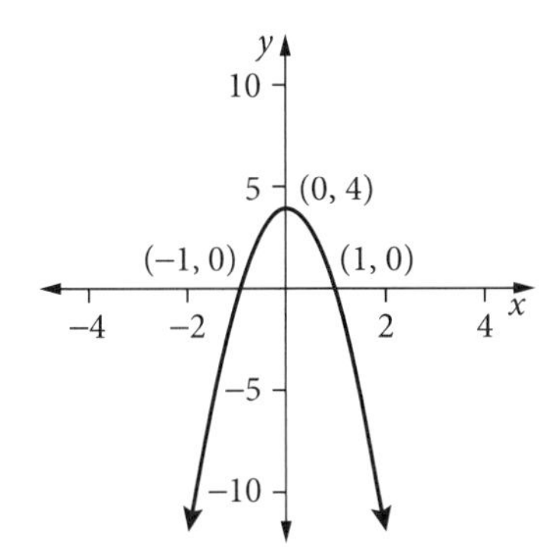


**a** the value of $A$ so that the maximal domain, in the form of $[A, \infty)$, will allow the inverse function $y^{-1}$ to exist — 1 mark

**b** the rule that defines the function $y^{-1}$. — 2 marks

**Question 3** (1 mark)

Find the remainder if the polynomial $P(x) = 55 - 18x + 2x^2 - 3x^3 + x^4$ is divided by $x + 2$.

**Question 4** (2 marks)

If $\log_e(y) = \log_e(x) + \log_e(p)$, write an equation relating $x$, $y$ and $p$ that does not involve logarithms.

**Question 5** (1 mark)

Evaluate $\dfrac{\log_3(27)}{\log_3(9)}$.

**Question 6** (2 marks)

Solve the equation $\sin(x°) = \dfrac{1}{\sqrt{2}}$ for $0° \le x° \le 180°$.

**Question 7** (3 marks)

The height of water in a dam can be represented by a curve of the form $y = a\cos(kt) + c$, where $t$ is in days after some rainfall. At $t = 0$, the height of water in the dam is 20 metres. At $t = 10$, the height of water in the dam is 60 metres. At $t = 20$, the height of water in the dam is back to 20 metres for the first time since $t = 0$.

**a** What is the value of $k$? — 1 mark

**b** What is the value of $a$? — 1 mark

**c** What is the value of $c$? — 1 mark

**Question 8** (1 marks)

Solve for $x$ in the equation $\log_e(x^2) = 2$.

**Question 9** (2 marks)

Find the rule that defines the inverse of the function $f(x) = 3 + \log_e(x + 2)$.

**Question 10** (4 marks)

For $g: [D, \infty) \to R, g(x) = (x - 2)^2 + 3$,

**a** find the smallest value of $D$ such that $g^{-1}$ exists 2 marks

**b** define $g^{-1}(x)$. 2 marks

**Question 11** (4 marks)

If $f(x) = x^2$ and $g(x) = \log_e(x - 2)$, state, with reasons, whether or not the functions $f(g(x))$ and $g(f(x))$ exist.

**Question 12** (2 marks)

Solve the equation $x^3 - x^2 - x + 1 = 0$.

**Question 13** (3 marks)

Find the value of $x$ such that $9^x - 2(3^{x+1}) + 9 = 0$.

**Question 14** (3 marks)

If $f(x) = x^3 + 1$ and $g(x) = \sqrt{x - 1}$, define $g(f(x))$ and $f(g(x))$.

**Question 15** (3 marks)

Find the exact values of $x$ in the domain $0 \le x \le 2\pi$ for which $2\sin^2(x) - \sin(x) - 1 = 0$.

**Question 16** (3 marks)

A sinusoidal curve of the form $T = a\sin(nt) + b$ could be used to model the rise and fall of temperature on a newly-discovered planet. The temperature hits a maximum of 100°F after 1.5 months, and reaches the minimum of 40°F three months later. Find the rule that best models this data, where $t$ is the time in months.

**Question 17** (6 marks)

**a** Find the general solution to the equation $\sin\left(2\left(x - \frac{\pi}{2}\right)\right) = \frac{1}{\sqrt{2}}$. 3 marks

**b** Hence, find all the exact solutions in the interval $\left(-\frac{\pi}{2}, \frac{\pi}{2}\right)$. 3 marks

**Question 18** (6 marks)

Find the value(s) of $k$ for which these simultaneous linear equations have the following solutions:

$$(k - 1)x - 2y = 6$$
$$-x + ky = 3$$

**a** a unique solution 2 marks

**b** no solutions 2 marks

**c** infinitely many solutions 2 marks

**Question 19** (2 marks)

Find the quotient and remainder when the polynomial $Q(x) = x^3 - 9x^2 + 7x - 3$ is divided by $x - 3$.

**Question 20** (3 marks)

If $3x^2 - 4x + 1 = a(x + b)^2 + c$, find the values of $a$, $b$ and $c$.

**Question 21** (3 marks)

Find the value(s) of $k$ for which the following simultaneous linear equations have a unique solution.

$$kx - 9y = 5$$
$$x - ky = k$$

**Question 22** (3 marks)

Find the value(s) of $k$ for which the following simultaneous linear equations have infinitely many solutions.

$$kx - 9y = 5$$
$$x - ky = k$$

## Multiple-choice questions

### Technology active: 48 questions

Solutions to this section start on page 218.

**Question 1**

The linear function shown could have the equation

**A** $y = 2x + 4$

**B** $y = 2x - 4$

**C** $y = -2x + 4$

**D** $y = -3x + 3$

**E** $y = 3x + 3$

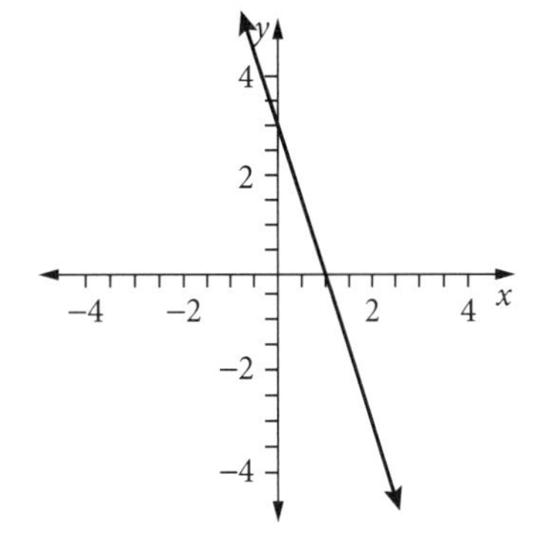


**Question 2**

The quadratic function shown could have the equation

**A** $y = 2x^2 + 3$

**B** $y = (x - 0.5)(x + 1.5)$

**C** $y = 4(x - 1)^2 - 1$

**D** $y = (x - 1)^2 - 1$

**E** $y = x^2 - 1$

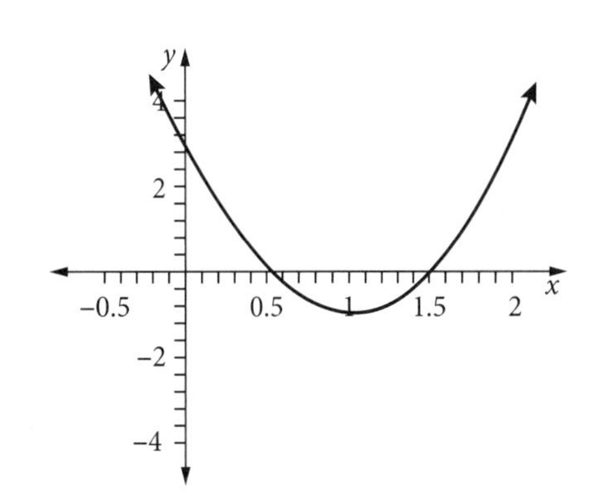

**Question 3**

Which of the following is **not** a polynomial?

**A** $f(x) = 1 + 2x$

**B** $f(x) = -3 + 4x + 2x^2$

**C** $f(x) = 4x + 2x^2$

**D** $f(x) = 2x^2 - 8x^3$

**E** $f(x) = -\frac{1}{x} + x - 2$

**Question 4**

If $P(x) = -2x^3 + 7x - 5$, the following **correct** statement is

**A** $P(1) = 0$

**B** $(x + 1)$ is a factor

**C** $P(-1) = 0$

**D** $P(x) = -2(1)^3 + 7(1) - 5$

**E** $(x - 5)$ is a factor

**Question 5**

Let $Q(x) = -x^4 - 2x^3 + 3x^2 + 4x - 4$.

The graph that best represents this polynomial is

**A**

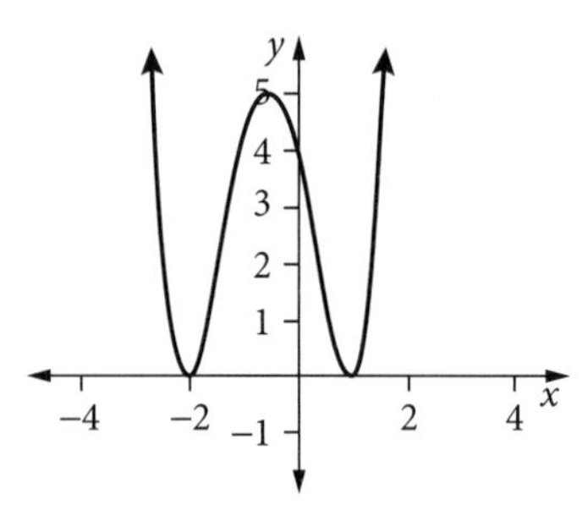


**B**

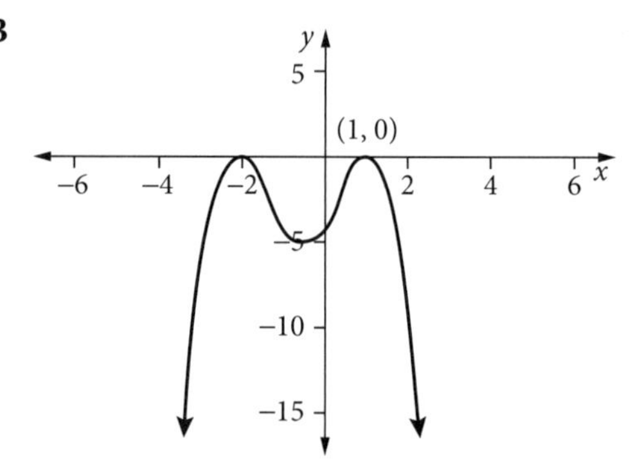


**C**

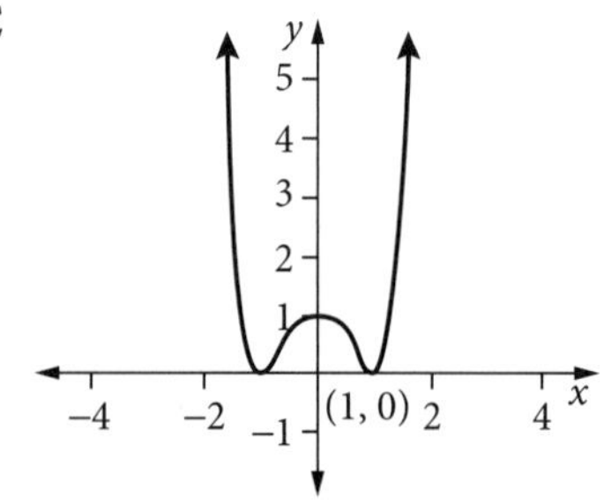


**D**

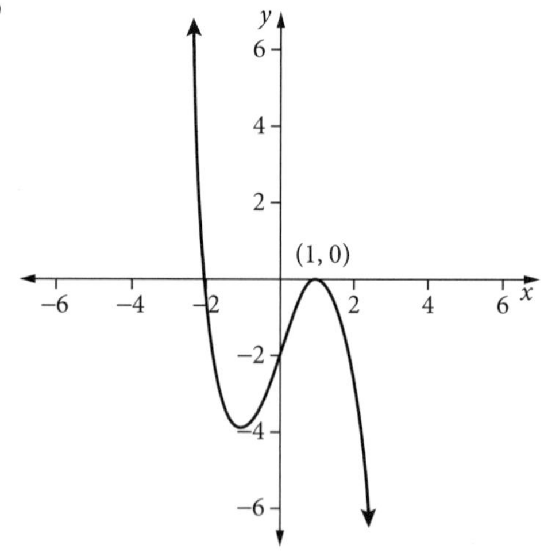


**E**

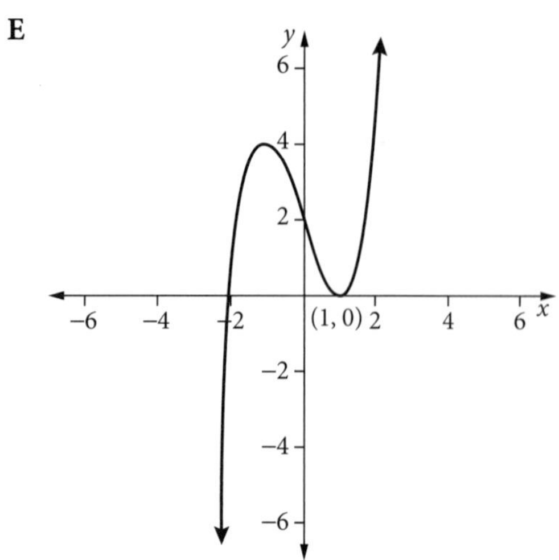

**Question 6**

One factor of the polynomial $Q(x) = -x^4 - 2x^3 + 3x^2 + 4x - 4$ is $(1 - x)$. The other factors are

**A** $(x-1)(x+2)$ **B** $(x-1)(1-x)(x+2)(x+2)$ **C** $(x-1)(1-x)(x-1)$

**D** $(x-1)(x+2)(x+2)$ **E** $-(x+2)$

**Question 7**

If $P(x) = -2(x-3)(x+2)(x-5)$, the $y$-intercept is at

**A** $(0,-60)$ **B** $(-60,0)$ **C** $(0,-30)$ **D** $(0,30)$ **E** $(0,60)$

**Question 8**

If $4x^2 - 3x + 1 = a(x-b)^2 + c$, the values of $a$, $b$ and $c$ are

**A** $a = 4, b = 3, c = 1$ **B** $a = 4, b = -\frac{3}{8}, c = \frac{7}{16}$ **C** $a = -4, b = \frac{3}{8}, c = -\frac{7}{16}$

**D** $a = 4, b = \frac{3}{16}, c = \frac{7}{16}$ **E** $a = 4, b = \frac{3}{8}, c = \frac{7}{16}$

**Question 9**

If $(ax + b)$ is a factor of $P(x)$, then

**A** $P(-b) = 0$ **B** $P\left(-\frac{b}{a}\right) = 0$ **C** $P\left(\frac{b}{a}\right) = 0$ **D** $P(a) = 0$ **E** $P\left(\frac{a}{b}\right) = 0$

**Question 10**

If we divide the polynomial $P(x) = x^3 + x^2 - 5x + 7$ by $(x + 1)$, the remainder is

**A** 0 **B** 4 **C** 7 **D** 10 **E** 12

**Question 11**

If $f(x) + f(y) = f(xy)$, the rule for $f(x)$ could be

**A** $f(x) = x$ **B** $f(x) = x^2$ **C** $f(x) = e^x$

**D** $f(x) = \log_e(x)$ **E** $f(x) = \frac{1}{x}$

**Question 12**

If $f(x) \times f(y) = f(x + y)$, the rule for $f(x)$ could be

**A** $f(x) = x$ **B** $f(x) = x^2$ **C** $f(x) = e^x$

**D** $f(x) = \log_e(x)$ **E** $f(x) = \frac{1}{x}$

**Question 13**

Consider $f: (-\infty, A] \to R, f(x) = x^2 + 4x$. The maximal value of $A$ for the inverse $f^{-1}$ to exist is

**A** $-4$ **B** $-2$ **C** 0 **D** 2 **E** 4

**Question 14**

Consider $f: (-\infty, 1) \to R$, $f(x) = x^2 - 2x$. The rule and domain for the inverse function $f^{-1}$ is

**A** $f^{-1}(x) = 1 + \sqrt{x+1}, x \in (-\infty, 1]$ **B** $f^{-1}(x) = 1 \pm \sqrt{x+1}, x \in (-\infty, 1]$

**C** $f^{-1}(x) = 1 - \sqrt{x+1}, x \in (-\infty, -1)$ **D** $f^{-1}(x) = 1 - \sqrt{x+1}, x \in (-1, \infty)$

**E** $f^{-1}(x) = 1 + \sqrt{x+1}, x \in (-1, \infty)$

**Question 15**

For the function $f: (-\infty, 0] \to R, f(x) = x^2 - 4$, the point(s) of intersection, correct to two decimal places, between $f$ and $f^{-1}$ is/are

**A** $(-1.56, -1.56)$ **B** $(1.56, 1.56)$ **C** $(-1.56, -1.56)$ and $(1.56, 1.56)$

**D** $(-2.30, 1.30)$ **E** $(2.56, 2.56)$

**Question 16**

The graph shown has the equation $y = 2a^2(x-1)^2$ where $a \in R$. The point(s) of intersection between this graph and the graph of $y = 2$ is/are

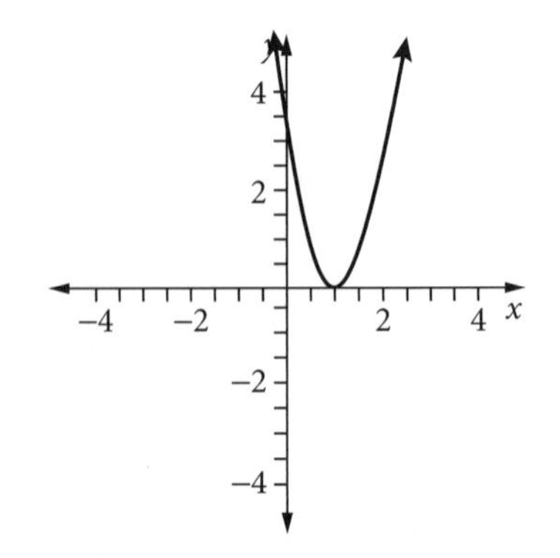


**A** $\left(\frac{a-1}{a}, 2\right), \left(\frac{a+1}{a}, 2\right)$ **B** $\left(\frac{a-1}{a}, 4\right), \left(\frac{a+1}{a}, 2\right)$

**C** $\left(\frac{a-1}{a}, \frac{a+1}{a}\right)$ **D** $\left(0, \frac{a+1}{a}\right), a+1, \left(0, \frac{a-1}{a}\right)$

**E** $\left(2, \frac{a-1}{a}\right), \left(2, \frac{a+1}{a}\right)$

**Question 17**

Written in radians, 25° is

**A** $\frac{\pi}{36}$ **B** $\frac{5\pi}{36}$ **C** 0.436 353 **D** 0.44 **E** $\frac{4500}{\pi}$

**Question 18**

Written in degrees, $\frac{\pi}{12}$ is

**A** $\frac{\pi^2}{2160}$ **B** 15° **C** 25° **D** 45° **E** 90°

**Question 19**

$\sin\left(\frac{7\pi}{2}\right) =$

**A** −2 **B** −1 **C** 0 **D** $\frac{1}{2}$ **E** 1

**Question 20**

$\cos\left(\frac{7\pi}{2}\right) =$

**A** −2 **B** −1 **C** 0 **D** $\frac{1}{2}$ **E** 1

**Question 21**

If $f(x) = -2\sqrt{x+2} - 1$, then the point(s) of intersection between $f(x)$ and the function $g(x) = x + 1$ is/are

**A** $(-2, 0)$ **B** $(2, 0)$ **C** $(-2, -1)$

**D** $(-2, -1)$ and $(-2, 0)$ **E** $(-2, -1)$ and $(2, 3)$

**Question 22**

Consider the graph of $y = 2(2^{x-1}) - 8$. Its inverse graph has the rule

**A** $y^{-1} = \dfrac{\log_e(x+8)}{\log_e(2)}$ **B** $y^{-1} = \log_2(x+8)$ **C** $y^{-1} = \dfrac{\log_2(x+8)}{\log_2(2)}$

**D** $y^{-1} = \log_e\left(\dfrac{x}{2} + 4\right)$ **E** $y^{-1} = \log_2(x-8)$

**Question 23**

Let $f(x) = x^2 - 2x$ and $g(x) = \dfrac{1}{x^2}$ for a maximal domain.

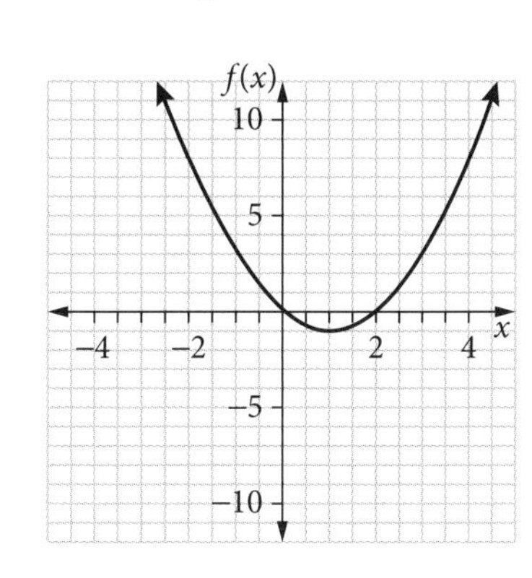


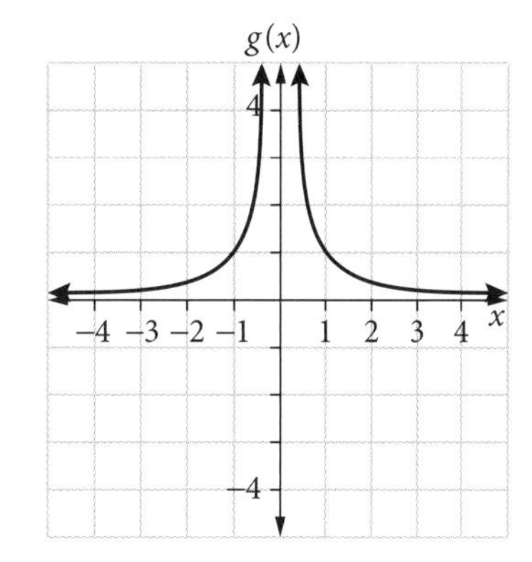


$g(f(x))$ is

**A** $g(f(x)) = \dfrac{1}{(x^2 - 2x)^2}$ for $x \in (0, \infty)$ **B** $g(f(x)) = \dfrac{1}{(x^2 - 2x)^2}$ for $x \in R \setminus \{0, 2\}$

**C** $g(f(x)) = \dfrac{1}{(x^2 - 2x)^2}$ for $x \in R$ **D** $g(f(x)) = \dfrac{1}{x^2 - 2x}$ for $x \in R$

**E** $g(f(x)) = x^4 - 2x^2$ for $x \in (1, \infty)$

**Question 24**

The line $2y - 6x = 3$ has an $x$-intercept of

**A** $x = -3$ **B** $x = -\dfrac{1}{2}$ **C** $x = -\dfrac{1}{6}$ **D** $x = \dfrac{9}{2}$ **E** $x = 3$

**Question 25**

An equivalent expression for $\sin\left(\dfrac{7\pi}{6}\right)$ is

**A** $\sin\left(\dfrac{\pi}{6}\right)$ **B** $\sin\left(\dfrac{5\pi}{6}\right)$ **C** $\cos\left(\dfrac{\pi}{6}\right)$ **D** $\cos\left(\dfrac{\pi}{3}\right)$ **E** $-\sin\left(\dfrac{\pi}{6}\right)$

**Question 26**

An equivalent expression for $\log_x(25) = 2$ is

**A** $x^2 = 25$ **B** $2^x = 25$ **C** $\log_5(25) = 2$

**D** $\log_2(25) = x$ **E** $5^2 = 25$

**Question 27**

If $9^{x-2} \times 3^x = 9$, then $x$ equals

**A** $x = 2$ **B** $x = 2.5$ **C** $x = 3$ **D** $x = 9$ **E** $x = 25$

**Question 28**

The maximal domain for the function $f(x) = -\frac{2}{\sqrt{x}}$ is

**A** $x > 0$ **B** $x < -2$ **C** $x \geq -2$ **D** $x \leq 0$ **E** $x \geq 0$

**Question 29**

The graph of $y = \sin(x) - \cos(2x)$ is shown.

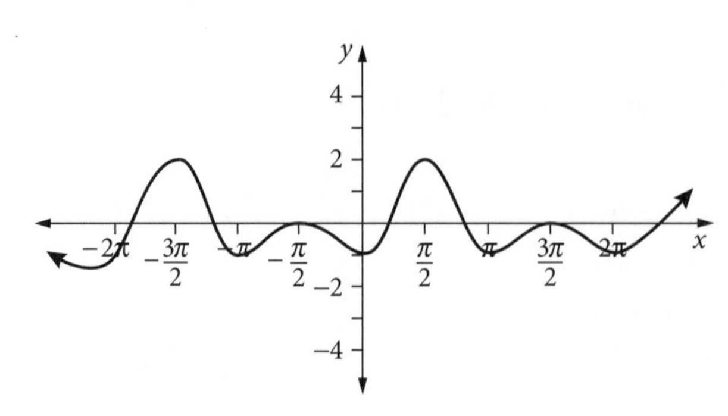


The solution(s) for the equation $y = 0$ for the domain $(\pi, 2\pi)$ is/are

**A** $x = \frac{\pi}{2}, \frac{3\pi}{2}$ **B** $x = \frac{\pi}{2}$ **C** $x = \frac{27\pi}{25}$ **D** $x = -\frac{\pi}{2}$ **E** $x = \frac{3\pi}{2}$

**Question 30**

A simplified version of the expression $2\log_3(x) - \log_3(x+1) + 2\log_3(x+4)$ is

**A** $\log_3\left(\frac{x^2+4x}{x+1}\right)$ **B** $\log_3\left(\frac{x^2(x+4)^2}{x+1}\right)$ **C** $\frac{x^2(x+4)^2}{x+1}$

**D** $\log_e\left(\frac{x^2+4x}{x+1}\right)$ **E** $\log_3\left(\frac{8x}{x+1}\right)$

**Question 31**

To make $d$ the subject in the literal equation $ax^2 + b = cd$, the answer is

**A** $d = ax^2 + b$ **B** $d = ax^2 + b - c$ **C** $d = \frac{ax^2+b}{c}$

**D** $d = \frac{ax^2-b}{c}$ **E** $b = cd - ax^2$

**Question 32**

The solutions to the simultaneous equations

$$x + y = 5$$
$$2x - y = 1$$

are

**A** $(2, 3)$ **B** $\left(\frac{1}{2}, \frac{9}{2}\right)$ **C** $(-2, 7)$ and $\left(\frac{3}{2}, \frac{7}{2}\right)$

**D** $(-2, 7)$ only **E** $\left(\frac{3}{2}, \frac{7}{2}\right)$ only

**Question 33**

The graph shown has the equation $f(x) = -2\sin(3(x - \pi))$.

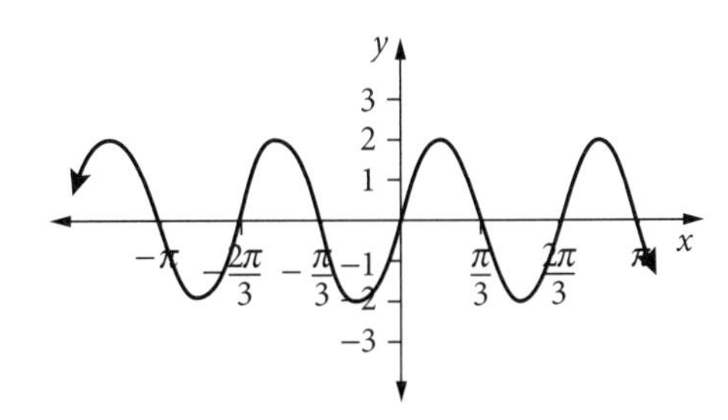


The number of solutions for the equation $f(x) = 2$ in the domain $(-\pi, \pi)$ is

**A** 0 **B** 1 **C** 2 **D** 3 **E** 4

**Question 34**

The inverse of the function $f: R \to R, f(x) = -e^{x+2} - 3$ is

**A** $f^{-1}: R \to R, f^{-1}(x) = \log_e(3 - x) - 2$

**B** $f^{-1}: (3, \infty) \to R, f^{-1}(x) = \log_e(x - 3) - 2$

**C** $f^{-1}: (-\infty, -3) \to R, f^{-1}(x) = \log_e(-3 - x) - 2$

**D** $f^{-1}: R \to R, f^{-1}(x) = e^{x-2} - 3$

**E** $f^{-1}: (-\infty, 3) \to R, f^{-1}(x) = -\log_e(x - 3) - 2$

**Question 35**

For $f(x) = e^x$ which of the following is correct?

**A** $f(xy) = f(x) + f(y)$

**B** $f(x)f(y) = f(x + y)$

**C** $f(x + y) = f(x) + f(y)$

**D** $f(x - y) = f(x) - f(y)$

**E** $f(x) = f(y)$

**Question 36**

If $\log_a(y) = \frac{1}{2}$, then $2\log_a(y^3)$ equals

**A** $\frac{1}{8}$ **B** $\frac{1}{2}$ **C** 2 **D** 3 **E** 4

**Question 37**

If $(\log_5(x))^2 = \log_5(x^2)$, then $x$ is equal to

**A** 1 or 25 **B** 0 or 3 **C** 1 only **D** 1 or 125 **E** 1 or 5

**Question 38**

If $N = Ae^{-kt}$ and $N = 4.12$ when $t = 2$, and $N = 2.62$ when $t = 5$, the values of $A$ and $k$, respectively (correct to one decimal place), are

**A** $A = 5.6, k = 0.2$

**B** $A = 5.6, k = -0.2$

**C** $A = 5.4, k = 0.1$

**D** $A = 5.4, k = 0.2$

**E** $A = 1.5, k = 0.5$

**Question 39**

The solutions to the equation $\sin^2(2\pi x) = \frac{3}{4}$ for $x \in \left[0, \frac{1}{2}\right]$ are

**A** $x = \frac{\sqrt{3}}{2}$ **B** $x = 0.13, 0.37$ **C** $x = \frac{1}{2}$ **D** $x = \frac{1}{6}, \frac{1}{3}$ **E** $x = \frac{1}{3}$

**Question 40**

$x^{-3} + x^{-1}$ simplifies to

**A** $\frac{1}{x^2}$ **B** $\frac{1+x^2}{x^3}$ **C** $\frac{2x}{x^3}$ **D** $x^2$ **E** $\frac{1+x}{x^2}$

**Question 41**

A system of linear equations is

$$3x + 4y + 1 = 0$$
$$6x + 8y + 2 = 0$$

The graph of these equations has

**A** no points of intersection

**B** one point of intersection

**C** an infinite number of points of intersection

**D** two points of intersection

**E** three points of intersection.

**Question 42**

For what value(s) of $k$ does the system of equations

$$kx + 4y = 1$$
$$(k-1)x + ky = 2k$$

have no solutions?

**A** $k \in R \setminus \{2\}$ **B** $k \in R \setminus \{-2\}$ **C** $k = 4$

**D** $k = 2$ **E** $k \in R \setminus \{\pm 2\}$

**Question 43**

If $f(x) = 3\log_e(x-b)$ and $f(2) = 6$, then $b$ equals

**A** $6 - e^{\frac{4}{3}}$ **B** $e^2 - 4$ **C** $e^{\frac{4}{3}} - 6$ **D** $2 - e^2$ **E** $4 - e^2$

**Question 44**

If $5^{x-2} = 10$, then $x$ is equal to

**A** $\frac{1}{\log_{10}(5)} + 2$ **B** $\frac{1}{\log_e(5)} + 2$ **C** $\log_5(10) - 2$

**D** $2 + \log_{10}(5)$ **E** $\log_e(5) + \log_e(10)$

**Question 45**

The simultaneous linear equations

$$(k-3)x - 5y = -2$$
$$2x - (2k+2)y = -4$$

have infinitely many solutions for

**A** $k = 4$ **B** $k = -2$ **C** $k \in R \setminus \{4\}$

**D** $k \in R \setminus \{-2\}$ **E** $k \in R \setminus \{4, -2\}$

**Question 46**

The general solution to the equation $\sin(x) = 1$ is

**A** $\frac{\pi}{2}$

**B** $\frac{\pi}{2} + 2n\pi, n \in Z$

**C** $\frac{\pi}{2}, \frac{3\pi}{2}$

**D** $\frac{\pi}{2} + n\pi, n \in Z$

**E** $\frac{\pi}{2} + 2n\pi, n \in R$

**Question 47**

The equation of the inverse of the graph shown could be

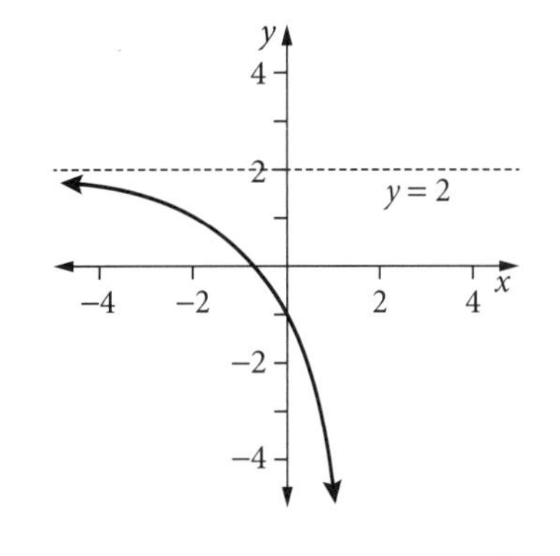


**A** $y = \log_e\left(\frac{2-x}{3}\right)$

**B** $y = -3\log_e(x) + 2$

**C** $y = 3e^x + 2$

**D** $y = -3e^x + 2$

**E** $y = 3e^{-x} + 2$

**Question 48**

The temperature of a hot chocolate cools according to the rule $T = T_0 \times 2^{-kt}$, where $T$ is the temperature in degrees Celsius and $t$ is the time in minutes. If it takes 20 minutes for the temperature to halve, what fraction of the original temperature is the temperature of the hot chocolate after 60 minutes?

**A** $\frac{1}{4}$

**B** $\frac{1}{5}$

**C** $\frac{3}{4}$

**D** $\frac{1}{16}$

**E** $\frac{1}{8}$

# Extended-answer questions

## Technology active: 6 questions

Solutions to this section start on page 222.

**Question 1** (10 marks)

Two lines are defined by $y = 3x - k$ and $y = 3kx + k$ for their maximal domains, where $k$ is a real constant.

**a** What are the value(s) of $k$ for the two lines to have no point of intersection? 3 marks

**b** If the lines $y = 3x - k$ and $y = 3kx + k$ are now perpendicular, show that the value of $k$ is equal to $-\frac{1}{9}$. 1 mark

**c** Hence, using $k = -\frac{1}{9}$, find the coordinates of the point of intersection of the lines. 2 marks

**d** Sketch the graphs of the lines $y = 3x + \frac{1}{9}$ and $y = -\frac{1}{3}x - \frac{1}{9}$, labelling the point of intersection found in part **c**. 2 marks

**e** Find the area of the shape bounded by the line with the –ve gradient, the line with the +ve gradient and the $x$-axis. 2 marks

**Question 2** (15 marks)

Consider $f: [-2, 2] \rightarrow R, f(x) = ax^4 - 2bx^2 + c$

**a** It is known that both $x - 2$ and $2x + 1$ are factors of $f(x)$, and when $f(x)$ is divided by $x + 1$, the remainder is 6. Find $a$, $b$ and $c$. 4 marks

**b** Hence, fully factorise $f(x)$. 1 mark

**c** Sketch the graph of $f(x)$ for $x \in [-2, 2]$, labelling exact $x$-intercepts as well as the turning points, correct to two decimal places. 3 marks

**d** Find a new function, $g(x)$, if it is the image of $f(x)$ after it is translated 2 units in the +ve direction of the $x$-axis, $\frac{8}{3}$ units in the +ve direction of the $y$-axis, and then dilated by 2 units from the $y$-axis. 3 marks

**e** Express this new function, $g(x)$, in fully factorised form, hence showing that one quadratic factor has rational roots and the other quadratic factor has irrational roots. 2 marks

**f** Sketch a graph of $g(x)$, labelling all axial intercepts. 2 marks

**Question 3** (7 marks)

Sue is throwing a javelin in an athletics event. She starts with a long run up before she throws her javelin, modelled by the quadratic function below.

$y = -0.25x^2 + 14.75x - 187$

She runs for 19 metres with the javelin held above her head at a height of 3 metres from the ground.

Sue's run, then throw, can be described by the piecewise function below.

$$f(x) = \begin{cases} 3 & 0 \le x < 19 \\ -0.25x^2 + 14.75x - 187 & 19 \le x \le A \end{cases}$$

**a** Why is it important to use the signs $< 19$ and $19 \le$ in the written piecewise function? 1 mark

**b** Find $f(10)$. 1 mark

**c** Find $f(20)$. 1 mark

**d** Find the point $A$, correct to two decimal places. 1 mark

**e** Sketch the piecewise function that represents this run and throw labelling with coordinates, correct to two decimal places, all significant points: turning point, cusps, end points and axis intercepts. 3 marks

**Question 4** (7 marks) ©VCAA 2022 2BQ2abcd ●●●

On a remote island, there are only two species of animals: foxes and rabbits. The foxes are the predators and the rabbits are their prey.

The populations of foxes and rabbits increase and decrease in a periodic pattern, with the period of both populations being the same, as shown in the graph below, for all $t \geq 0$, where time $t$ is measured in weeks.

One point of minimum fox population, $(20, 700)$, and one point of maximum fox population, $(100, 2500)$, are also shown on the graph.

The graph has been drawn to scale.

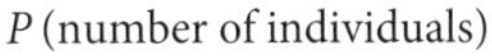


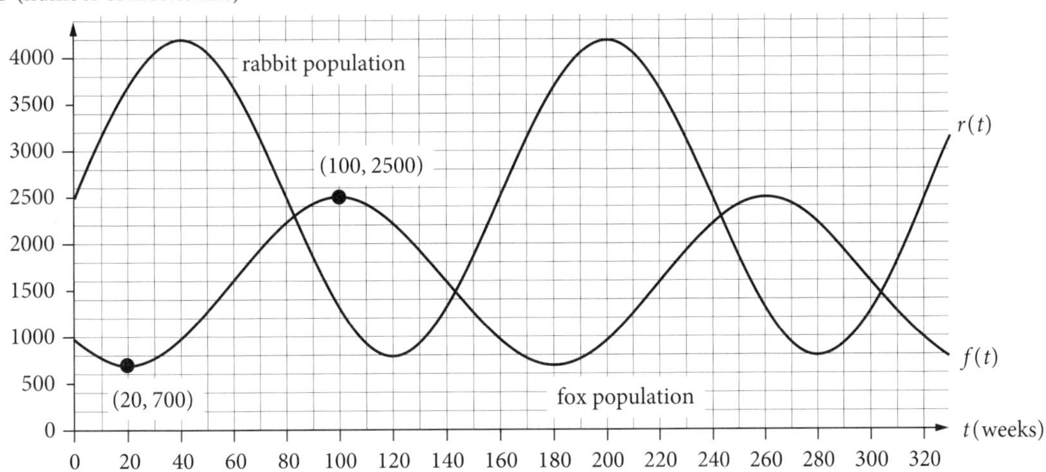


The population of rabbits can be modelled by the rule $r(t) = 1700 \sin\left(\frac{\pi t}{80}\right) + 2500$.

**a** **i** 97% State the initial population of rabbits. 1 mark

**ii** 89% State the minimum and maximum population of rabbits. 1 mark

**iii** 84% State the number of weeks between maximum populations of rabbits. 1 mark

The population of foxes can be modelled by the rule $f(t) = a \sin(b(t - 60)) + 1600$.

**b** 75% Show that $a = 900$ and $b = \frac{\pi}{80}$. 2 marks

**c** 38% Find the maximum combined population of foxes and rabbits. Give your answer correct to the nearest whole number. 1 mark

**d** 57% What is the number of weeks between the periods when the combined population of foxes and rabbits is a maximum? 1 mark

**Question 5** (8 marks) ©VCAA 2022 2BQ4abcd

Consider the function $f$, where $f: \left(-\frac{1}{2}, \frac{1}{2}\right) \to R,\ f(x) = \log_e\left(x + \frac{1}{2}\right) - \log_e\left(\frac{1}{2} - x\right)$.

Part of the graph of $y = f(x)$ is shown below.

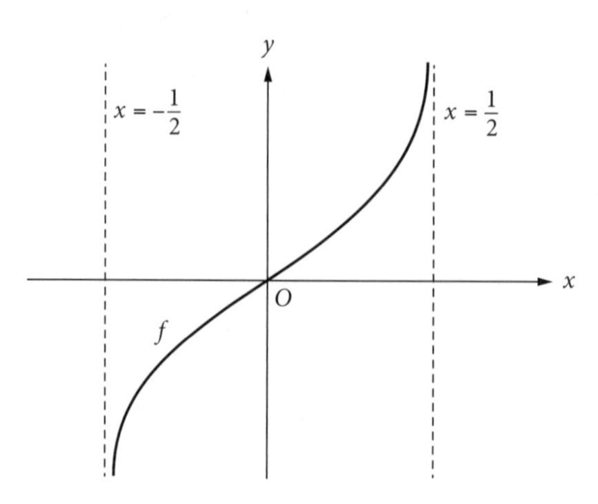


**a** 82% State the range of $f(x)$. 1 mark

**b** **i** 88% Find $f'(0)$. 2 marks

**ii** 57% State the maximal domain over which $f$ is strictly increasing. 1 mark

**c** 72% Show that $f(x) + f(-x) = 0$. 1 mark

**d** 71% Find the domain and the rule of $f^{-1}$, the inverse of $f$. 3 marks

**Question 6** (4 marks) ©VCAA 2021 2BQ5abcd

Part of the graph of $f: R \to R,\ f(x) = \sin\left(\frac{x}{2}\right) + \cos(2x)$ is shown below.

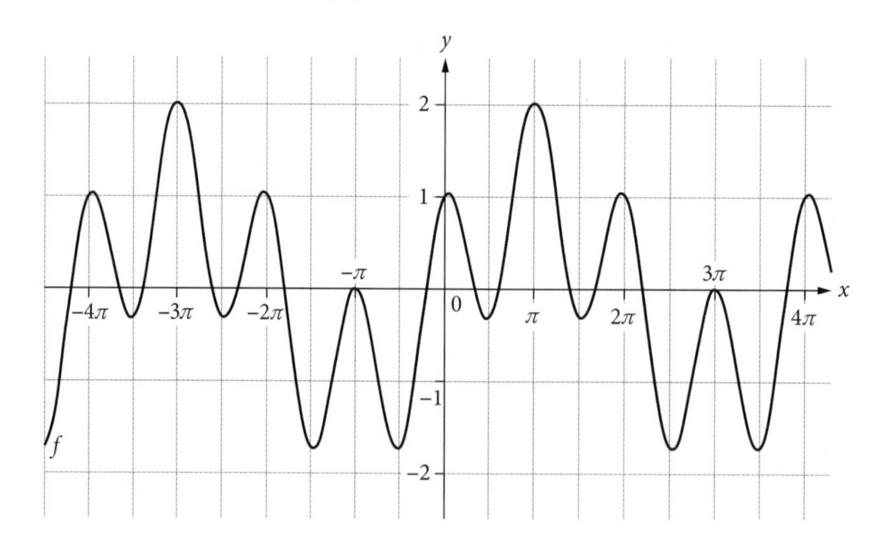


**a** 71% State the period of $f$. 1 mark

**b** 61% State the minimum value of $f$, correct to three decimal places. 1 mark

**c** 21% Find the smallest positive value of $h$ for which $f(h - x) = f(x)$. 1 mark

Consider the set of functions of the form $g_a : R \to R,\ g_a(x) = \sin\left(\frac{x}{a}\right) + \cos(ax)$, where $a$ is a positive integer.

**d** 68% State the value of $a$ such that $g_a(x) = f(x)$ for all $x$. 1 mark

# Chapter 3
# Area of Study 3: Calculus

## Content summary notes

**Calculus**

### Differentiability

- Limits
- Continuity
- Differentiability

### Differentiation

- Gradient
- Average and instantaneous rate of change
- Graphs of derivatives
- Derivatives of basic functions

### Applications of differentiation

- Stationary points
- Strictly increasing and decreasing
- Points of inflection
- Equations of tangents
- Maxima and minima

### Derivative rules

- Properties of derivatives
- Product rule
- Quotient rule
- Chain rule

### Anti-differentiation

- Anti-derivatives of basic functions
- Graphs of anti-derivatives
- Approximation of areas: LEA, REA, Trapezium rule
- Anti-differentiation by recognition

### Integration of functions

- Indefinite integral
- Definite integral
- Properties of definite integrals
- Fundamental theorem of calculus

### Modelling calculus

- Finding + $c$
- Areas under curves
- Area between two curves
- Average value
- Kinematics
- Practical situations

In this area of study students cover graphical treatment of limits, continuity and differentiability of functions of a single real variable, and differentiation, anti-differentiation and integration of these functions. This material is to be linked to applications in practical situations.

This area of study includes:

- deducing the graph of the derivative function from the graph of a given function and deducing the graph of an anti-derivative function from the graph of a given function
- derivatives of $x^n$ for $n \in Q$, $e^x$, $\log_e(x)$, $\sin(x)$, $\cos(x)$ and $\tan(x)$

- derivatives of $f(x) \pm g(x), f(x) \times g(x), \frac{f(x)}{g(x)}$, and $(f \circ g)(x)$ where $f$ and $g$ are polynomial, exponential, circular, logarithmic or power functions and transformations or simple combinations of these functions
- application of differentiation to graph sketching and identification of key features of graphs, including stationary points and points of inflection, and intervals over which a function is strictly increasing or strictly decreasing
- identification of local maximum/minimum values over an interval and application to solving optimisation problems in context, including identification of interval endpoint maximum and minimum values
- anti-derivatives of polynomial functions and functions of the form $f(ax + b)$ where $f$ is $x^n$ for $n \in Q$, $e^x$, $\sin(x)$, $\cos(x)$ and linear combinations of these
- informal consideration of the definite integral as a limiting value of a sum involving quantities such as area under a curve and approximation of definite integrals using the trapezium rule
- anti-differentiation by recognition that $F'(x) = f(x)$ implies $\int f(x)\,dx = F(x) + c$ and informal treatment of the fundamental theorem of calculus, $\int_a^b f(x)\,dx = F(b) - F(a)$
- properties of anti-derivatives and definite integrals
- application of integration to problems involving finding a function from a known rate of change given a boundary condition, calculation of the area of a region under a curve and simple cases of areas between curves, average value of a function and other situations.

# 3.1 Average and instantaneous rates of change

This section reviews the concepts of average and instantaneous rates of change, tangents to a graph leading to the concept of the **derivative function**.

## 3.1.1 Gradients

Consider the parabola with the equation $y = x^2 + 1$.

The tangent to the graph at the point $(1, 2)$ has the equation $y = 2x$.

This means that the gradient of the curve $y = x^2 + 1$ at $x = 1$ is 2.

The gradient of a curve constantly changes, as the curve changes.

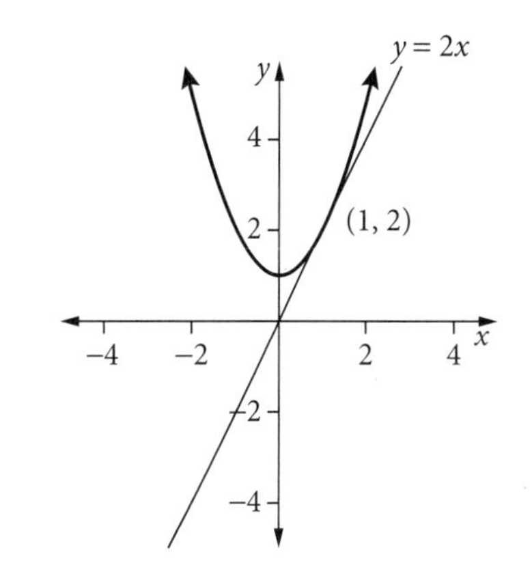


The graph shows the gradient, $m$, at $x = 0, 1, 2$.

This is called the **instantaneous rate of change**.

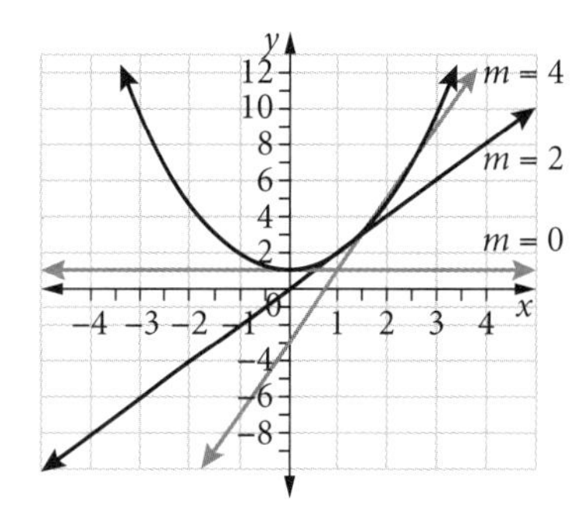


## 3.1.2 Average rate of change

Consider again the graph of the function $f(x) = x^2 + 1$.

The **average rate of change** between the points $(0, 1)$ and $(2, 5)$ is the gradient of the chord/secant shown on the graph.

$$\text{average rate of change} = \frac{\text{rise}}{\text{run}} = \frac{y_2 - y_1}{x_2 - x_1}$$

Average rate of change between the points $(0, 1)$ and $(2, 5)$ is $\frac{5-1}{2-0} = \frac{4}{2} = 2$.

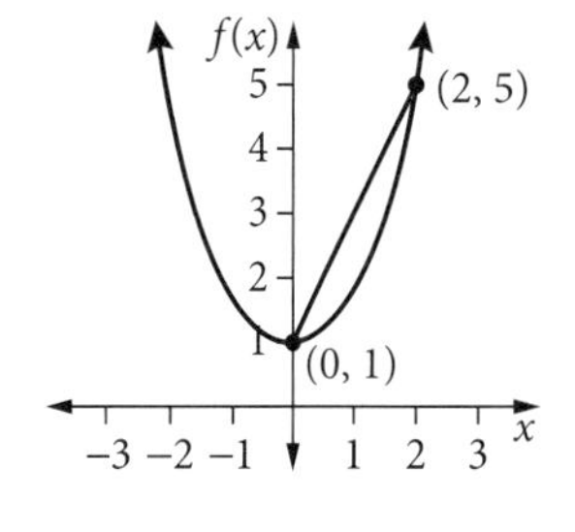


### 3.1.2.1 Developing first principles

Average rate of change between the points $(x, f(x))$ and $(x + h, f(x + h))$ equals $\frac{f(x+h) - f(x)}{x + h - x}$.

$$\text{average rate of change} = \frac{f(x+h) - f(x)}{h}$$

f(x)

A

(x + h, f(x + h))

B

(x, f(x))

x

Instantaneous rate of change uses the concept of average rate of change, where $h$ becomes continually smaller: $h \to 0$

$$\text{instantaneous rate of change} = \lim_{h \to 0} \frac{f(x+h) - f(x)}{h}$$

**Example 1**

For the curve $f(x) = x^2 + 1$, find $f'(x)$, using first principles formula $f'(x) = \lim_{h \to 0} \frac{f(x+h) - f(x)}{h}$.

**Solution**

$$f'(x) = \lim_{h \to 0} \frac{(x+h)^2 + 1 - (x^2 + 1)}{h}$$

Continue to simplify, cancelling terms:

$$\begin{aligned} f'(x) &= \lim_{h \to 0} \frac{x^2 + 2xh + h^2 + 1 - x^2 - 1}{h} \\ &= \lim_{h \to 0} \frac{2xh + h^2}{h} \\ &= \lim_{h \to 0} \frac{h(2x+h)}{h} \\ &= \lim_{h \to 0} (2x + h) \end{aligned}$$

Let $h \to 0$: $= 2x$

**Hint**
$f(x) = x^2 + 1$ and $f(x+h) = (x+h)^2 + 1$ which gives the coordinates $(x, x^2 + 1)$ and $(x + h, (x + h)^2 + 1)$.

## 3.1.3 Summary

The derivative of a function gives the instantaneous rate of change of a function. This is also known as the gradient of the function or the gradient of the tangent to a curve at a point.

- Constant rate of change — the gradient does not change, so the function is a straight line.
- Average rate of change — the gradient of a chord between two points $(x_1, y_1)$ and $(x_2, y_2)$. In this case, do not differentiate.
- Instantaneous rate of change — the gradient at a particular point. Find the derivative of the function when substituting a particular point.

Find $f'(x)$ if $f(x) = x^2 + 1$. This gives $f'(x) = 2x$.

Then at a point find $f'(3)$ if $f(x) = x^2 + 1$.
This gives $f'(3) = 2 \times 3 = 6$.

```
d/dx (x^2+1)                2·x
diff(x^2+1,x,1,3)           6
```

The derivative or instantaneous rate of change of a polynomial function is given in two separate notations as

$f(x) = ax^n$ then $f'(x) = anx^{n-1}$ or $y = ax^n$ then $\frac{dy}{dx} = anx^{n-1}$

**Example 2**

Find the derivative of $f(x) = 2x^3 + x$ at $x = 3$, using gradient formula $f'(x) = anx^{n-1}$.

**Solution**

$$f(x) = 2x^3 + x$$
$$\Rightarrow f'(x) = 6x^2 + 1$$

Substitute $x = 3$:

$$\begin{aligned} f'(3) &= 6 \times 3^2 + 1 \\ &= 55 \end{aligned}$$

**Hint**
Apply the gradient formula to each separate term.

# 3.2 Graphs of derivatives

This section will cover

- deducing the graph of the derivative function from the graph of a given function and deducing the graph of an anti-derivative function from the graph of a given function.

VCE Mathematics Study Design 2023–2027 p. 99, © VCAA 2022

## 3.2.1 Graphs of the gradient function

The **gradient function** can be sketched when given the graph of the original function. If you know the equation of the function, you can use CAS to sketch the graph of the gradient function.

### Example 3

Define the quadratic function $f(x) = x^2 - 4x - 10$ and then graph the linear derivative function $f'(x) = 2x - 4$.

**Hint**

Note the point $x = 2$ is the turning point of the parabola and also the root of the gradient graph where the gradient is 0.

### Solution

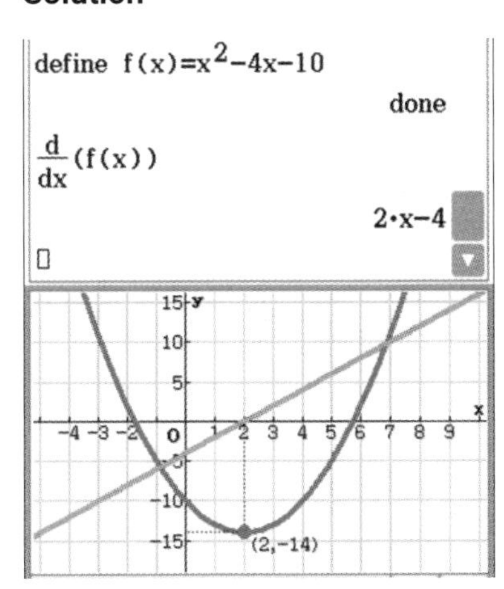


We can sketch graphs of the gradient function without knowing the equation of the original function by using information about the gradient.

Consider the graph of $f(x)$ below. The graph of $f'(x)$ is shown in the graph on the right-hand side.

The graph shown looks like a +ve cubic.

The gradient graph will therefore be a +ve quadratic.

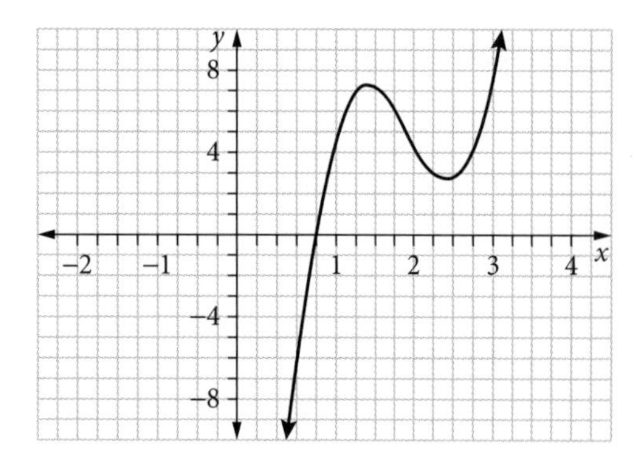


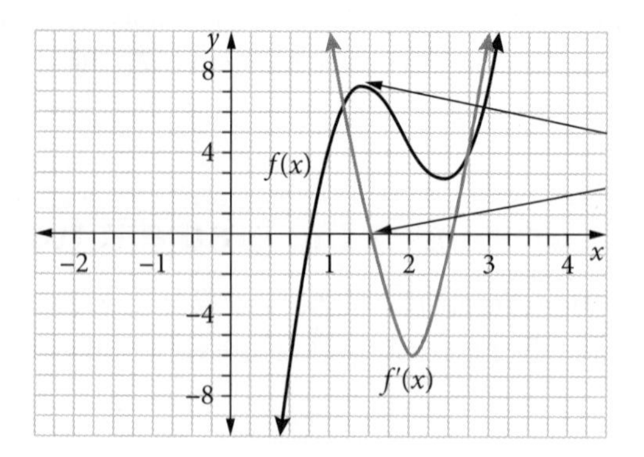


The stationary points on $f(x)$ match the $x$-intercepts on $f'(x)$.

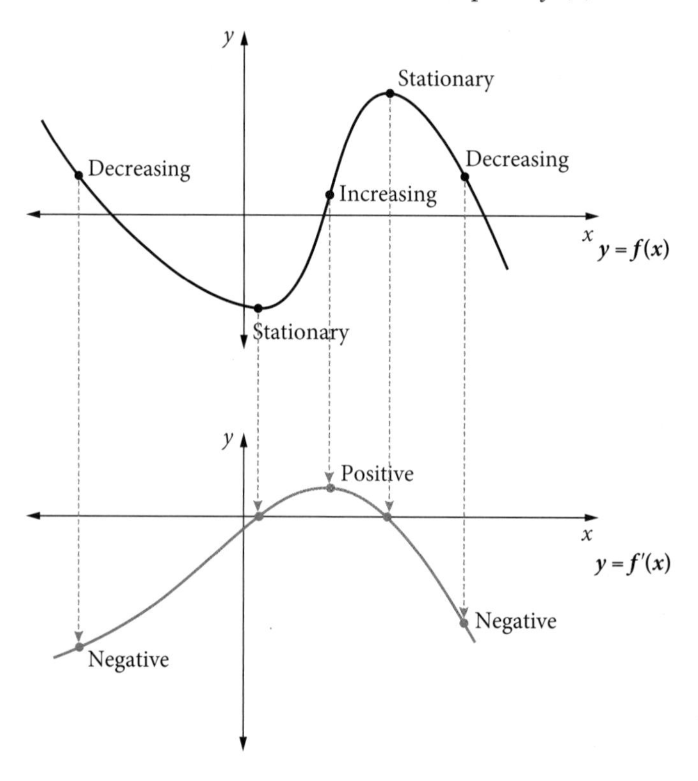


## 3.2.2 Graphs of the anti-derivative function

**Anti-differentiation** is the reverse process of differentiation, so we reverse the process described previously.

We can sketch the graph of the **anti-derivative** function when given the graph of a gradient function. If the equation of the gradient function is known, use CAS to sketch the graph of the anti-derivative function.

## Example 4

Define $f'(x) = 4x^2 + 4x + 2$ and then graph the anti-derivative function $f(x) = \frac{4}{3}x^3 + 2x^2 + 2x + c$.

### Solution

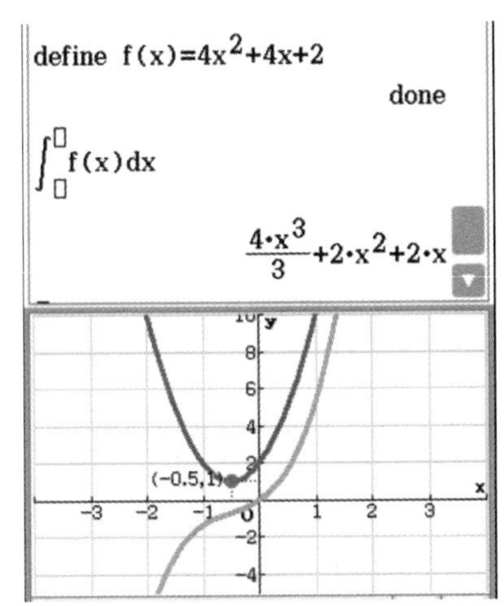


**Hint**

Note the point (−0.5, 1) is the turning point of the original parabola and also the $x$ value on the anti-derivative cubic graph where the gradient would be equal to 1.

The + $c$ term reminds us that $f'(x)$ has an infinite number of anti-derivatives. In this case, sketch the general shape of the function without the specific value of the + $c$. Note that CAS does not include the + $c$.

We can sketch graphs of the anti-derivative function without knowing the equation of the gradient function by using information about the anti-derivative.

Consider the graph of $f'(x)$ below. Use it to sketch the graph of $f(x)$.

The graph shown looks like a +ve cubic.

The anti-derivative graph will therefore be a +ve quartic.

The stationary point on $f(x)$ matches the $x$-intercept on $f'(x)$.

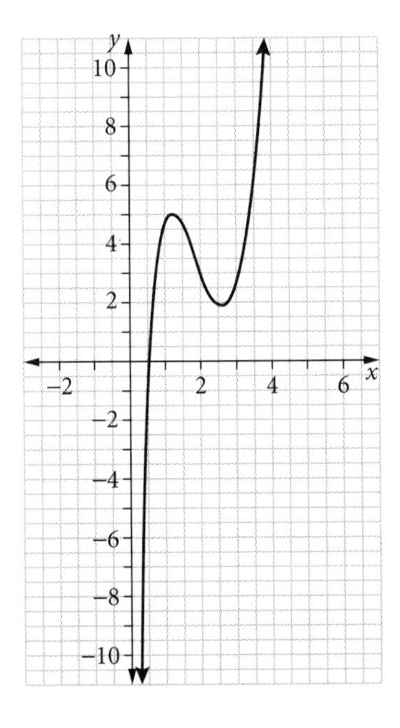

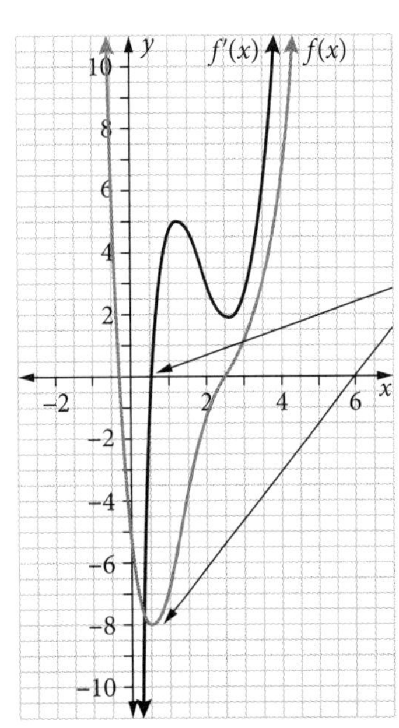

# 3.3 Derivatives of basic functions

This section will cover

- derivatives of $x^n$ for $n \in Q$, $e^x$, $\log_e(x)$, $\sin(x)$, $\cos(x)$ and $\tan(x)$.

VCE Mathematics Study Design 2023–2027 p. 99, © VCAA 2022

## 3.3.1 Summary of basic derivatives

| | |
|---|---|
| $y = x^n$ | $\frac{dy}{dx} = nx^{n-1}$ |
| $y = ax^n$ | $\frac{dy}{dx} = anx^{n-1}$ |
| $y = e^x$ | $\frac{dy}{dx} = e^x$ |
| $y = e^{kx}$ | $\frac{dy}{dx} = ke^{kx}$ |
| $y = \log_e(x)$ | $\frac{dy}{dx} = \frac{1}{x}$ |
| $y = \log_e(kx)$ | $\frac{dy}{dx} = \frac{k}{kx} = \frac{1}{x}$ |
| $y = \sin(x)$ | $\frac{dy}{dx} = \cos(x)$ |
| $y = \sin(kx)$ | $\frac{dy}{dx} = k\cos(kx)$ |
| $y = \cos(x)$ | $\frac{dy}{dx} = -\sin(x)$ |
| $y = \cos(kx)$ | $\frac{dy}{dx} = -k\sin(kx)$ |
| $y = \tan(x)$ | $\frac{dy}{dx} = \sec^2(x)$ <br> **Note** <br> $\sec(x) = \frac{1}{\cos(x)}$ and $\sec^2(x) = \frac{1}{\cos^2(x)}$ |
| $y = \tan(kx)$ | $\frac{dy}{dx} = k\sec^2(kx)$ |

Using the table above, we can find the instantaneous rate of change (the derivative) of a variety of different functions.

Given the function $y = 3x^6$:

Using the rule $y = ax^n$, $\frac{dy}{dx} = anx^{n-1}$, we get $\frac{dy}{dx} = 18x^5$.

**Example 5**

a For the function $f(x) = 2\cos(4x)$, find $f'(x)$.

b For the function $y = \log_e(4x)$, find $\frac{dy}{dx}$.

c For the function $f(x) = -2e^{3x}$, find $f'(x)$.

CAS can be used to find derivatives.

**Solution**

a Using the rule $f'(x) = -k\sin(kx)$, we get $f'(x) = -8\sin(4x)$.

b Using the rule $\frac{dy}{dx} = \frac{k}{kx} = \frac{1}{x}$, we get $\frac{dy}{dx} = \frac{4}{4x} = \frac{1}{x}$.

**Note**

In this case, we can go straight to the answer $\frac{1}{x}$.

c Using the rule $f'(x) = ke^{kx}$, we get $f'(x) = -6e^{3x}$.

$$\frac{d}{dx}(2\cdot\cos(4\cdot x))$$
$$-8\cdot\sin(4\cdot x)$$
$$\frac{d}{dx}(\ln(4\cdot x))$$
$$\frac{1}{x}$$
$$\frac{d}{dx}(-2e^{3x})$$
$$-6\cdot e^{3\cdot x}$$

## 3.4 Derivative rules

This section will cover

- derivatives of $f(x) \pm g(x)$, $f(x) \times g(x)$, $\frac{f(x)}{g(x)}$, and $(f \circ g)(x)$ where $f$ and $g$ are polynomial, exponential, circular, logarithmic or power functions and transformations or simple combinations of these functions.

VCE Mathematics Study Design 2023–2027 p. 100, © VCAA 2022

We use the **product rule**, **quotient rule** and **chain rules** to derive more complicated functions.

### 3.4.1 Properties of derivatives

- The derivative of the scalar multiple of a function equals the scalar multiple of the derivative.

$$(a\,f(x))' = a\,f'(x)$$

- The derivative of the sum, or difference, equals the sum, or difference, of the derivatives.

  Considering the sum and difference of functions: $f(x) \pm g(x)$,

$$(a\,f(x) \pm b\,g(x))' = a\,f'(x) \pm b\,g'(x)$$

### 3.4.1.1 Product rule

Considering the product of functions: $f(x) \times g(x)$.

For $y = f(x) \times g(x)$, we get

$$\frac{dy}{dx} = f(x)\, g'(x) + g(x)\, f'(x)$$

Or for $y = u \times v$:

$$\frac{dy}{dx} = u \times \frac{dv}{dx} + v \times \frac{du}{dx}$$

---

**Example 6**

For the function $y = e^{3x} \sin(2x)$, find $\frac{dy}{dx}$.

**Solution**

Using the rule $\frac{dy}{dx} = u \times \frac{dv}{dx} + v \times \frac{du}{dx}$:

$u = e^{3x}$ and $v = \sin(2x)$

$\frac{du}{dx} = 3e^{3x}$ and $\frac{dv}{dx} = 2\cos(2x)$

$$\frac{dy}{dx} = e^{3x} \times 2\cos(2x) + \sin(2x) \times 3e^{3x}$$

Simplify: $\frac{dy}{dx} = 2e^{3x}\cos(2x) + 3e^{3x}\sin(2x)$

**Note**

Be aware that CAS may give quite a different version of the answer you get 'by-hand'.

simplify$\left(\frac{d}{dx}\left(e^{3x}\sin(2x)\right)\right)$

$(2\cdot\cos(2\cdot x)+3\cdot\sin(2\cdot x))\cdot e^{3\cdot x}$

### 3.4.1.2 Quotient rule

Considering the quotient of functions $\frac{f(x)}{g(x)}$.

For $y = \frac{f(x)}{g(x)}$, we get

$$\frac{dy}{dx} = \frac{g(x)f'(x) - f(x)g'(x)}{(g(x))^2}$$

Or for $y = \frac{u}{v}$:

$$\frac{dy}{dx} = \frac{v\frac{du}{dx} - u\frac{dv}{dx}}{v^2}$$

---

**Example 7**

For the function $y = \frac{e^{3x}}{\sin(2x)}$, find $\frac{dy}{dx}$.

**Solution**

Using the rule $\frac{dy}{dx} = \frac{v\frac{du}{dx} - u\frac{dv}{dx}}{v^2}$:

$u = e^{3x}$ and $v = \sin(2x)$

$\frac{du}{dx} = 3e^{3x}$ and $\frac{dv}{dx} = 2\cos(2x)$

$$\frac{dy}{dx} = \frac{\sin(2x) \times 3e^{3x} - e^{3x} \times 2\cos(2x)}{\sin^2(2x)}$$

Simplify: $\frac{dy}{dx} = \frac{3e^{3x}\sin(2x) - 2e^{3x}\cos(2x)}{\sin^2(2x)}$

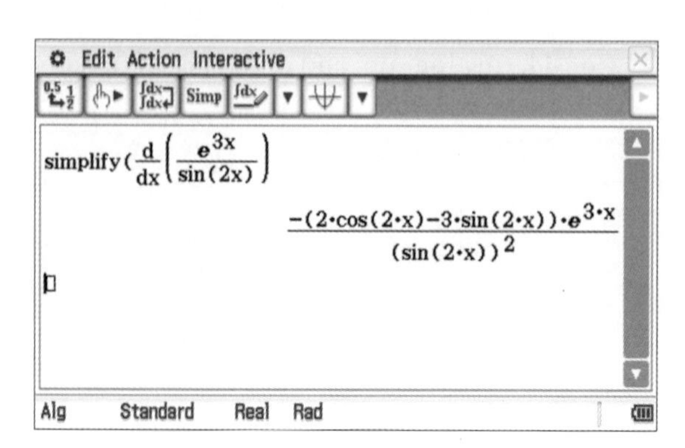

### 3.4.1.3 Chain rule

Considering combinations of functions $y = f(g(x))$ or $(f \circ g)(x)$, we use the **chain rule**.

For $y = f(g(x))$, we get

$$\frac{dy}{dx} = f'(g(x)) \times g'(x)$$

Or in terms of $u$ and $v$:

$$\frac{dy}{dx} = \frac{dy}{du} \times \frac{du}{dx}$$

This rule can be seen as the derivative of the 'outer' ($y$) times the derivative of the 'inner' ($u$).



---

**Example 8**

For the function $y = \cos(3x^2 + 1)$, find $\frac{dy}{dx}$.

**Solution**

Using the rule $\frac{dy}{dx} = \frac{dy}{du} \times \frac{du}{dx}$:

$u = 3x^2 + 1$ and $y = \cos(u)$

$\frac{du}{dx} = 6x$ and $\frac{dy}{du} = -\sin(u)$

$$\frac{dy}{dx} = -\sin(u) \times 6x$$

Simplify, replacing $u = 3x^2 + 1$: $\frac{dy}{dx} = -6x\sin(3x^2 + 1)$

simplify$\left(\frac{d}{dx}(\cos(3x^2+1))\right)$

$-6 \cdot x \cdot \sin(3 \cdot x^2+1)$

You will need to differentiate a function that is a mixture of product and chain rule or quotient and chain rule.

---

**Example 9**

For the function $y = \sin(x) \times \log_e(x + x^4)$, find $f'(x)$.

**Solution**

Using the **Product rule:** $\frac{dy}{dx} = f(x)g'(x) + g(x)f'(x)$

$$\frac{dy}{dx} = \sin(x) \times \frac{d}{dx}(\log_e(x + x^4)) + \log_e(x + x^4) \times \cos(x)$$

For the **Chain rule:** $u = x + x^4$ and $y = \log_e(u)$

$$\therefore \frac{du}{dx} = 1 + 4x^3 \text{ and } \frac{dy}{du} = \frac{1}{u}$$

Find $\frac{d}{dx}(\log_e(x + x^4))$:

$$\frac{d}{dx}(\log_e(x + x^4)) = (1 + 4x^3) \times \frac{1}{u}$$

giving $\frac{d}{dx}(\log_e(x + x^4)) = (1 + 4x^3) \times \frac{1}{x + x^4} = \frac{1 + 4x^3}{x + x^4}$.

Combine Product and Chain rule and simplify: $\frac{dy}{dx} = \sin(x) \times \frac{1 + 4x^3}{x + x^4} + \log_e(x + x^4) \times \cos(x)$

$$\frac{dy}{dx} = \frac{1 + 4x^3}{x + x^4}\sin(x) + \cos(x)\log_e(x + x^4)$$

# 3.5 Applications of differentiation

This section will cover

- application of differentiation to graph sketching and identification of key features of graphs, including stationary points and points of inflection, and intervals over which a function is strictly increasing or strictly decreasing.



## 3.5.1 Stationary points

A **stationary point** is any point on the graph of $y = f(x)$ where $f'(x) = 0$.

There are three types of stationary points:

**local minimum turning point**

$x < a, f'(x) < 0$
$x = a, f'(x) = 0$
$x > a, f'(x) > 0$

**local maximum turning point**

$x < b, f'(x) > 0$
$x = b, f'(x) = 0$
$x > b, f'(x) < 0$

**stationary point of inflection**

$x < c, f'(x) > 0$
$x = c, f'(x) = 0$
$x > c, f'(x) > 0$

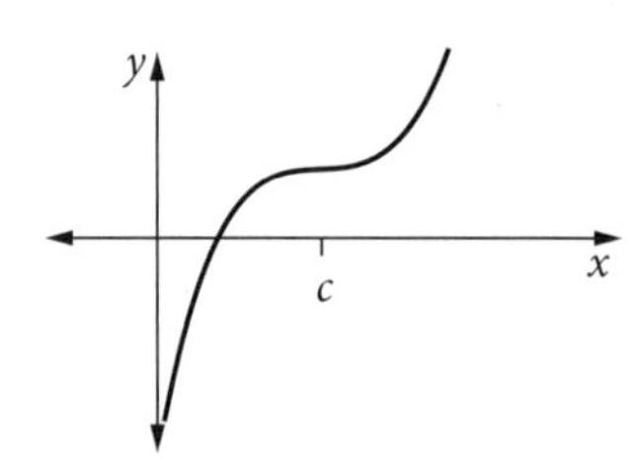


or

$x < c, f'(x) < 0$
$x = c, f'(x) = 0$
$x > c, f'(x) < 0$

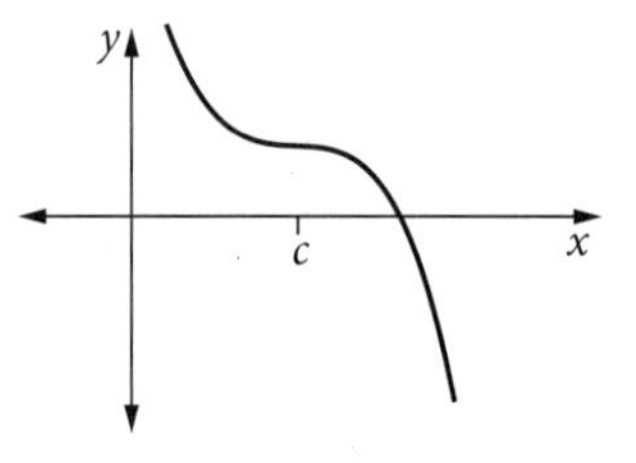


### 3.5.1.1 Coordinates and nature of stationary points

When asked to find the coordinates and nature of stationary points,

1 find where $f'(x) = 0$.

2 decide which of the three types of stationary points are being used.

This can be done either by

- using the sign diagram

  Test $x$ values on either side of the point found, where $f'(x) = 0$, to find the sign of the gradient.

- looking at the shape of the graph

  Write 'From the graph it can be seen that …'

- or by using the 2nd derivative test.
  - Where $f''(x) < 0$, the graph may have a local maximum.
  - Where $f''(x) > 0$, the graph may have a local minimum.
  - Where $f''(x) = 0$, the graph may have a stationary point of inflection.

9780170465366

- Further investigation is required to distinguish between such cases. In particular, where $f''(x) = 0$, if the graph has a stationary point of inflection, the concavity of the graph must necessarily change either side of the stationary point.

**Note**

Some ambiguities arise when using the 2nd derivative test when repeated factors occur at the same point. Consideration of the 2nd derivative is not required, but can be used, in Maths Methods exams.



### Example 10

Find the coordinates and nature of the stationary points of $f(x) = 2x^3 + x^2 + 2$.

Solve $f'(x) = 0$.

### Solution

Find where $f'(x) = 0$:

$f(x) = 2x^3 + x^2 + 2$ gives $f'(x) = 6x^2 + 2x$

$$6x^2 + 2x = 0$$

$$2x(3x + 1) = 0$$

Thus, $x = 0$ and $x = -\frac{1}{3}$.

```
Define f(x)=2·x^3+x^2+2
                                done
solve(d/dx(f(x))=0, x)
                      {x=0, x=-1/3}
```

To find the coordinates, substitute these $x$ values back into the original function.

Coordinates of stationary points of the graph are $(0, 2)$ and $\left(-\frac{1}{3}, \frac{55}{27}\right)$.

```
f(0)
                                  2
f(-1/3)
                              55/27
```

Decide which of the three types of stationary points are being used.

Use the sign diagram.

Test $x$ values on either side of the point found, where $f'(x) = 0$, to find the sign of the gradient.

Test: $x = 0$

$x < 0$ $\left(\text{say } x = -\frac{1}{4}\right)$ gives $f'\left(-\frac{1}{4}\right) = -\frac{1}{8} < 0$

$x > 0$ (say $x = 1$) gives $f'(1) = 8 > 0$

So $x = 0$ is a local minimum. ⋃

And testing the other point:

Test: $x = -\frac{1}{3}$

$x < -\frac{1}{3}$ (say $x = -1$) gives $f'(-1) = 4 > 0$

$x > -\frac{1}{3}$ $\left(\text{say } x = -\frac{1}{4}\right)$ gives $f'\left(-\frac{1}{4}\right) = -\frac{1}{8} < 0$

So $x = -\frac{1}{3}$ is a local maximum. ⋂

**Stationary points of inflection** will have the same $x$ value where $f'(x) = 0$ and $f''(x) = 0$, necessarily with the concavity of the graph changing either side of the stationary point.

Consider $f(x) = x^4 + 2x^3 - 2x - 1$.

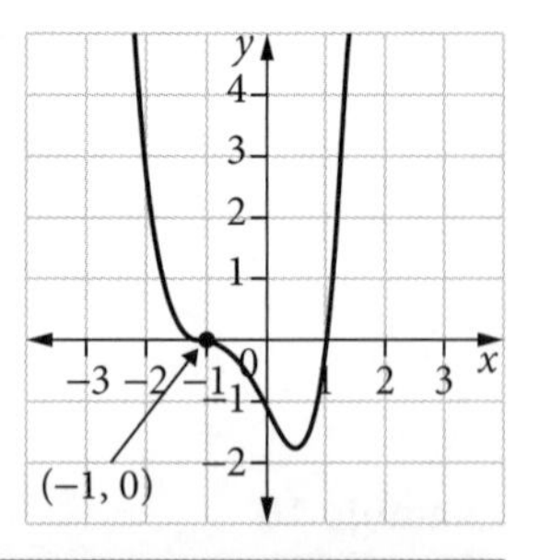


Solving $f'(x) = 4x^3 + 6x^2 - 2 = 0$ for stationary points gives $x = -1$ and $x = \frac{1}{2}$.

Solving $f''(x) = 12x^2 + 12x = 0$ for the point of inflection gives $x = -1$ and $x = 0$.

The solution $x = -1$ is the same $x$ value where $f'(x) = 0$ and $f''(x) = 0$.

Stationary point of inflection at $x = -1$.

For the function where $f(x) = x^4 + 2x^3 - 2x - 1$, it is found that $(-1, 0)$ is a stationary point of inflection.

**Hint**

Check that concavity changes in the graph either side of $x = -1$.

## 3.5.2 Strictly increasing and strictly decreasing

- A function $f(x)$ is said to be **strictly increasing** on an interval if whenever $a$, $b$ are in the interval and $a < b$, then $f(b) > f(a)$.
- A function is **strictly increasing** where the values of $y$ are getting larger. If $a < b$, then $f(a) < f(b)$.

**Note**

This is a '$y$ value' discussion rather than a 'gradient' discussion. Don't confuse this for a +ve and −ve gradient question. A function can be strictly increasing, or decreasing, even at a point where the gradient is zero.

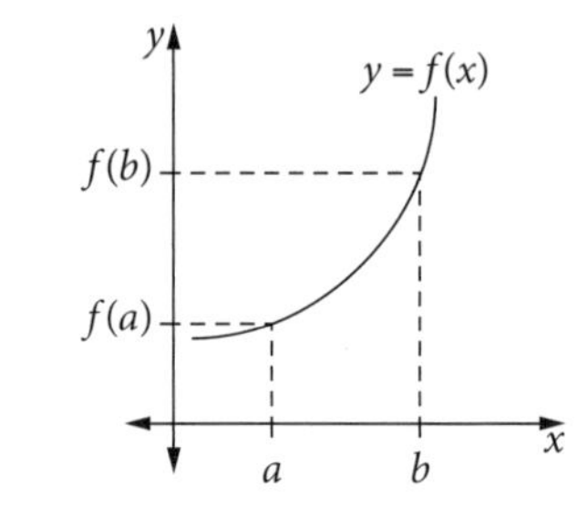


When $f(x)$ is strictly increasing $f(a) < f(b)$.

- A function $f(x)$ is said to be **strictly decreasing** on an interval if whenever $a$, $b$ are in the interval and $a < b$ then $f(a) > f(b)$.
- A function is **strictly decreasing** where the values of $y$ are getting smaller. If $a < b$, then $f(a) > f(b)$.

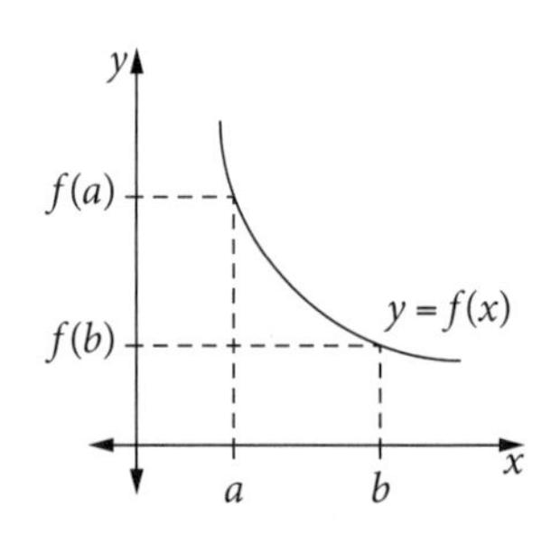


When $f(x)$ is strictly decreasing $f(a) > f(b)$.

Consider the graph of $f(x) = \frac{1}{9}x^3 - \frac{1}{3}x^2 - x$ where the $x$-coordinates of the stationary points are $x = -1$ and $x = 3$.

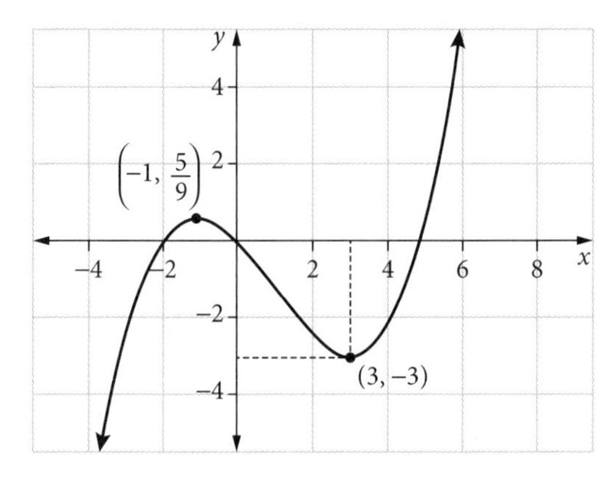


**Note**
Note the square brackets except for ∞ which we approach but never reach.

- The function is **strictly increasing** for $x \in (-\infty, -1]$ and for $x \in [3, \infty)$.
- The function is **strictly decreasing for** $x \in [-1, 3]$.

---

**Example 11**

For $f(x) = x^4 + 2x^3 - 2x - 1$, find where the graph is

- strictly increasing
- strictly decreasing

**Solution**

Find the stationary points where $f'(x) = 0$.

$f'(x) = 4x^3 + 6x^2 - 2 = 0$ gives $x = -1$ and $x = \frac{1}{2}$.

**strictly increasing** for $x \in \left[\frac{1}{2}, \infty\right)$.

**strictly decreasing** for $x \in (-\infty, -1)$ and for $(-1, \frac{1}{2}]$.

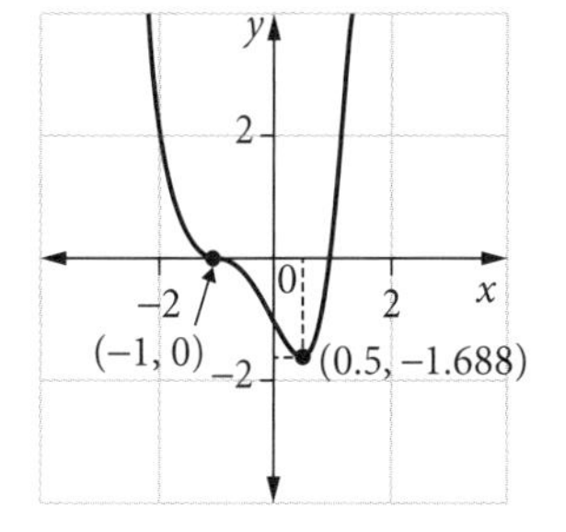


**Note**
Note the square brackets and the inclusion of $x = -1$ in the strictly decreasing section, even though the gradient at $x = -1$ is zero.

## 3.5.2.1 Maximum rate of increase or decrease

For the graph of $f(x) = \frac{1}{3}x^3 - x^2 - 3x$, the **maximum rate of decrease** occurs in the interval $x \in (-1, 3)$.

To find the maximum rate of decrease, the gradient should be maximised (negative).

Differentiating the function gives $f'(x) = x^2 - 2x - 3$ and differentiating the gradient function gives $f''(x) = 2x - 2$.

The maximum rate of decrease occurs when $2x - 2 = 0$.

So at $x = 1$, the rate is $f'(1) = -4$.

That is, the maximum rate of decrease is 4.

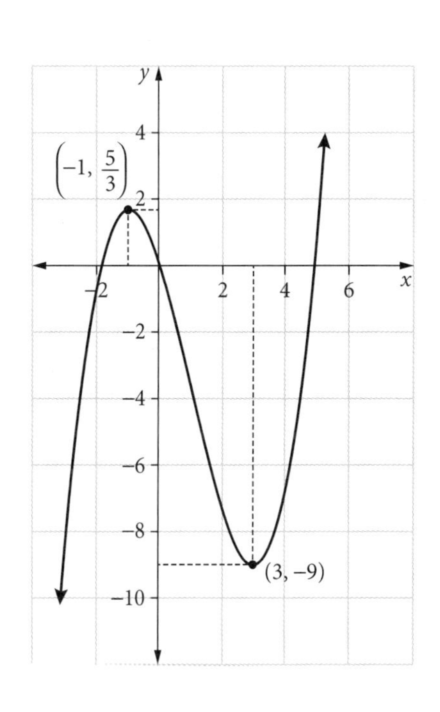

The point $\left(1, -\frac{11}{3}\right)$ is called a **non-stationary point of inflection**.

The **maximum rate of increase** or decrease is the point on the graph where the gradient is the steepest.

CAS finds the 2nd derivative. In this case:

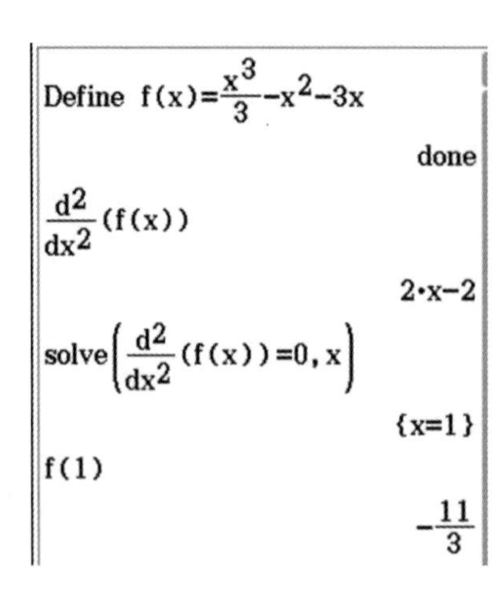


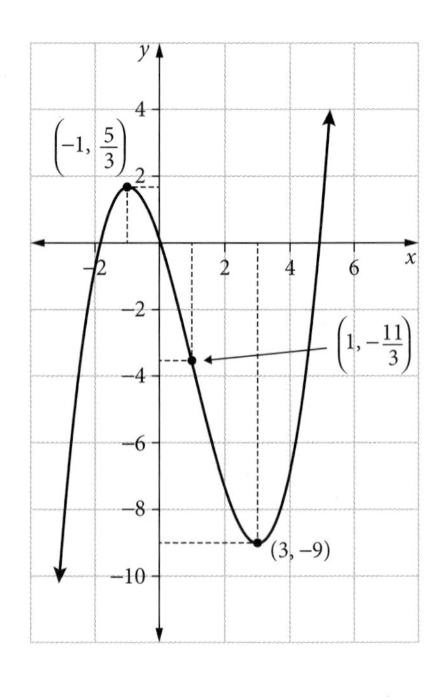


## 3.5.3 Equations of tangents

To find the equation of a **tangent** to a curve $y = f(x)$ at the point $(x_1, y_1)$, use the equation

$$y - y_1 = m(x - x_1)$$

where $m$ = gradient of the curve at $x = x_1$.

We can use CAS to find the equation of the tangent at a given point.

Consider the equation of the tangent to the graph of $y = \sin(x + \pi)$ at $x = \frac{\pi}{3}$.

define f(x)=sin(x+π)
done
tanLine(f(x), x, π/3)
$\frac{-x}{2}+\frac{\pi}{6}-\frac{\sqrt{3}}{2}$

(1.0472,−0.866)

The equation of the tangent is $y = -\frac{x}{2} + \frac{\pi}{6} - \frac{\sqrt{3}}{2}$.

**Hint**
Make sure to write $y =$ to make this an equation. CAS doesn't include $y =$ in this case.

**Example 12**

For $f(x) = x^2 - 2x - 3$, find the equation of the

**a** tangent line

**b** line perpendicular to the tangent at $x = 3$.

**Solution**

**a** Find the gradient at $x = 3$.

$f'(x) = 2x - 2$
$f'(3) = 4$
Point $(3, 0)$

To find the tangent line at $x = 3$, use the equation $y - y_1 = m(x - x_1)$.

$y - y_1 = m(x - x_1)$
$y - 0 = 4(x - 3)$
$y = 4x - 12$

**b** Find the perpendicular gradient at $x = 3$.

perpendicular gradient $= -\frac{1}{4}$

$y - 0 = -\frac{1}{4}(x - 3)$

To find the line perpendicular to the curve at $x = 3$, use the equation $y - y_1 = m(x - x_1)$

$y = -\frac{1}{4}x + \frac{3}{4}$

## 3.6 Maxima and minima

This section will cover

- identification of local maximum/minimum values over an interval and application to solving optimisation problems in context, including identification of interval endpoint maximum and minimum values.

VCE Mathematics Study Design 2023–2027 p. 100, © VCAA 2022

Differentiation can be used to help solve practical applications such as the minimum surface area of a material needed to construct a container or the maximum profit a company can make.

### 3.6.1 Local and absolute maxima and minima

A **local maximum and minimum point** is found where $f'(x) = 0$.

An **absolute** or **global maximum and minimum point** is the highest or lowest point over the whole interval. These are not necessarily the same point as the local maximum or minimum.

Consider the graph of $f: [-2, 5] \to R, f(x) = -(x - 1)(x + 1)(x - 4)$.

The graph below shows an absolute maximum at the point $(-2, 18)$ and an absolute minimum of $(5, -24)$.

However, the local maximum within its locality is at $x = \frac{4 + \sqrt{19}}{3}$

and the local minimum within its locality is at $x = \frac{4 - \sqrt{19}}{3}$. These are different points.

So particularly when a graph is restricted, make sure to check the endpoints.

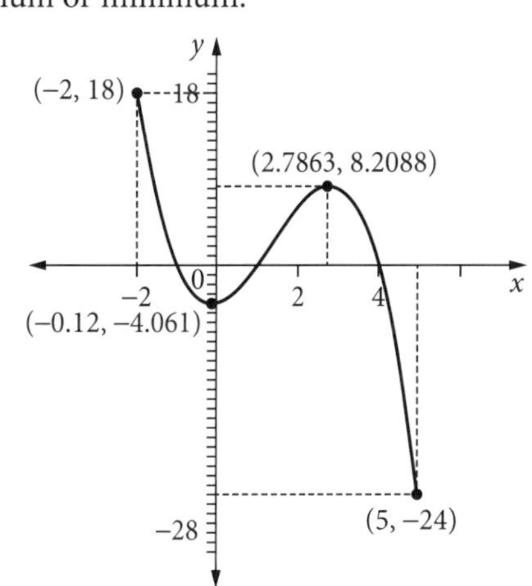

## 3.6.2 Finding the endpoint maximum and minimum values over an interval

Sometimes the maximum or minimum values do not correspond to the point where there is a local maximum turning point or local minimum turning point within the domain. We need to check the ends of the domain.

---

### Example 13

A model for the amount of medication in mg/L after $x$ days is $y = 2\sqrt{x}e^{-0.1x} + 4$ for the domain $x \in [0, 15]$.

Find the maximum and minimum doses.

### Solution

Sketch the graph of $y = 2\sqrt{x}e^{-0.1x} + 4$ for $x \in [0, 15]$.

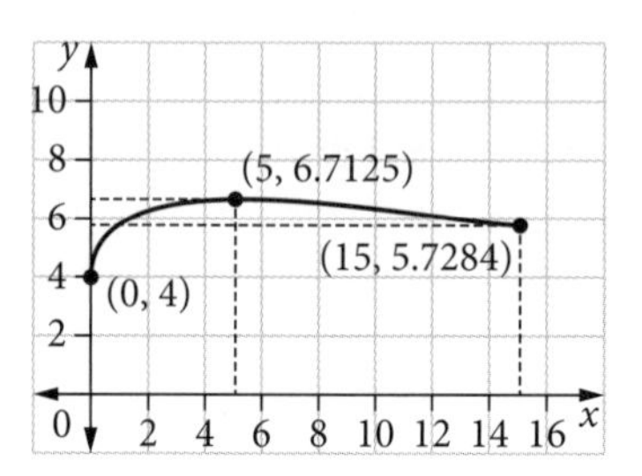


Find maximum and minimum doses, checking endpoints as well.

Maximum dose of 6.71 mg/L occurs at the local maximum at $x = 5$.

Minimum dose of 4 mg/L occurs at one of the endpoints at $x = 0$.

## 3.6.3 Maximum and minimum problems

Useful steps for solving maximum and minimum problems are:

1 always draw a diagram

2 decide on and define the variables

3 establish which variable is to be maximised or minimised. Set up an equation for this variable (sometimes this equation is given).

4 calculate the derivative, equate it to zero and solve for the unknown variable

5 clarify the nature of the stationary points

6 reread the question to determine which value has to be calculated

7 check that the value found is the actual maximum or minimum required. Check the graph of the function, noting the end points of the domain. If an endpoint gives the maximum or minimum, this is called the **absolute** or **global maximum** or **minimum**.

---

### Example 14

A rectangular piece of cardboard, measuring 10 cm by 6 cm, has squares of length $x$ units cut out at each corner. The remaining faces are folded up to make a small packing box for jewellery. Find the maximum possible volume, in cm$^3$ correct to one decimal place, for these jewellery boxes.

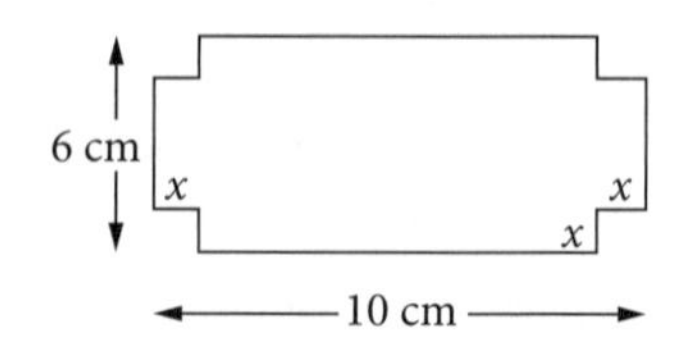

**Solution**

Sketch a diagram.

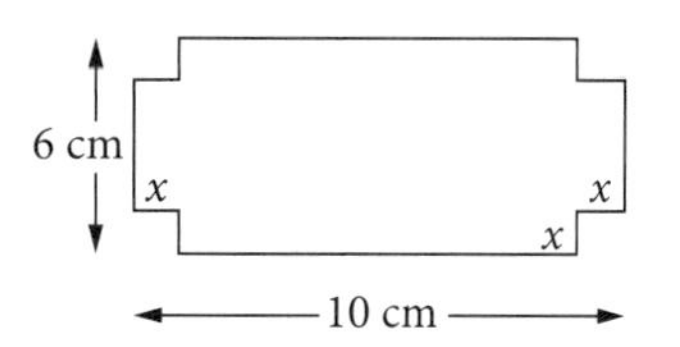


Set up an equation to describe the volume of the box.

height = $x$

length = $10 - 2x$

width = $6 - 2x$

$V(x) = x(10 - 2x)(6 - 2x)$

Solve $V'(x) = 0$ for local maximum and minimum values.

$$V(x) = x(10 - 2x)(6 - 2x) = 4x^3 - 32x^2 + 60x$$

$$V'(x) = 12x^2 - 64x + 60 = 0$$

$$x = \frac{8 \pm \sqrt{19}}{3}$$

Identify the domain for this problem.

$x > 0$ as it is a length.

$(10 - 2x) > 0$ so $x < 5$.

Also $(6 - 2x) > 0$ so $x < 3$.

The domain is $(0, 3)$.

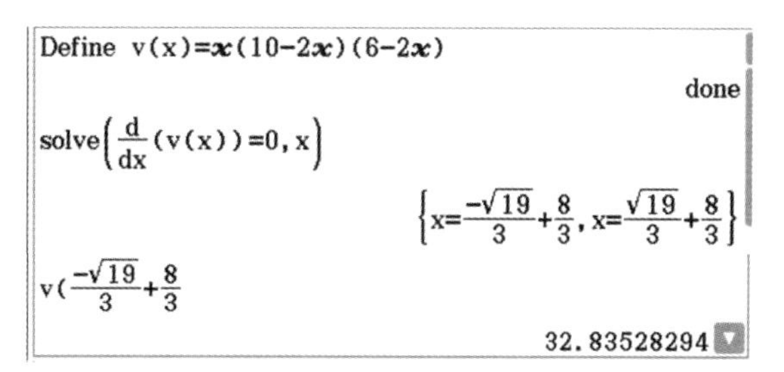


Consider the graph shape to decide which value of $x$ gives the maximum value and whether it lies in the domain $(0, 3)$.

The maximum point is at

$$x = \frac{8 - \sqrt{19}}{3} \approx 1.2137$$

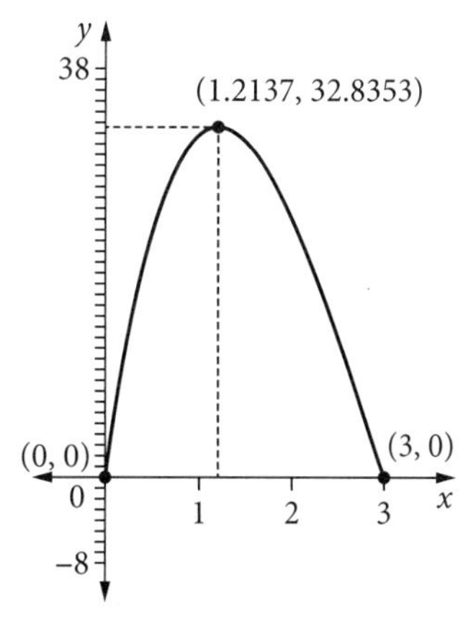


Substitute this value of $x$ into $V(x)$ to find the maximum volume of the jewellery box.

$V(1.2137) = 32.835$

The maximum possible volume of the box is $32.8\,\text{cm}^3$.

**Note**

After solving maximum and minimum problems, re-read the question to make sure you have answered it. One common mistake students make in exams is to forget to answer the question, even though all their working is correct. Do you need the $x$ or the $V$ value?

# 3.7 Anti-derivative of basic functions

This section will cover

- anti-derivatives of polynomial functions and functions of the form $f(ax+b)$, where $f$ is $x^n$ for $n \in Q$, $e^x$, $\sin(x)$, $\cos(x)$ and linear combinations of these.

VCE Mathematics Study Design 2023–2027 p. 100, © VCAA 2022

Integration (anti-differentiation) is the reverse of the differentiation process.

## 3.7.1 Summary of basic anti-derivatives

| | |
|---|---|
| $y = x^n$ | $\int x^n \, dx = \frac{x^{n+1}}{n+1} + c$ <br> **Note** Add 1 to the power, divide by the new power. |
| $y = (ax+b)^n$ | $\int (ax+b)^n \, dx = \frac{(ax+b)^{n+1}}{a(n+1)} + c$ |
| $y = e^x$ | $\int e^x \, dx = e^x + c$ |
| $y = e^{kx}$ | $\int e^{kx} \, dx = \frac{1}{k} e^{kx} + c$ |
| $y = \frac{1}{x}$ | $\int \frac{1}{x} \, dx = \log_e(x) + c$ |
| $y = \frac{1}{ax+b}$ | $\int \frac{1}{ax+b} \, dx = \frac{1}{a} \log_e(ax+b) + c$ |
| $y = \frac{f'(x)}{f(x)}$ | $\int \frac{f'(x)}{f(x)} \, dx = \log_e(f(x)) + c$ |
| $y = \sin(x)$ | $\int \sin(x) \, dx = -\cos(x) + c$ |
| $y = \sin(kx)$ | $\int \sin(kx) \, dx = -\frac{1}{k} \cos(kx) + c$ |
| $y = \cos(x)$ | $\int \cos(x) dx = \sin(x) + c$ |
| $y = \cos(kx)$ | $\int \cos(kx) dx = \frac{1}{k} \sin(kx) + c$ |
| $y = \sec^2(x)$ | $\int \sec^2(x) dx = \tan(x) + c$ |
| $y = \sec^2(kx)$ | $\int \sec^2(x) dx = \frac{1}{k} \tan(kx) + c$ |

**Example 15**

Find the anti-derivative function $f(x)$ for $f'(x) = 2\cos(4x)$.

**Solution**

Using the rule $\int \cos(kx)dx = \frac{1}{k}\sin(kx) + c$,

we get $f(x) = 2 \times \frac{1}{4}\sin(4x) + c$

$= \frac{1}{2}\sin(4x) + c$

$$\int_{\square}^{\square} 2\cdot\cos(4\cdot x)dx \qquad \frac{\sin(4\cdot x)}{2}$$

Anti-derivatives can be found on CAS.

Note that we have to add the $+ c$ (**constant of integration**).

## 3.8 Definite integral: properties and fundamental theorem of calculus

This section will cover

- informal consideration of the definite integral as a limiting value of a sum involving quantities such as area under a curve and approximation of definite integrals using the trapezium rule.
- informal treatment of the fundamental theorem of calculus, $\int_a^b f(x)\,dx = F(b) - F(a)$
- properties of anti-derivatives and definite integrals.

VCE Mathematics Study Design 2023–2027 p. 100, © VCAA 2022

### 3.8.1 Approximations leading to definite integrals

The area under a curve between $x = a$ and $x = b$ can be calculated approximately by dividing this area into several rectangles, then adding the areas of these rectangles. This method leads to the definition of a **definite integral** as the sum of these rectangles over the interval $[a, b]$ as their width tends to zero.

### 3.8.2 Approximation methods for finding areas under curves

#### 3.8.2.1 Left endpoint approximation or area (called LEA)

Draw rectangles from the left-hand $x$ value of the interval.

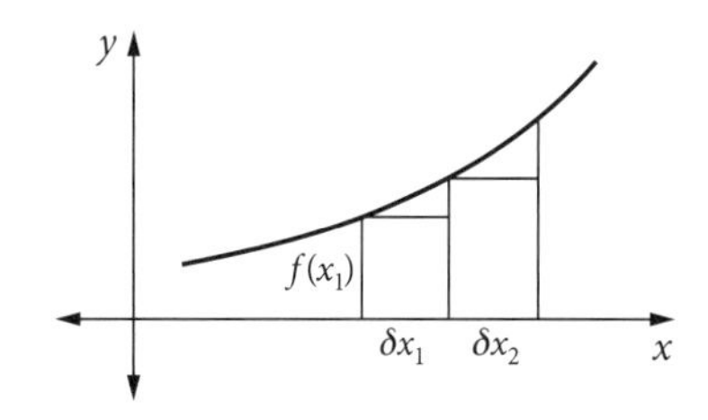


#### 3.8.2.2 Right endpoint approximation or area (called REA)

Draw rectangles from the right-hand $x$ value of the interval.

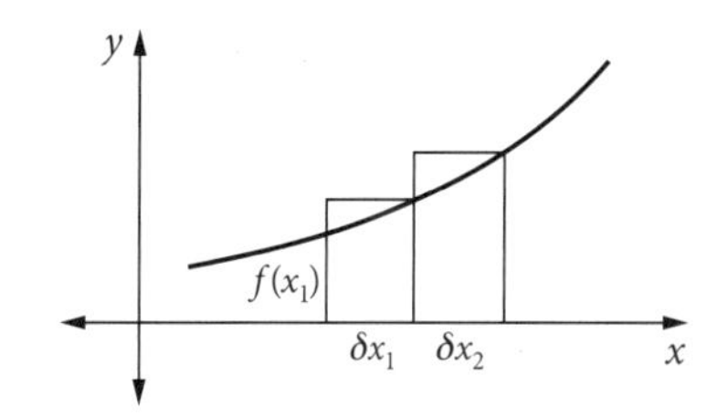


Adding the areas of these rectangles gives an estimate of the area under the curve. If we make these rectangles thinner and thinner, we get the expression $\sum_{i=1}^{n} f(x_i)\,\delta x_i$, where $\Sigma$ means 'the sum of'.

The area of the first rectangle is $f(x_1)\,\delta x_1$, the area of the second rectangle is $f(x_2)\,\delta x_2$ and so on.

The smaller the value of $\delta x_i$, the more accurate the approximation.

9780170465366

### 3.8.2.3 The trapezium rule

The average of the right endpoint rectangles and the left endpoint rectangles gives a closer approximation to the actual area under a curve and is called the **trapezium rule.**

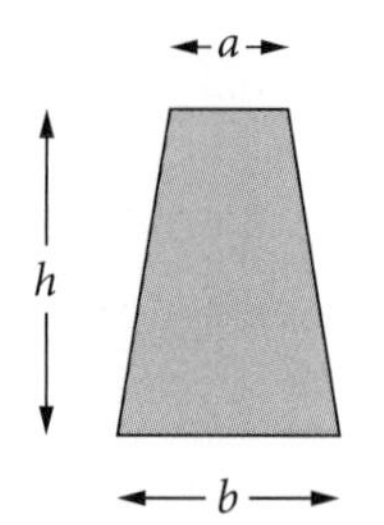


The trapezium rule uses the formula for the area of a trapezium,

$$\text{area} = \frac{(a+b)}{2}h$$

When summing multiple trapeziums, this can be written as

$$\text{area} = \frac{w}{2}(1\text{st } y_0 + y_{\text{last}} + y_1 + y_2),$$

where $w$ = width of strip and the end of each trapezium becomes the beginning of the next trapezium.

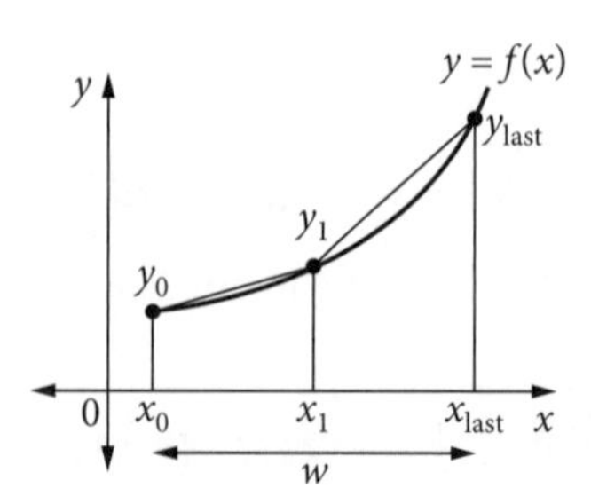


**Example 16**

Consider $f(x) = 2\sin(x)$.

Use LEA and REA and the trapezium rule to approximate the area under the curve $f$, between $x = 0$ and $x = \frac{\pi}{2}$ with the width of strip $\frac{\pi}{4}$.

**Solution**

Sketch the diagram of LEA width of strip $\frac{\pi}{4}$.

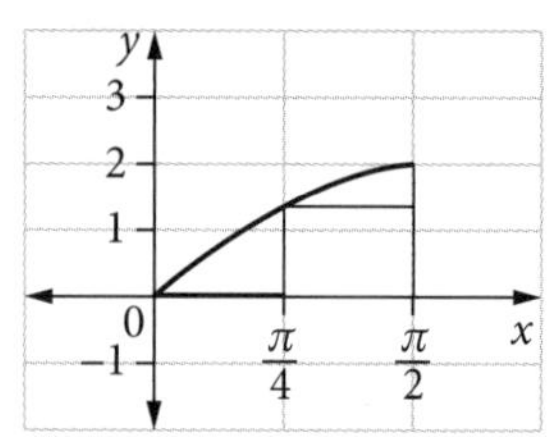


Sketch the diagram of REA width of strip $\frac{\pi}{4}$.

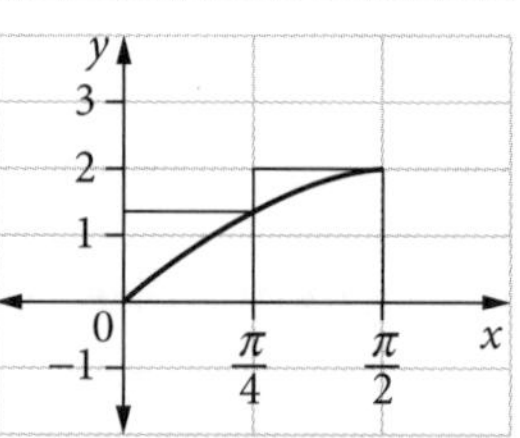


Calculate LEA and REA.

$$\text{LEA} = \frac{\pi}{4}\left(f(0) + f\left(\frac{\pi}{4}\right)\right) = \frac{\pi\sqrt{2}}{4}$$

$$\text{REA} = \frac{\pi}{4}\left(f\left(\frac{\pi}{4}\right) + f\left(\frac{\pi}{2}\right)\right) = \frac{\pi(\sqrt{2}+2)}{4}$$

Calculate the average of LEA and REA.

$$\frac{\text{LEA and REA}}{2} = \frac{\frac{\pi\sqrt{2}}{4} + \frac{\pi(\sqrt{2}+2)}{4}}{2}$$

$$= \frac{\pi(\sqrt{2}+1)}{4} \approx 1.896$$

Use trapezium rule.

$$\text{trapezium rule} = \frac{w}{2}(1\text{st } y_0 + y_{\text{last}} + y_1 + y_2),$$

$$= \left(\frac{\pi}{8}\right)\left(0 + 2 + 2\left(\frac{\sqrt{2}}{2}\right)\right)$$

$$= 1.896$$

In the 17th century, mathematician Wilhelm Leibnitz defined the definite integral so that

$$\lim_{\delta x_i \to \infty} \sum_{i=1}^{n} f(x_i)\delta x_i = \int_a^b f(x)\,dx$$

## 3.8.3 The fundamental theorem of calculus

The area under a curve between $x = a$ and $x = b$ is given by:

$$\text{area} = \int_a^b f(x)\,dx = [F(x)]_a^b = F(b) - F(a)$$

where $F(x)$ is the anti-derivative of $f(x)$.

Consider the function $f(x) = -4x^3 + 3x^2 + x$ between $x = 0$ and $x = 1$.

**a** Evaluate the integral $\int_0^1 f(x)\,dx$

$$\int_0^1 f(x)\,dx = \left[-x^4 + x^3 + \frac{x^2}{2}\right]_0^1 = \frac{1}{2}$$

$\int_0^1 -4x^3+3x^2+x\,dx$ $\quad \frac{1}{2}$

**b** Find the area under the curve of the graph of $f(x) = -4x^3 + 3x^2 + x$ between $x = 0$ and $x = 1$.

First, sketch a graph to see where the required area is.

The area required is above the $x$-axis.

Area above the $x$-axis is +ve and below the $x$-axis is –ve.

$$\text{area} = \frac{1}{2} \text{ sq. units}$$

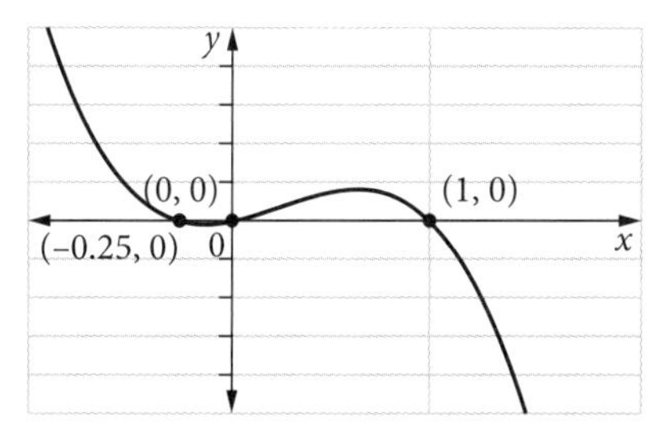


The CAS screen shows the area as 0.5.

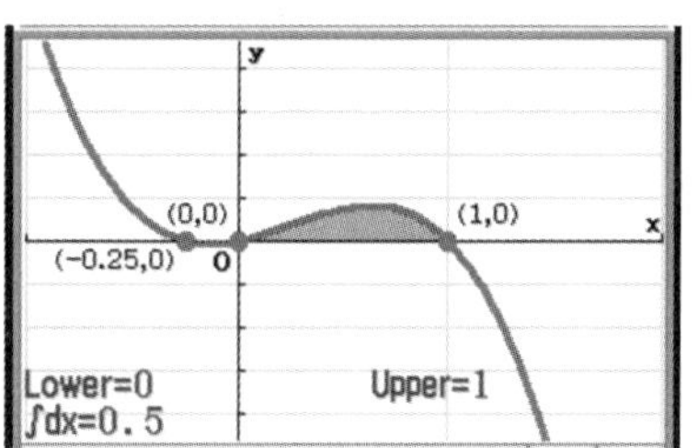


### Example 17

Consider $f(x) = 2\sin(x)$.

Find the area under the curve $f$ between $x = 0$ and $x = \frac{\pi}{2}$ using the fundamental theorem of calculus.

### Solution

Area is above the $x$-axis.

$$\text{area} = \int_0^{\frac{\pi}{2}} f(x)\,dx$$

$$\int_0^{\frac{\pi}{2}} 2\sin(x)\,dx = [-2\cos(x)]_0^{\frac{\pi}{2}}$$

$$= \left(-2\cos\left(\frac{\pi}{2}\right)\right) - (-2\cos(0))$$

$$= 2$$

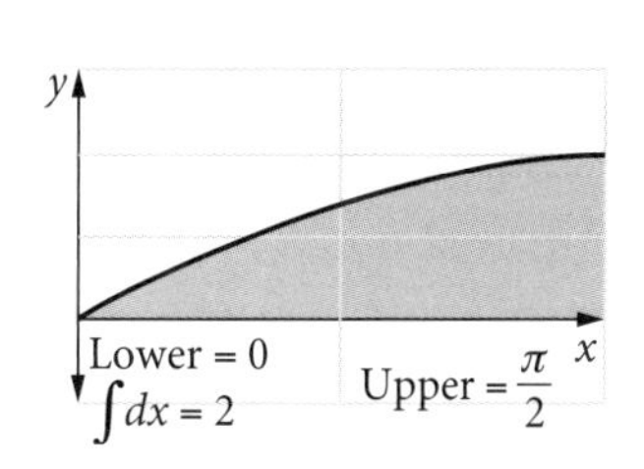

### 3.8.4 Properties of definite integrals

The following properties apply to definite integrals.

1 $\int_a^b kf(x)\,dx = k\int_a^b f(x)\,dx$

2 $\int_a^b f(x) + g(x)\,dx = \int_a^b f(x)\,dx + \int_a^b g(x)\,dx$

3 $\int_a^c f(x)\,dx = \int_a^b f(x)\,dx + \int_b^c f(x)\,dx$, where $a < b < c$

4 $\int_a^b f(x)\,dx = -\int_b^a f(x)\,dx$

5 $\int_a^a f(x)\,dx = 0$

Consider the integral $\int_1^5 (3f(x) - 2)\,dx$, where $\int_1^5 f(x)\,dx = 3$.

First, simplify using the rules above to get $\int_1^5 f(x)\,dx$ by itself.

Using property 2: $\int_1^5 3f(x) - 2\,dx = \int_1^5 3f(x)\,dx - \int_1^5 2\,dx$

Using property 1: $\int_1^5 3f(x)\,dx - \int_1^5 2\,dx = 3\int_1^5 f(x)\,dx - \int_1^5 2\,dx$

Using $\int_1^5 f(x)\,dx = 3$, we get $3 \times 3 - [2x]_1^5 = 9 - (10 - 2) = 9 - 8 = 1$.

## 3.9 Anti-differentiation by recognition

This section introduces a particular technique where we can use the inverse relationship of derivatives and anti-derivatives to solve problems.

- anti-differentiation by recognition that $F'(x) = f(x)$ implies that $\int f(x)\,dx = F(x) + c$.



Anti-differentiation by recognition requires us to differentiate first and then anti-differentiate a related function that is required in the answer. The questions often come in two steps.

**Example 18**

If $y = x\cos(x)$,

**a** find $\frac{dy}{dx}$.

**b** Hence, find an anti-derivative for $\int 2x\sin(x)dx$.

**Solution**

**a** Use the product rule.
$$\frac{dy}{dx} = -x\sin(x) + \cos(x)$$

**b** Using the answer to part **a**, it follows that
$$\int -x\sin(x) + \cos(x)dx = x\cos(x)(+ c)$$

Split the integral.
$$\int -x\sin(x)dx + \int \cos(x)dx = x\cos(x)$$

Rearrange.
$$\int -x\sin(x)dx = x\cos(x) - \int \cos(x)dx$$

Dividing by $-1$, we nearly get what we want on the LHS.
$$\int x\sin(x)dx = -x\cos(x) + \int \cos(x)dx$$

Multiplying both sides by 2.
$$\int 2x\sin(x)dx = -2x\cos(x) + 2\int \cos(x)dx$$

So
$$\int 2x\sin(x)dx = -2x\cos(x) + 2\sin(x)$$

**Note**

The question asked for **an** anti-derivative so we do not include the $+ c$ in the answer.

# 3.10 Modelling calculus

This last section will cover

- application of integration to problems involving finding a function from a known rate of change given a boundary condition, calculation of the area of a region under a curve and simple cases of areas between curves, average value of a function and other situations.

VCE Mathematics Study Design 2023–2027 p. 100, © VCAA 2022

## 3.10.1 Integrating to find a function

Up to now we have added the $+ c$ to most integration statements unless it was a definite integral. We can now determine the value of $c$ using given conditions.

To find the equation of a curve if the gradient curve is defined by $f'(x) = 3e^{2x} + 1$ and the curve of $y = f(x)$ goes through the point $(0, 1)$.

First, integrate.
$$\int 3e^{2x} + 1\,dx = \frac{3}{2}e^{2x} + x + c$$

This means that the original curve is
$$f(x) = \frac{3}{2}e^{2x} + x + c$$

Substitute $(0, 1)$.
$$1 = \frac{3}{2}e^{0} + 0 + c \Rightarrow c = -\frac{1}{2}$$

So
$$f(x) = \frac{3}{2}e^{2x} + x - \frac{1}{2}.$$

## 3.10.2 Areas under curves

If the area is underneath the $x$-axis, the definite integral will be negative, so we need to change things so that the answer to the area is positive.

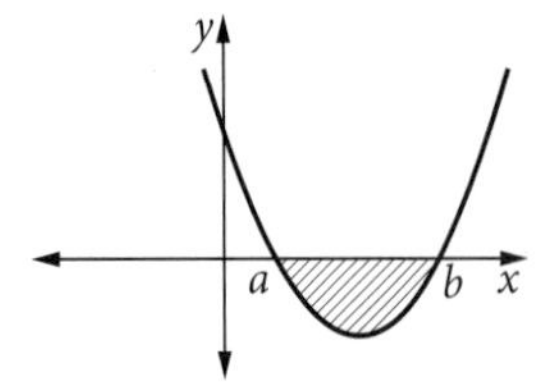


**Hint**

Remember, the area above the $x$-axis is +ve and the area below the $x$-axis is −ve.

This shaded area can be expressed as:

1 area $= -\int_a^b f(x)\,dx$ (making the −ve area +ve)

or

2 area $= \left|\int_a^b f(x)\,dx\right|$ (making the −ve integral +ve)

or

3 area $= \int_b^a f(x)\,dx$ (swapping limits changes the −ve integral to a +ve integral)

---

**Example 19**

Find the area between the graph of $y = x^2 - 2x - 8$ and the $x$-axis.

**Solution**

First, sketch the graph to see where the required area is.

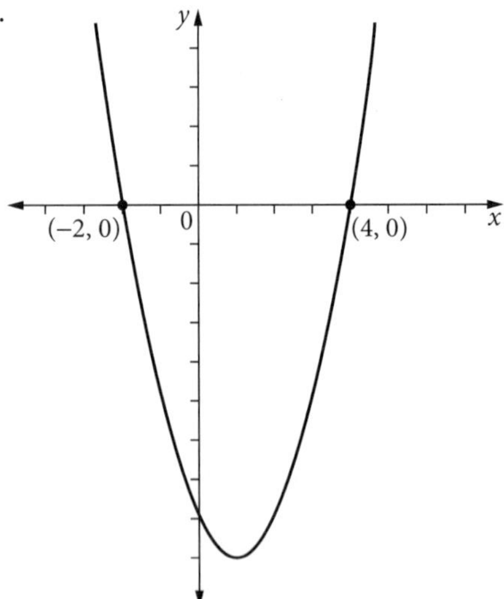


The area required is wholly below the $x$-axis.

Use the 1st method above.

$$\text{area} = -\int_{-2}^{4} (x^2 - 2x - 8)\,dx$$

$$= -\left[\frac{x^3}{3} - \frac{2x^2}{2} - 8x\right]_{-2}^{4}$$

$$= -\left[\left(\frac{4^3}{3} - 4^2 - 32\right) - \left(\frac{(-2)^3}{3} - (-2)^2 + 16\right)\right]$$

Evaluate the answer.

area = 36 sq. units

On CAS, using the 1st, 2nd and 3rd methods of making areas +ve.

$$-\int_{-2}^{4} x^2-2x-8\,dx \qquad 36$$

$$\left|\int_{-2}^{4} x^2-2x-8\,dx\right| \qquad 36$$

$$\int_{4}^{-2} x^2-2x-8\,dx \qquad 36$$

Graphically on CAS.

Here we are expecting the area to come up as –ve as it is below the $x$-axis.

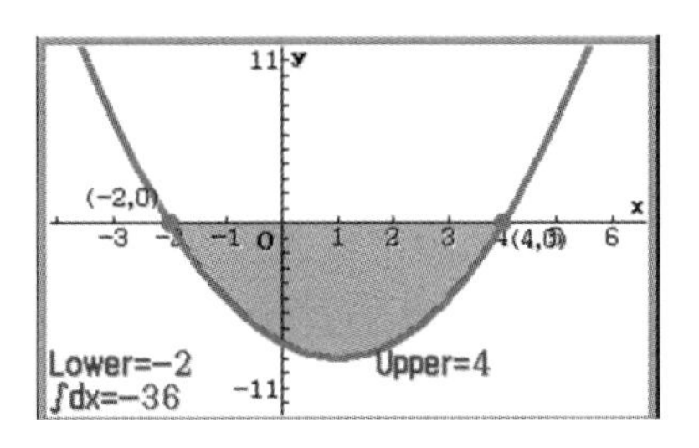


Sometimes you have to split the curve at the $x$ values where it goes above and below the axis.

Consider the area enclosed by the curve of $y = x^2 - 2x$, the $x$-axis and $x = 0$ and $x = 3$.

First, sketch the graph to see where the required area is.

The graph shows that the area required is partly above and partly below the $x$-axis.

First, find where the graph cuts the $x$-axis.

Solving $x^2 - 2x = 0$ gives $x = 0$ and $x = 2$.

Recognise that the shaded section of the area that is below the $x$-axis will produce a negative area.

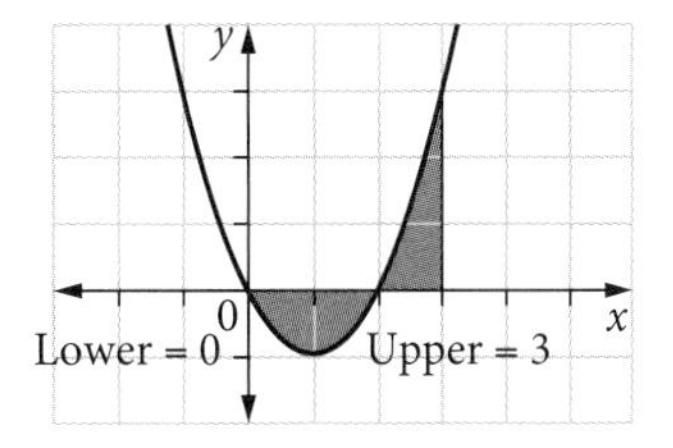


$$\begin{aligned}\text{area} &= -\int_0^2 (x^2 - 2x)\,dx + \int_2^3 (x^2 - 2x)\,dx \\ &= -\left[\frac{x^3}{3} - x^2\right]_0^2 + \left[\frac{x^3}{3} - x^2\right]_2^3 \\ &= \frac{8}{3}\text{ sq. units}\end{aligned}$$

## 3.10.3 Area between two curves

The rule for finding the area between curves is

$$\text{area} = \int (\text{upper} - \text{lower})\,dx$$

The area in the graph shown is found by

$$\text{area} = \int_a^b f(x) - g(x)\,dx + \int_b^c g(x) - f(x)\,dx$$

where, from $a$ to $b$, $f(x)$ is the upper curve, and from $b$ to $c$, $g(x)$ is the upper curve.

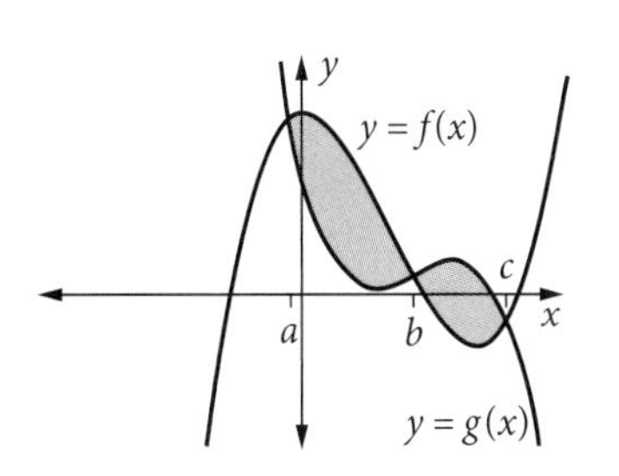

### Example 20

Find the area between the curves of $y = x^2 - 2x - 8$ and $y = x + 2$.

#### Solution

Sketch the graph to see where the upper and lower curves are.

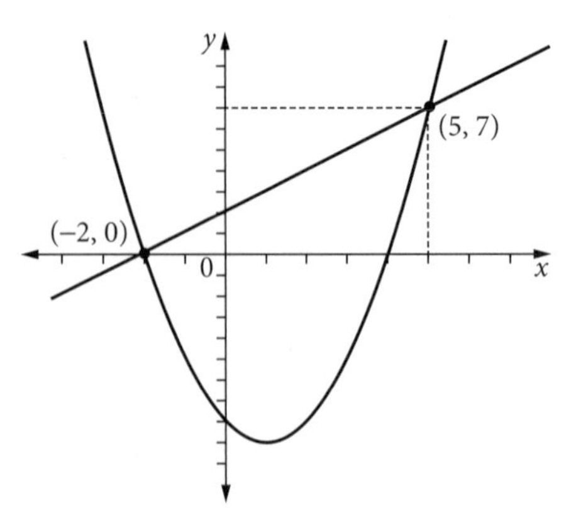


**Hint**
When using upper – lower, don't worry about which graph is above or below the axis as the formula deals with this.

Upper curve is $y = x + 2$.

Lower curve is $y = x^2 - 2x - 8$.

Find the points of intersection $(-2, 0)$ and $(5, 7)$.

Use the formula:

$$\text{area} = \int_{-2}^{5} (\text{upper} - \text{lower})\,dx$$

**Hint**
It is easier to simplify before you integrate.

Evaluate the answer.

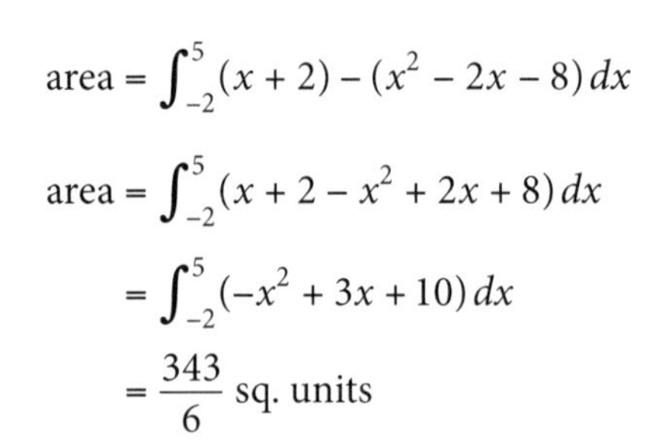

$$\text{area} = \int_{-2}^{5} (x + 2) - (x^2 - 2x - 8)\,dx$$

$$\text{area} = \int_{-2}^{5} (x + 2 - x^2 + 2x + 8)\,dx$$

$$= \int_{-2}^{5} (-x^2 + 3x + 10)\,dx$$

$$= \frac{343}{6} \text{ sq. units}$$

Area is approximately equal to 57.167 sq. units.

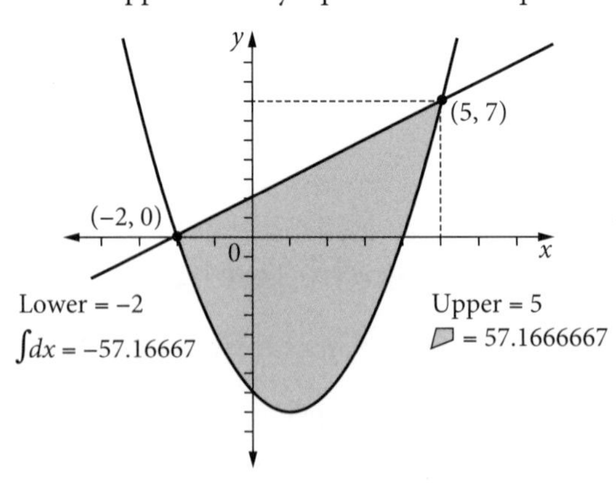

## 3.10.4 Kinematics

Differential and integral calculus is used in the topic of kinematics. **Displacement, velocity and acceleration** can be found by differentiating or anti-differentiating as required.

Suppose the velocity ($v$ m/s) of a particle, at time $t$ seconds ($t \geq 0$) is given by $v(t) = \frac{1}{2}t^2 - 3t + 4$.

**a** Calculate the distance travelled between 2 and 4 seconds.

The distance in the velocity-time graph shown is represented by the shaded area:

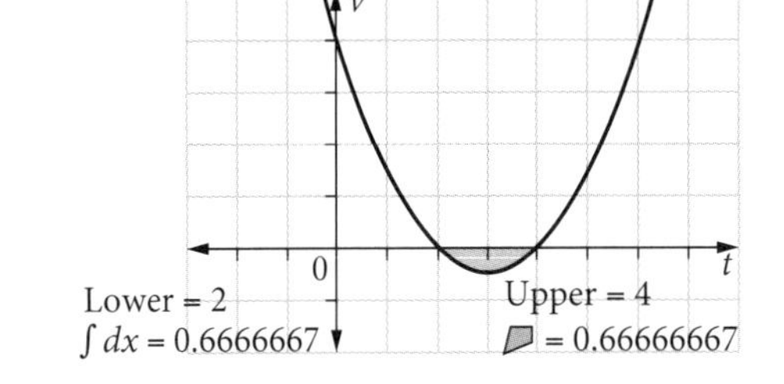


$$\text{area} = -\int_2^4 \left(\frac{1}{2}t^2 - 3t + 4\right) dt$$

$$= \frac{2}{3}$$

Thus from $t = 2$ to $t = 4$, the body travels $\frac{2}{3}$ metres.

**b** Calculate the body's displacement at any time, $t$.

$$\text{displacement } x(t) = \int v(t)\, dt$$

$$= \int \left(\frac{1}{2}t^2 - 3t + 4\right) dt$$

$$= \frac{t^3}{6} - \frac{3t^2}{2} + 4t + c$$

**c** If the body starts at the origin, find its displacement at $t = 2$.

$$x(t) = \frac{t^3}{6} - \frac{3t^2}{2} + 4t$$

Displacement is presumed zero initially, giving $c = 0$.

$$x(2) = \frac{10}{3} \text{ metres}$$

## 3.10.5 Average value of a function

$$\text{average value} = \frac{1}{b-a}\int_a^b f(x)\,dx$$

This is the **average value of a function** $f$ over the interval $[a, b]$.

---

**Example 21**

The height of rainfall, in mm, collected in a barrel over a period of time, $t$ minutes, is measured by the function $h(t) = \sin(3t)$.

Find the average height of rainfall collected from $t = 0$ to $t = \frac{\pi}{3}$.

**Solution**

Use the formula:

$$\text{average value} = \frac{1}{b-a}\int_a^b f(t)\,dt$$

$$\text{average value} = \frac{1}{\frac{\pi}{3}-0}\int_0^{\frac{\pi}{3}} \sin(3t)\,dt$$

$$= \frac{2}{\pi}\text{ mm}$$

Find the average height of rainfall collected from $t = 0$ to $t = \frac{\pi}{3}$.

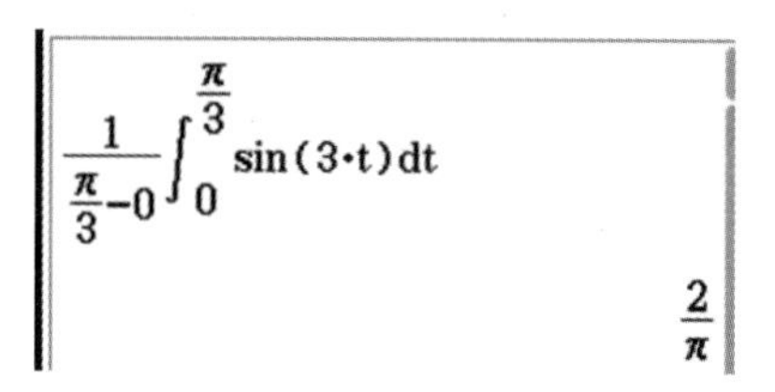

**Hint**

The average value of a simple trig. graph over one cycle is the mean of the function. For example, in $y = \sin(x) + 2$ the average value over one cycle, between $x = 0$ to $x = 2\pi$, is 2.

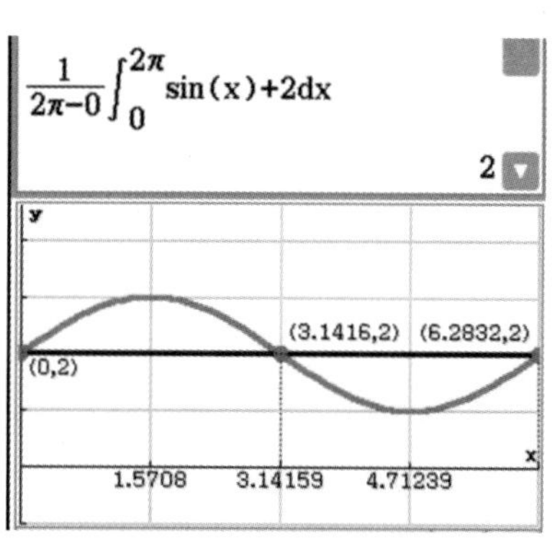

# Glossary

**A+ DIGITAL FLASHCARDS**
Revise this topic's key terms and concepts by scanning the QR code or typing the URL into your browser.

https://get.ga/aplus-vce-maths-methods

**absolute maximum/minimum point** The maximum/minimum point of a function over a given interval. It is either a local maximum/minimum point or one of the endpoints of the interval. For the function graphed over the given interval below, the absolute maximum point is $(1.6667, 9.4815)$ and the absolute minimum point is $(4, -25)$.

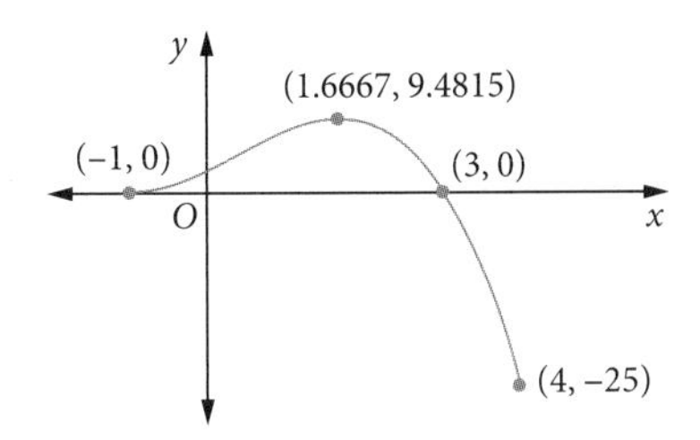


**acceleration** The rate of change of velocity of a moving object, represented by the function $a = \frac{dv}{dt}$, where $v$ is the velocity ('signed speed'). Conversely, $v = \int a(t)\,dt$.

**anti-derivative** (or **integral** or **primitive**) The opposite of the derivative. The anti-derivative of $f(x)$ is a function $F(x)$ whose derivative is $f(x)$: $F'(x) = f(x)$.

**anti-differentiation** (or **integration**) The opposite of differentiation, that is, the process of finding the anti-derivative or integral.

**average rate of change** The rate of change between point $A(x_1, y_1)$ and point $B(x_2, y_2)$ on the graph of a function, given by the gradient $m = \frac{y_2 - y_1}{x_2 - x_1}$ of the straight line connecting the two points.

**average value of a function** The average $y$ value of a function, or the average height of its graph, over an interval $[a, b]$, calculated by the formula $\frac{1}{b-a}\int_a^b f(x)\,dx$.

**chain rule** A formula for finding the derivative of a composite function $f(g(x))$. If $y = f(u)$ and $u = g(x)$, then $\frac{dy}{dx} = \frac{dy}{du} \times \frac{du}{dx}$.

**composite function rule** *See* **chain rule**.

**concave** The shape of the curve of a graph in an upward or downward direction.

**concave downwards** Upside-down U-shape, describing the shape of a curve, especially a parabola.

**concave upwards** U-shaped, describing the shape of a curve, especially a parabola.

**constant of integration** A constant value ('$+ c$') added to the anti-derivative of a function.

**continuous function** A function without a break in its graph. Its graph can be drawn without taking pen off paper.

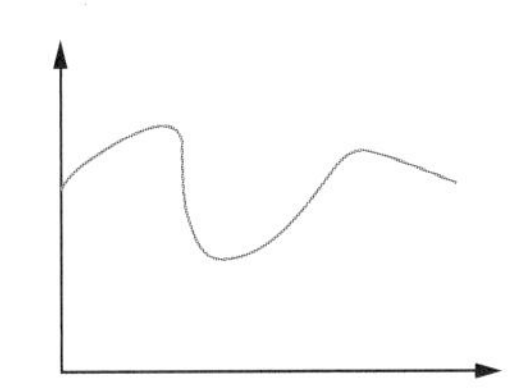

**definite integral** An integral of the form $\int_a^b f(x)\,dx$, whose value is the (signed) area under a curve (shaded $A$ in the diagram below) and is read 'the integral of $f(x)$ between $a$ and $b$ with respect to $x$'.

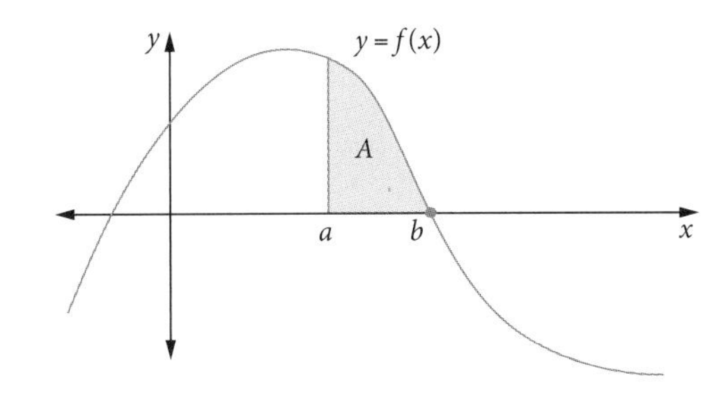


**derivative** For the function $y = f(x)$, the derivative of $y$, $\frac{dy}{dx}$ or $f'(x)$, is the function of the instantaneous rate of change of $f(x)$, also called the gradient function.

**differentiation** The process of finding the derivative of a function.

**differentiation by first principles** Finding the derivative of a function using the formula $\frac{dy}{dx} = f'(x) = \lim_{h \to 0} \frac{f(x+h) - f(x)}{h}$.

**discontinuous function** A function with a break in its graph. Its graph cannot be drawn without taking the pen off the paper.

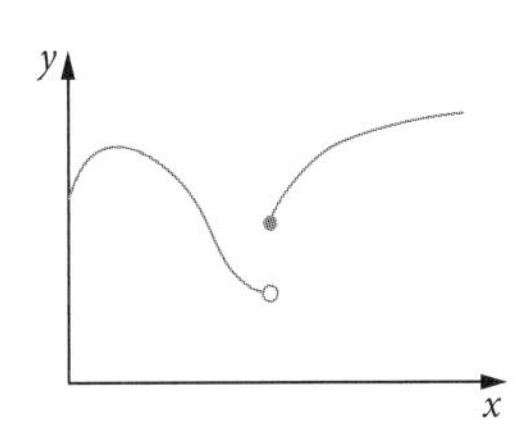


**displacement** The 'signed distance' from the origin of a moving object, represented by the function $x(t)$, where displacement is a function of time.

**global maximum/minimum point** The maximum/minimum point of a function over a given interval. It is either a local maximum/minimum point or one of the endpoints of the interval. For the function graphed over the given interval below, the absolute maximum point is $(1.6667, 9.4815)$ and the absolute minimum point is $(4, -25)$.

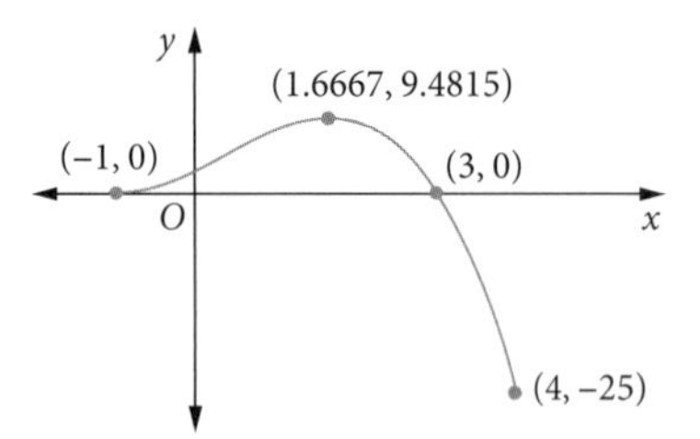


*See also* **absolute maximum/minimum point**.

**gradient** The slope of a line, given by $m = \frac{\text{rise}}{\text{run}} = \frac{y_2 - y_1}{x_2 - x_1}$. For a curve, the gradient is the slope of the tangent to the curve at a particular point.

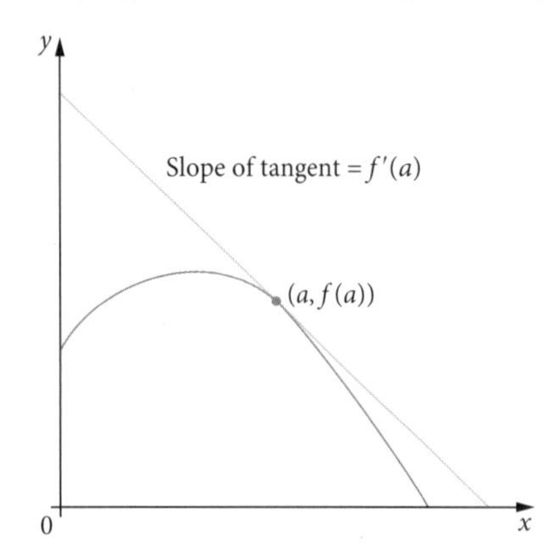


**gradient function** *See* **derivative**.

**indefinite integral** An integral of the form $\int f(x)dx$, which is an anti-derivative function and is read 'the integral of $f(x)$ with respect to $x$'.

**instantaneous rate of change** The rate of change at a point $(x_1, y_1)$ on the graph of a function $f(x)$, given by the gradient of the tangent to the graph at that point, which is the derivative of the function at that point, $f'(x_1)$. *See* **gradient** for diagram.

**integral** (or **anti-derivative** or **primitive**) The opposite of the derivative. The anti-derivative of $f(x)$ is a function $F(x)$ whose derivative is $f(x)$: $F'(x) = f(x)$. *See also* **definite integral** and **indefinite integral**.

**integrand** The function being integrated; the $f(x)$ in $\int f(x)dx$.

**integration** (or **anti-differentiation**) The limiting sum of rectangles under a curve; the opposite of differentiation, the process of finding the anti-derivative or integral.

**limits of integration** The lower and upper bound (interval) over which the anti-derivative applies. It represents the area of the region under the curve from the lower bound to the upper bound.

**local maximum/minimum point** *See* **turning point** and **global maximum/minimum point**.

**maximum/minimum point** *See* **turning point**.

**maximum rate of increase/decrease** The rate of increase or decrease where a function is increasing or decreasing most rapidly; where the gradient of its graph is steepest in the positive and negative directions.

**primitive** *See* **integral**.

**product rule** A formula for finding the derivative of the product of two functions: $\frac{d}{dx}(uv) = u'v + uv'$.

**quotient rule** A formula for finding the derivative of the ratio of two functions: $\frac{d}{dx}\left(\frac{u}{v}\right) = \frac{vu' - uv'}{v^2}$.

**stationary point** A point on a graph where the gradient equals 0. The graph is flat at the stationary point, neither increasing nor decreasing. $\frac{dy}{dx} = f'(x) = 0$. A stationary point is either a turning point (local maximum or minimum point) or a stationary point of inflection.

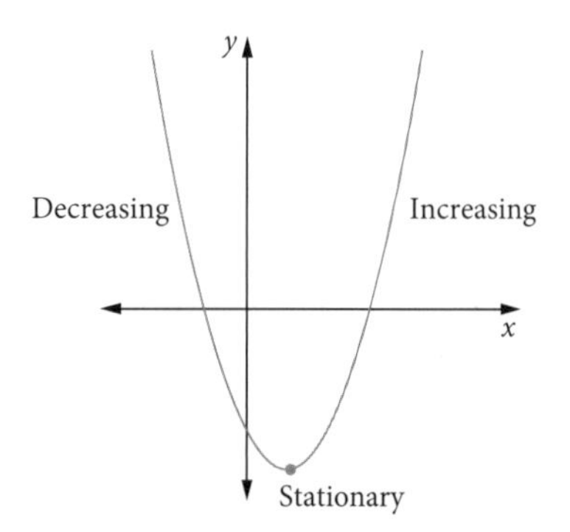


**stationary point of inflection** A stationary point on a graph where the sign of the gradient stays the same on both sides.

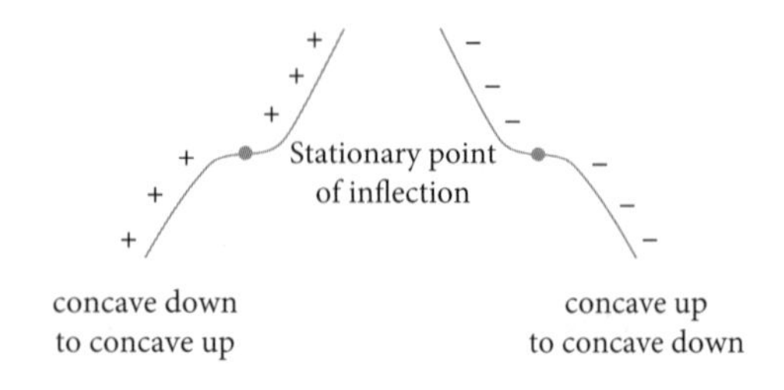

**strictly decreasing** A function is strictly decreasing when its $y$ values are getting smaller and its graph is moving downwards to the right. If $a < b$, then $f(a) > f(b)$.

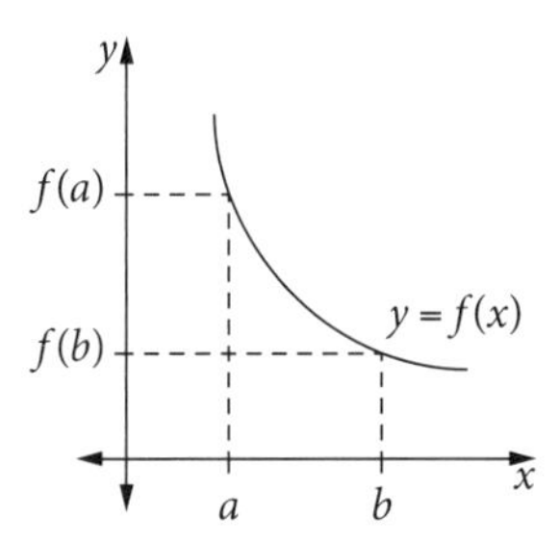


When $f(x)$ is strictly decreasing $f(a) > f(b)$.

**strictly increasing** A function is strictly increasing when its $y$ values are getting larger and its graph is moving upwards to the right. If $a < b$, then $f(a) < f(b)$.

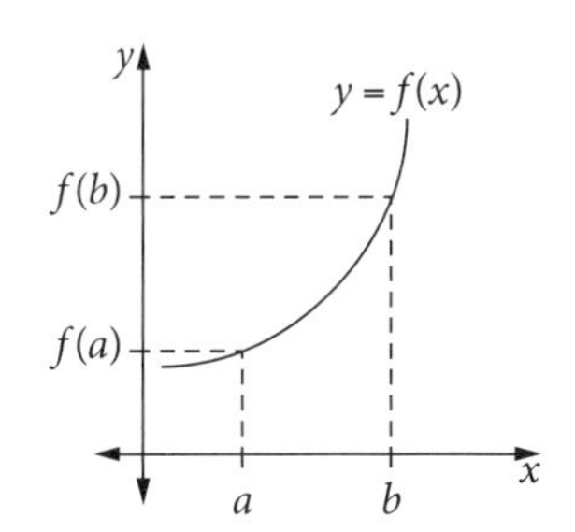


When $f(x)$ is strictly increasing $f(a) < f(b)$.

**tangent** A straight line that touches a curve at only one point. See **gradient** for diagram.

**trapezium rule** A method of approximating the area under the curve of a function by dividing it into a sequence of trapeziums.

**turning point** A stationary point on a graph where the sign of the gradient changes on either side. If it changes from negative (decreasing) to positive (increasing), it is a local minimum point. If it changes from positive (increasing) to negative (decreasing), it is a local maximum point.

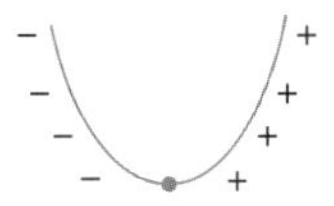


Local minimum point

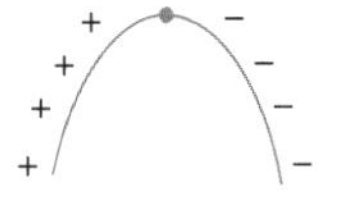


Local maximum point

**velocity** The 'signed speed' of a moving object, represented by the function $v = \frac{dx}{dt}$; the rate of change of the displacement, $x$. Conversely, $x = \int v(t)\,dt$.



CHAPTER 3 – GLOSSARY

# Exam practice

## Short-answer questions

### Technology free: 20 questions

Solutions to this section start on page 224.

**Question 1** (2 marks)

Find $\frac{dy}{dx}$ when $y = 2x\sin(3x)$.

**Question 2** (5 marks)

For the quadratic graph shown, find

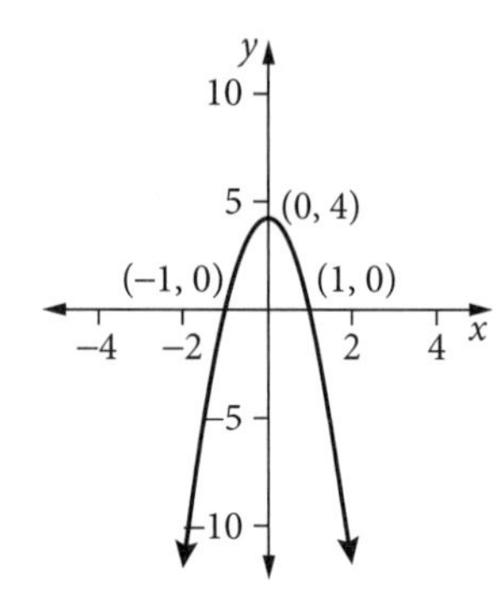


a the average rate of change between $x = -1$ and $x = 1$ — 2 marks

b the average rate of change between $x = 0$ and $x = 1$ — 2 marks

c the rate of change at $x = 1$. — 1 mark

**Question 3** (4 marks)

If $y = x\sin(x)$,

a find $\frac{dy}{dx}$. — 2 marks

b Hence, find an anti-derivative for $\int 2x\cos(x)dx$. — 2 marks

**Question 4** (4 marks)

If $y = x\log_e(x)$,

a find $\frac{dy}{dx}$. — 2 marks

b Hence, find $\int_1^2 2\log_e(x)dx$. — 2 marks

**Question 5** (3 marks)

Find $\frac{dy}{dx}$ when $y = \frac{\log_e(2x)}{x^2}$.

**Question 6** (2 marks)

Find the area enclosed by the curve and the $x$-axis for the graph $y = -x^2 + 2x$.

**Question 7** (3 marks)

Our biorhythms can be represented by sinusoidal curves of the form $y = \sin(kt)$, where $t$ is in days. What is the value of $k$ if the gradient of the curve equals $-\pi$ at $t = 1$?

**Question 8** (4 marks)

The area enclosed by the curve of $y = x^2 - 2x$, the $x$-axis and $x = 0$ and $x = 3$ is shown.

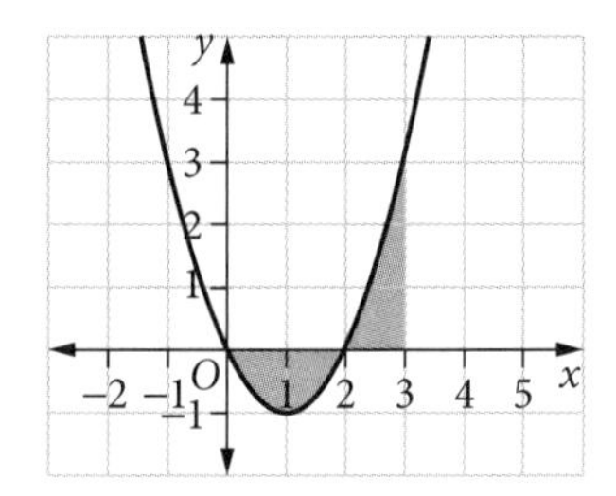


**a** Find the area as described. 2 marks

**b** Evaluate $\int_0^3 (x^2 - 2x)\,dx$. 2 marks

**Question 9** (3 marks)

Suppose that the velocity ($v$ m/s) of a particle, at time $t$ seconds ($t \geq 0$), of a body is given by $v(t) = t^2 - 6t + 8$. The particle is initially 2 metres to the right of a fixed point, 0.

Calculate the distance travelled between 1 and 3 seconds.

**Question 10** (4 marks)

If $g$: $[k, 4] \to R$, $g(x) = -(x - 2)^2 + 3$,

**a** find the smallest value of $k$ such that $g^{-1}$ exists. 2 marks

**b** find the derivative of $g^{-1}(x)$. 2 marks

**Question 11** (3 marks)

Find the derivative of $\sqrt{2x^2 + 4}$ at $x = 2$.

**Question 12** (4 marks)

Find the coordinates of any stationary points in the function $y = x^3 - x^2 - x - 2$.

**Question 13** (4 marks)

If $f(x) = x^3 + ax^2 + b$ has a stationary point at $(2, -3)$, calculate

**a** the values of $a$ and $b$. 2 marks

**b** the coordinates of any other stationary points. 2 marks

**Question 14** (2 marks)

If $f(x) = \sin\left(\frac{x}{5}\right)$, find the gradient of $f(x)$ at $x = \frac{5\pi}{4}$.

**Question 15** (4 marks)

If $f(x) = x^3 + 1$ and $g(x) = \sqrt{x - 1}$,

**a** define $g(f(x))$ and $f(g(x))$ if they exist. 2 marks

**b** find the derivative(s) of $g(f(x))$ and $f(g(x))$ if they exist. 2 marks

**Question 16** (4 marks)

If $y = (5x + 2)^4$, calculate

**a** the gradient of the tangent to the curve at $x = 0$. 2 marks

**b** the equation of the tangent to the curve at $x = 0$. 2 marks

**Question 17** (2 marks)

Sketch the graph of the gradient function $y = f'(x)$ if $y = f(x)$ is shown in the graph.

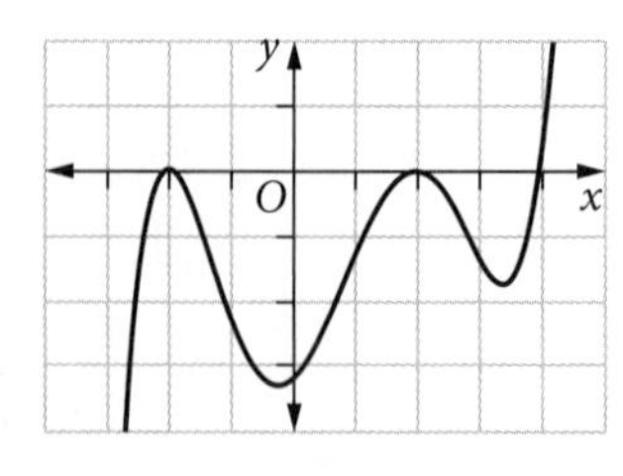


**Question 18** (3 marks)

**a** Find the gradient of the chord $PQ$ to the function $f(x) = 3x^2 + 2$ if the $x$-coordinates of $P$ and $Q$ are $-2$ and $-2 + h$, respectively. 2 marks

**b** Hence, find the gradient of the tangent at $P$. 1 mark

**Question 19** (3 marks)

The graph of $y = f'(x)$ is shown.

Sketch what the graph of the anti-derivative, $y = f(x)$, might look like.

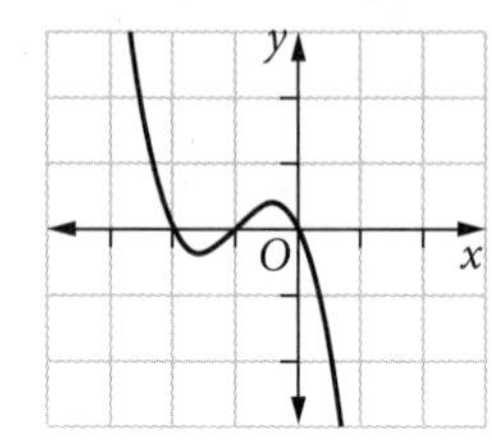


**Question 20** (4 marks)

The temperature, $T$°C, on a mountain is related to height ($h$ metres) by the rule

$$T = -0.002h^3 + 30, h \geq 0.$$

**a** Calculate the average rate of change of temperature over the first 10 metres. 2 marks

**b** Calculate the rate of change of temperature at $h = 10$. 2 marks

# Multiple-choice questions

## Technology active: 50 questions

Solutions to this section start on page 226.

**Question 1**

The derivative of the function $f(x) = 2x^2 + 4x$ is

**A** $f'(x) = 2x + 4$ **B** $f'(x) = 2x^2 + 4x$ **C** $f'(x) = 4x$

**D** $f'(x) = 2x + 4x$ **E** $f'(x) = 4x + 4$

**Question 2**

An anti-derivative of the function $f(x) = 2x^2 + 4x$ is

**A** $y = 2x + 4$ **B** $y = \frac{x^3}{3} + 4x^2$ **C** $y = \frac{2x^3}{3} + 2x^2$

**D** $y = \frac{2x^3}{3} + 4x$ **E** $y = 4x + 4$

**Question 3**

If $f(x) = 3\sqrt{x}$, then $f'(3)$ equals

**A** $\frac{\sqrt{3}}{2}$ **B** $2\sqrt{3}$ **C** $\sqrt{3}$ **D** $\frac{2}{\sqrt{3}}$ **E** $-\frac{1}{\sqrt{3}}$

**Question 4**

If $f(x) = 2\sqrt{x}$, then $f'(x)$ equals

**A** $\frac{1}{\sqrt{x}}$ **B** $2\sqrt{x}$ **C** $\sqrt{x}$ **D** $\frac{2}{\sqrt{x}}$ **E** $-\frac{1}{\sqrt{x}}$

**Question 5**

Let $g(x) = -\frac{1}{4}(x^4 - 6x^2 + 8x)$.

The graph that best represents $g'(x)$ is:

**A**

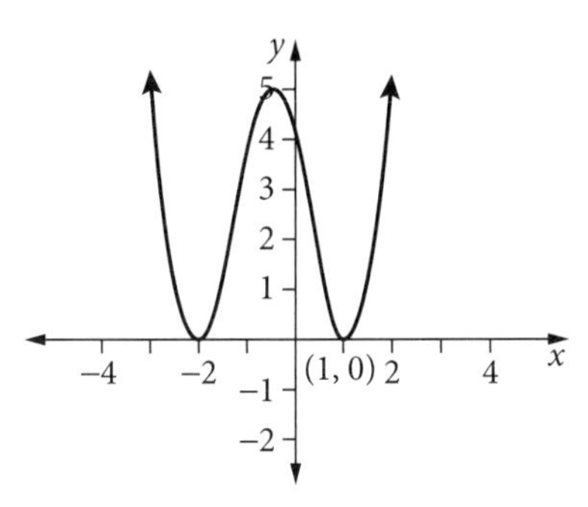


**B**

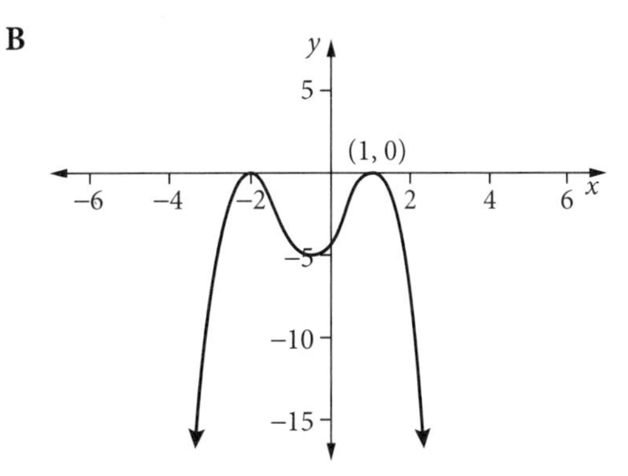


**C**

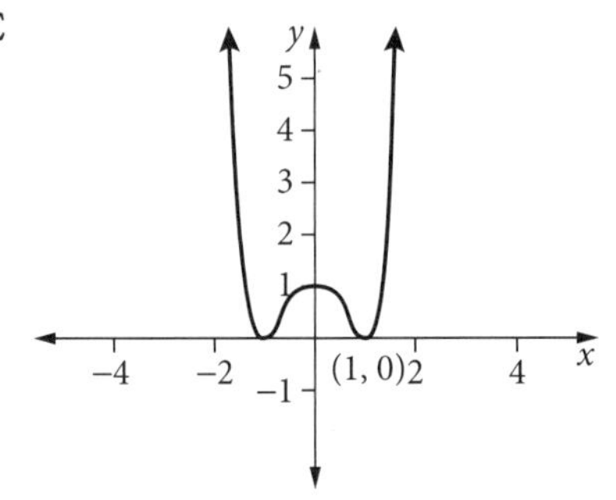


**D**

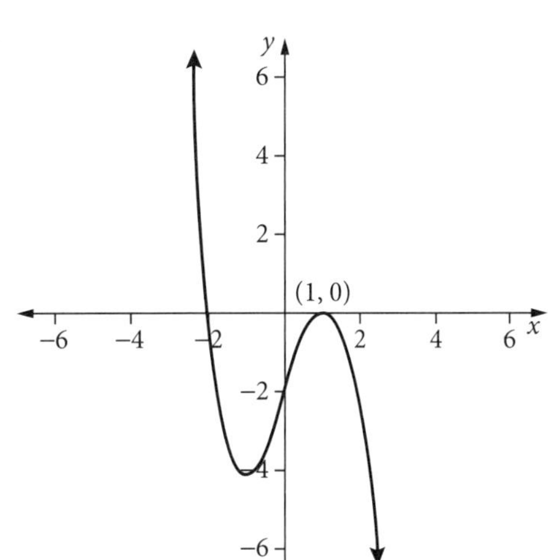


**E**

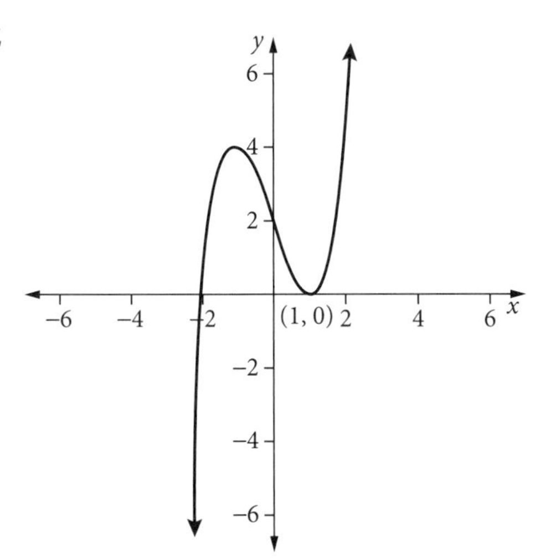


**Question 6**

The derivative of the function $f(x) = -x^4 - 2x^3 + 3x^2 + 4x - 4$ can be expressed as

**A** $(x - 1)(x + 2)$ **B** $(x - 1)(1 - x)(x + 2)(x + 2)$ **C** $(x - 1)(1 - x)(x - 1)$

**D** $2(x + 2)(x - 1)(2x + 1)$ **E** $-2(x + 2)(x - 1)(2x + 1)$

9780170465366

**Question 7**

If $g(x) = -2(x-3)(x+2)(x-5)$, an anti-derivative can be expressed as

**A** $-\frac{x^4}{2} + 4x^3 + x^2 - 60x + 10$ **B** $-\frac{x^4}{2} + 4x^3 + x^2$ **C** $-6x^2 + 24x + 2$

**D** $-6x^2 + 24x$ **E** $-12x + 24$

**Question 8**

If $f(x) = 2x^3 - 3x^2 + 4$, the $x$ value where the gradient of the curve is a minimum is

**A** $-\frac{3}{2}$ **B** 0 **C** $\frac{1}{2}$ **D** 1 **E** $\frac{3}{2}$

**Question 9**

$\frac{d}{dx}(-\sin(4x))$ equals

**A** $4\cos(4x)$ **B** $-4\cos(3x)$ **C** $-4x\cos(4x)$ **D** $-4\cos(4x)$ **E** $-\frac{1}{4}\cos(4x)$

**Question 10**

If $f(x) = e^{-3x}$, then $f'(x)$ equals

**A** $-3xe^{-3x}$ **B** $-3e^{-3x-1}$ **C** $-3e^{-3x}$ **D** $-3e^{-2x}$ **E** $-3e^{-4x}$

**Question 11**

If $f(x) = e^{-5x}$, then $f'(5)$ equals

**A** $-5xe^{-5}$ **B** $-5e^{-5x}$ **C** $-5e^{-25}$ **D** $-5e^{25}$ **E** $5e^{25}$

**Question 12**

The velocity ($v$ m/s) of a particle at time $t(s)$ is given by $v = t^3 - t^2 + 4$, $t \geq 0$. The rate of change of velocity at $t = 3$ is

**A** $1\,\text{m/s}^2$ **B** $8\,\text{m/s}^2$ **C** $16\,\text{m/s}^2$ **D** $21\,\text{m/s}^2$ **E** $40\,\text{m/s}^2$

**Question 13**

The graph shown has

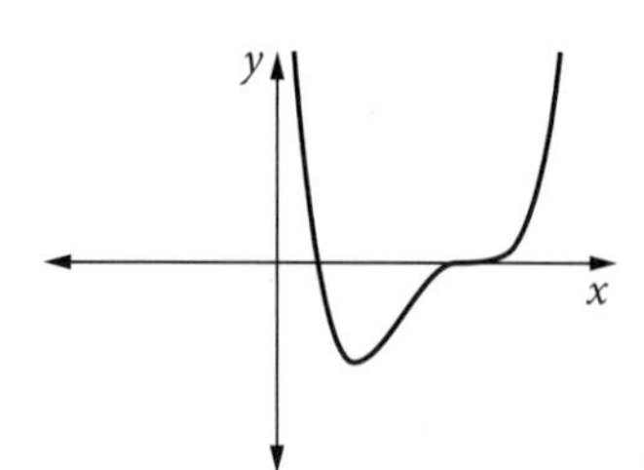


**A** one stationary point

**B** two stationary points

**C** the shape of a parabola

**D** a local maximum

**E** a negative gradient for $x \geq 0$

**Question 14**

Consider $f: (-4, 2] \to R, f(x) = x^2 + 4x$. The maximum value of $f$ is

**A** −4 **B** 0 **C** 2 **D** 12 **E** $\infty$

**Question 15**

Consider $f: (-4, 2] \to R, f(x) = x^2 + 4x$. The minimum value of $f$ is

**A** −4 **B** 0 **C** 2 **D** 12 **E** $\infty$

**Question 16**

The maximum height (in metres) of a ball thrown in the air with the height given by $h(t) = -t^2 + 2t + 1,\ t \geq 0$, is

**A** 1 m **B** 2 m **C** 12 m **D** 13 m **E** 14 m

**Question 17**

The graph shown has the equation $y = 4(x - 1)^2$.
The area bounded by the curve and the line $y = 1$ can be expressed as

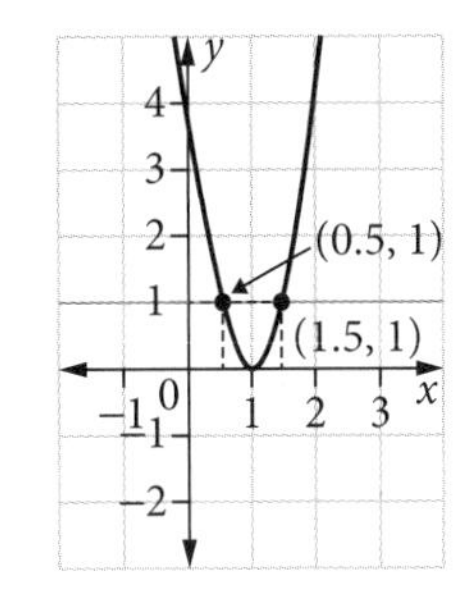


**A** $\int_0^2 4(x-1)^2 dx - 1$

**B** $\int_{\frac{1}{2}}^{\frac{3}{2}} 4(x-1)^2 dx - 1$

**C** $1 - \int_{\frac{1}{2}}^{\frac{3}{2}} 4(x-1)^2 dx$

**D** $\int_{\frac{1}{2}}^{\frac{3}{2}} 1 - 4(x-1)^2 dx$

**E** $\int_{\frac{1}{2}}^{\frac{3}{2}} 4(x-1)^2 - 1 dx$

**Question 18**

The population, $P$, of a certain town at $t$ years is given by $P(t) = 1000e^{1-t},\ t \geq 0$. The rate at which the population is decreasing after 10 years is approximated by

**A** −1000 **B** −0.123 **C** 0.123 **D** $1000e^{-9}$ **E** 1000

**Question 19**

If $y = 2\sin^2(3x)$, then $\frac{dy}{dx}$ is equal to

**A** $12\sin(3x)\cos(3x)$ **B** $2\sin(3x)\cos(3x)$ **C** $6x\sin(3x)\cos(3x)$

**D** $6\sin(3x)\cos(3x)$ **E** $3\cos^2(3x)$

**Question 20**

For the curve $y = 4x^3 - 3x^2 - 2$, the coordinates of the point(s) at which the gradient is zero are

**A** $(0,-2), \left(\frac{1}{2}, -\frac{9}{4}\right)$ **B** $(0,0), \left(\frac{1}{2}, 0\right)$ **C** $(0,-2)$

**D** $(0,-2), (2, 19)$ **E** $(0,-1), \left(\frac{1}{2}, -\frac{5}{4}\right)$

**Question 21**

Consider $f(x) = ax^2 - bx$. The gradient of the tangent to the curve at the point at $(1, 1)$ is equal to −1. The values of $a$ and $b$, respectively, are

**A** −3, −2 **B** −2, −3 **C** −3, 2 **D** −2, 3 **E** 1, −1

**Question 22**

If $f(x) = -2\sqrt{x+2} - 1$, then the gradient at $x = 2$ is

**A** −2 **B** −1 **C** −0.5 **D** 0.5 **E** 2

**Question 23**

If $y = \log_e(3x + 3)$, the value of $x$ where the gradient is equal to 1 is

**A** $-\frac{1}{3}$ **B** 0 **C** $\frac{1}{5}$ **D** $\frac{3}{5}$ **E** 1



CHAPTER 3 – EXAM PRACTICE

**Question 24**

Let $f(x) = x^2 - 2x$ and $g(x) = \frac{1}{x^2}$ for a maximal domain.

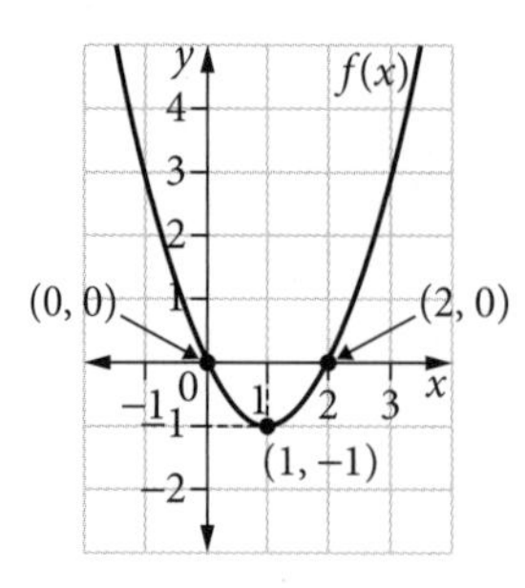


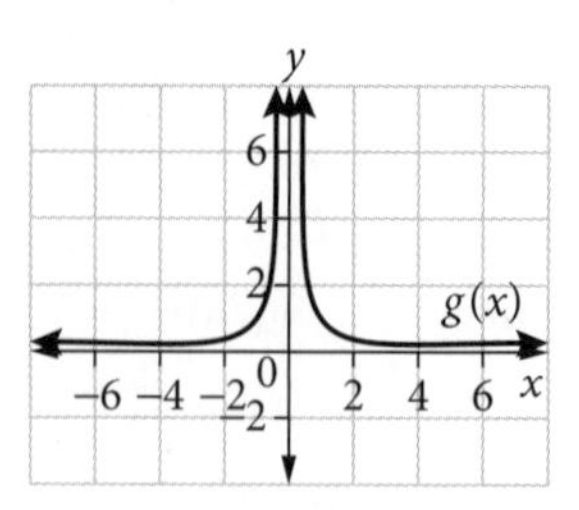


If $h = g(f(x))$, the derivative $h'(x)$ is equal to

**A** $h'(x) = \dfrac{-4x+4}{(x^2-2x)^2}$ for $x \in (-\infty, 0) \cup (0, 2) \cup (2, \infty)$

**B** $h'(x) = \dfrac{1}{(x^2-2x)^2}$ for $x \in (2, \infty)$

**C** $h'(x) = \dfrac{-4x+4}{(x^2-2x)^3}$ for $x \in R$

**D** $h'(x) = \dfrac{-4x+4}{(x^2-2x)^3}$ for $x \in (-\infty, 0) \cup (0, 2) \cup (2, \infty)$

**E** $h'(x) = x^4 - 2x^2$ for $x \in (1, \infty)$

**Question 25**

The line $2y - 6x = 3$ has a gradient of

**A** $-3$ **B** $-\frac{1}{2}$ **C** $-\frac{1}{6}$ **D** $3$ **E** $\frac{9}{2}$

**Question 26**

The average rate of change of $y = x^2 - 2x$ between $x = 1$ and $x = 3$ is

**A** $-1$ **B** $1$ **C** $2$ **D** $3$ **E** $4$

**Question 27**

The equation of the tangent to the curve $y = 2x^3$ at $x = 2$ is

**A** $x + 24y + 32 = 0$ **B** $y = 24x - 24$ **C** $y = 24$

**D** $y = 24x - 32$ **E** $y = 12x - 16$

**Question 28**

The gradient of the secant from $x = 2$ to $x = 2 + h$ in the function $y = 3x^2 + 2x - 1$ is found by

**A** $\lim\limits_{h \to 0} \dfrac{f(x+h) - f(x)}{2}$ **B** $\lim\limits_{h \to 0} \dfrac{f(2+h) - f(2)}{2}$ **C** $\dfrac{f(2+h) - f(2)}{h}$

**D** $\dfrac{f(2) - f(2-h)}{h}$ **E** $6x + 2$

**Question 29**

If $y = \tan(x)$, the rate of change of $y$ with respect to $x$ at $x = a$, where $\frac{\pi}{2} < a < \pi$, is:

**A** $\sec^2(a)$ **B** $-\tan(a)$ **C** $\tan(a)$ **D** $\sec^2(0)$ **E** $-\dfrac{1}{\cos^2(a)}$

**Question 30**

The graph of $y = \sin(x) - \cos(2x)$ is shown.

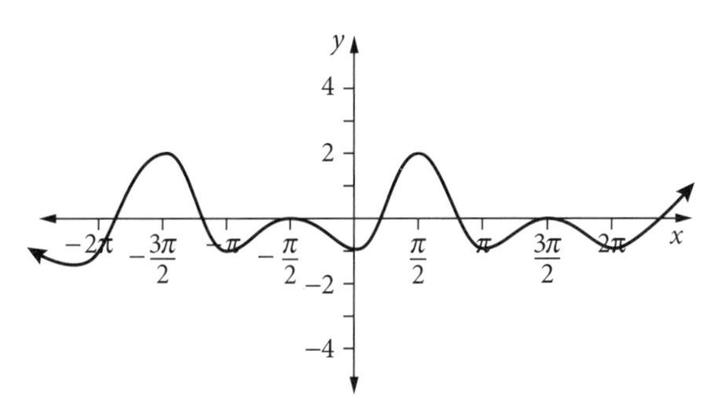


The absolute maximum points in the domain shown are at

**A** $x = \frac{\pi}{2}, \frac{3\pi}{2}$

**B** $x = \frac{\pi}{2}$

**C** $x = -\frac{3\pi}{2}, \frac{\pi}{2}$

**D** $x = -\frac{3\pi}{2}, -\frac{\pi}{2}, \frac{\pi}{2}, \frac{3\pi}{2}$

**E** $x = \frac{3\pi}{2}$

**Question 31**

The derivative of $g(x) = \sqrt{x}(x^p - 1)$ with respect to $x$ is

**A** $\frac{1}{2}x^{-\frac{1}{2}}(px^{p-1} - 1)$

**B** $\left(p + \frac{1}{2}\right)x^{p-\frac{1}{2}} - \frac{3\sqrt{x}}{2}$

**C** $\frac{p}{2}x^{p-\frac{3}{2}} + \frac{1}{2\sqrt{x}}$

**D** $\frac{1}{2}(2p+1)x^{p-\frac{1}{2}} - \frac{1}{2\sqrt{x}}$

**E** $\sqrt{x}(px^{p-1} - 1)$

**Question 32**

The derivative of $f(x) = \sin^n(x)$ is

**A** $n\sin^{n-1}(x)$

**B** $\frac{n\tan^n(x)}{\cos(x)}$

**C** $n\sin(x)\cos^{n-1}(x)$

**D** $n\cos(x)\sin^{n-1}(x)$

**E** $n\cos^2(x)\sin^n(x)$

**Question 33**

Which of the following is *not* true for the curve of $y = f(x)$, where $f(x) = x^{\frac{1}{8}}$?

**A** The domain is $[0, \infty)$.

**B** The range is $R^+ \cup \{0\}$.

**C** The curve passes through $(16, \sqrt{2})$.

**D** The function is strictly decreasing for all $x \in R^+$.

**E** The gradient is undefined at $x = 0$.

**Question 34**

If $y = x\log_{10}(2x^2 - 1)$, the gradient of the tangent at $x = 2$ is closest to

**A** 0.50 **B** 1.06 **C** 1.83 **D** 1.84 **E** undefined

### Question 35

The graph shown has the equation $f(x) = -2\sin(3(x - \pi))$.
The number of stationary points in the domain $(-\pi, \pi)$ is

**A** 1 **B** 2 **C** 3

**D** 4 **E** 6

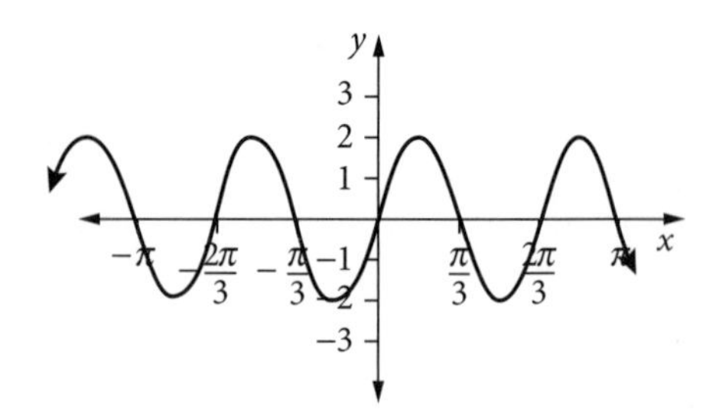


### Question 36

The parabola shown has the equation $y = x^2 + 1$.

The tangent to the graph at the point $(1, 2)$ has the equation $y = 2x$.

The instantaneous rate of change of the curve $y = x^2 + 1$ at $x = 1$ is

**A** 1 **B** 2 **C** 3

**D** 4 **E** 6

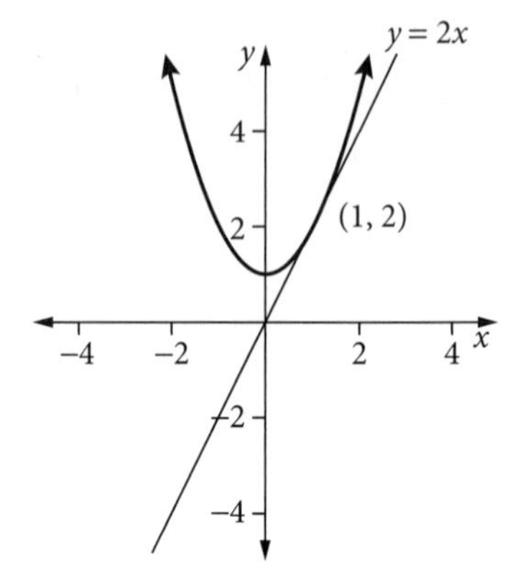


### Question 37

Consider the graph below of $y = f(x)$.

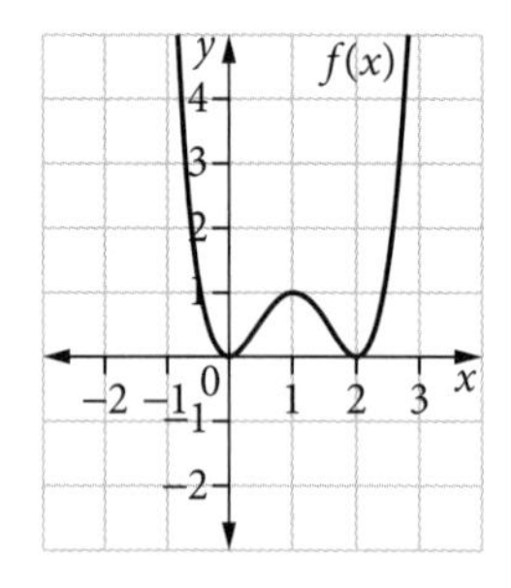


The graph of $f'(x)$ could look like

**A**

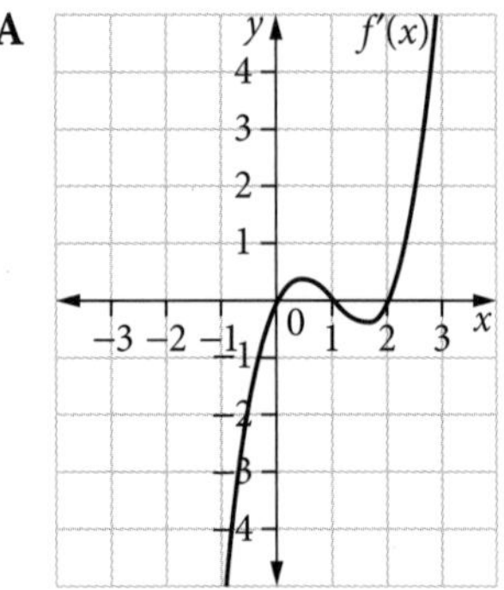


**B**

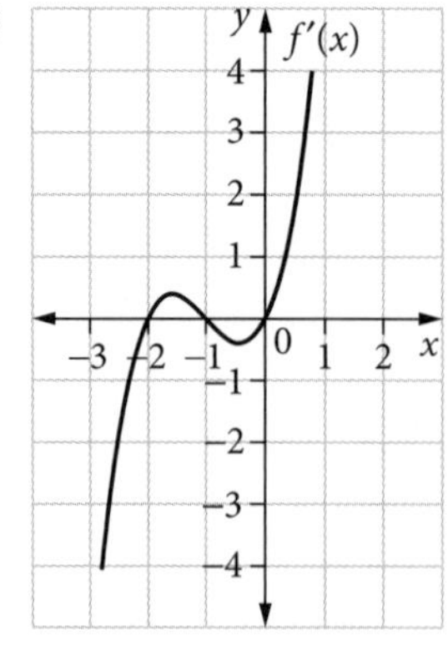


**C**

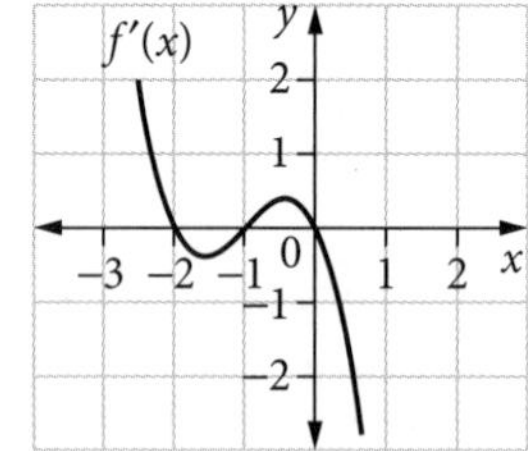


**D**

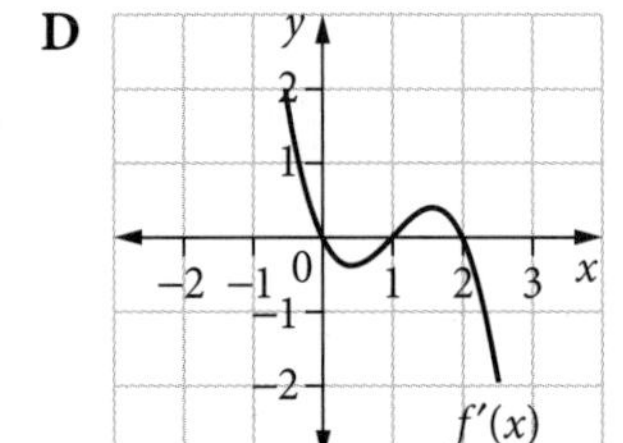


**E**

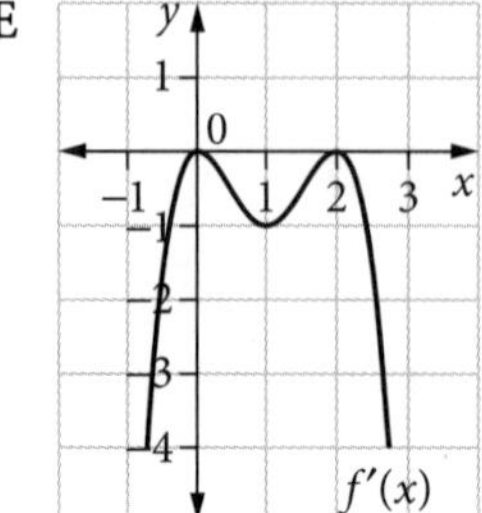

**Question 38**

The average rate of change between the points $(x, f(x))$ and $(x+h, f(x+h))$ in any function is found by

**A** $x+h$

**B** $\dfrac{f(x+h)-f(x)}{h}$

**C** $\lim\limits_{h\to 0}\dfrac{f(x+h)-f(x)}{h}$

**D** $\lim\limits_{h\to 0}\dfrac{h(2x+h)}{h}$

**E** $2x$

**Question 39**

If $N = Ae^{-kt}$, an anti-derivative of $N$ could be

**A** $Ae^{-kt}$

**B** $-Ake^{-kt}$

**C** $-Ake^{kt}$

**D** $-\dfrac{A}{k}e^{-kt}$

**E** $\dfrac{A}{k}e^{-kt}$

**Question 40**

The equation of the tangent to the function $h(x) = \tan(2\pi x) + 1$ at $x = 0$ is

**A** $y = 2\pi x + 1$

**B** $y = 2\pi x$

**C** $y = \pi x + 1$

**D** $y = 2\pi$

**E** $y = 2\pi x - 2\pi + 1$

**Question 41**

The gradient graph for the function $f(x) = x^{-2} + x^{-1}$ has

**A** an asymptote at $x = 0$

**B** an asymptote at $y = 0$

**C** asymptotes at both $x = 0$ and $y = 0$

**D** a defined value at $x = 0$

**E** a derivative at $x = 0$.

**Question 42**

The function $g(x) = 2x + x^2$ has

**A** a constant rate of change of 2

**B** an average rate of change of 3 between $x = 0$ and $x = 1$

**C** an instantaneous rate of change of 3 at $x = 1$

**D** a chord gradient of 2 between $x = 0$ and $x = 1$

**E** a derivative of 3 at $x = 0$.

**Question 43**

For the function $y = e^{3x}\sin(x)$, we get $\dfrac{dy}{dx} =$

**A** $3e^{3x}\cos(x)$

**B** $3e^{3x}$

**C** $e^{3x}\cos(x) + 3\sin(x)$

**D** $3e^{3x}(\cos(x) + 3\sin(x))$

**E** $e^{3x}(\cos(x) + 3\sin(x))$

**Question 44**

If $f(x) = 3\log_e(x-b)$, then $f'(4)$ equals

**A** $-\dfrac{3}{4-b}$

**B** $\dfrac{3}{4-b}$

**C** $3\log_e(4-b)$

**D** $\dfrac{3}{x-b}$

**E** $4-b$

**Question 45**

For the function $y = \dfrac{x}{\log_e(x)}$, $\dfrac{dy}{dx}$ equals

**A** $\dfrac{\log_e(x)}{\log_e(x-1)}$

**B** $\dfrac{\log_e(1)-1}{\log_e(x-1)}$

**C** $\dfrac{\log_e(x)-1}{(\log_e(x))^2}$

**D** $\dfrac{\log_e(x-1)}{(\log_e(x))^2}$

**E** $\log_e(5)+\log_e(10)$

**Question 46**

To find the gradient of the function $y = \cos(3x^2+1)$, what would $u$ be in the chain rule $\dfrac{dy}{dx} = \dfrac{dy}{du} \times \dfrac{du}{dx}$?

**A** $\cos(u)$ **B** $\cos(x)$ **C** $3x^2+1$ **D** $6x$ **E** $-\sin(u)$

**Question 47**

To find the gradient of the function $y = \cos(3x^2+1)$, what would $\dfrac{du}{dx}$ be in the chain rule $\dfrac{dy}{dx} = \dfrac{dy}{du} \times \dfrac{du}{dx}$?

**A** $\cos(u)$ **B** $\cos(x)$ **C** $3x^2+1$ **D** $6x$ **E** $-\sin(u)$

**Question 48**

Evaluate $\int_0^1 (-3f(x)+2x)\,dx$, where $\int_0^1 f(x)\,dx = 2$.

**A** −6 **B** −5 **C** 1 **D** 2 **E** 7

**Question 49**

The the graph of $f(x) = x\log_e(x+1)$ for all $x > -1$ has the general shape of

**A**

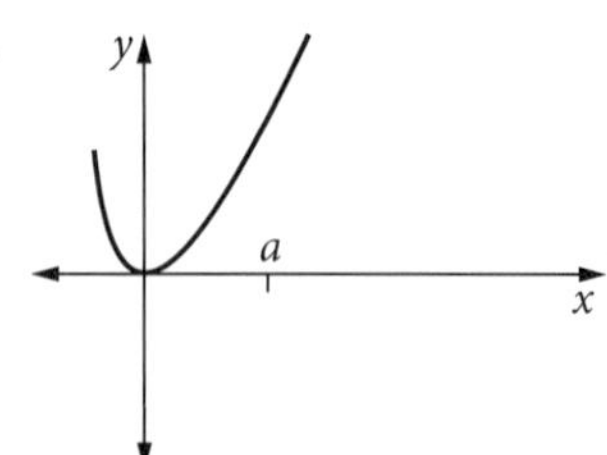


**B**

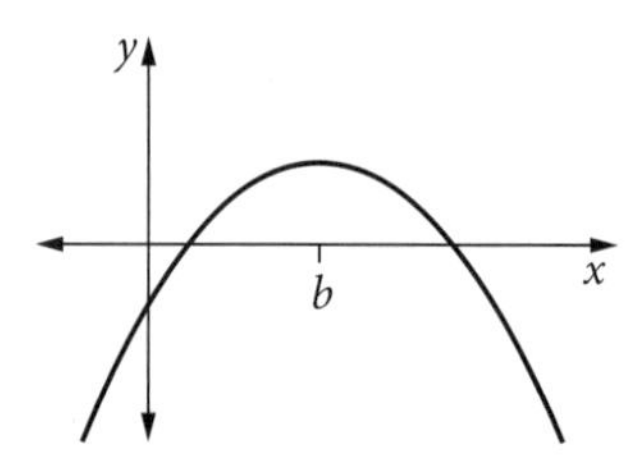


**C**

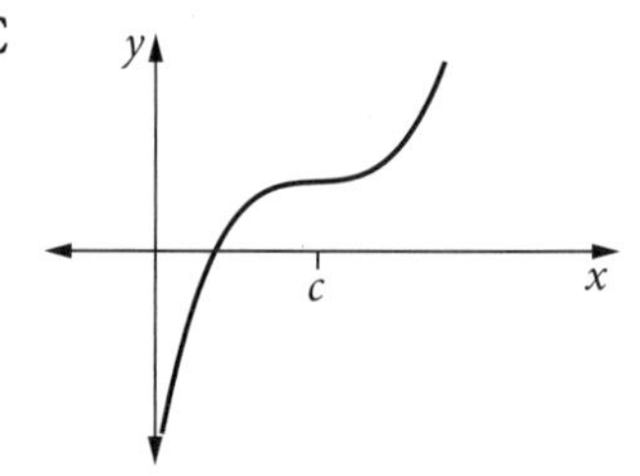


**D**

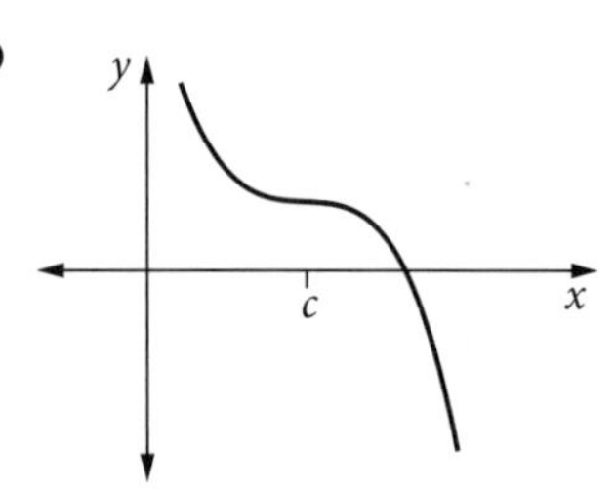


**E** none of the above

**Question 50**

$\lim\limits_{\delta x_i \to 0} \sum\limits_{i=1}^{n} f(x_i)\,\delta x_i$ can be expressed as

**A** $\lim\limits_{\delta x_i \to 0} \sum\limits_{0}^{n} f(x)\,\delta x$ **B** $\sum\limits_{0}^{n} f(x)\,\delta x$ **C** $\int_0^i f(x_i)\,dx$ **D** $\int_a^b f(x)\,dx$ **E** $\int_0^i f(x)\,dx$

## Extended-answer questions

### Technology active: 13 questions

Solutions to this section start on page 231.

**Question 1** (10 marks)

**a** Find $\frac{dy}{dx}$ if $y = 3x\cos(x)$. 1 mark

**b** Hence find an expression for $\int x\sin(x)\,dx$. 3 marks

**c** Evaluate $\int_0^{\pi} x\sin(x)\,dx$. 1 mark

Consider the curve of $f(x) = x\sin(x)$.

**d** Sketch the graph of $f: [0, \pi] \to R, f(x) = x\sin(x)$, labelling the coordinates of the endpoints and the local maximum point, correct to three decimal points. 2 marks

**e** Find the area under the curve $f(x) = x\sin(x)$ for $x \in [0, \pi]$. 1 mark

**f** Solve the equation $f'(x) = 0$ for $x \in [0, \pi]$, giving your answer correct to three decimal places. 2 marks

**Question 2** (14 marks)

A particular rock concert emits sound (measured in decibels) that ranges from 90 dB to quite a high level. The function that models the sound at the concert follows the function of $f(t) = 10\sin(\pi t) + 100$, where $t \geq 0$, $t$ hours after the start of the concert at 7 pm.

**a** What is the average noise level, in dB, at the rock concert from 7 pm to 11 pm? 1 mark

**b** The concert is so good that the band returns for encores, and the concert finishes at 11:45 pm. What is the average sound, in dB, correct to two decimal places, at the rock concert from 7 pm to 11:45 pm? 1 mark

**c** For $f(t) = 10\sin(\pi t) + 100$, what is the amplitude and period of the function? 2 marks

**d** Sketch the graph $f(t) = 10\sin(\pi t) + 100$ for the domain $x \in [0, 4]$, showing the coordinates of the endpoints and the maximum and minimum turning points. 3 marks

The level of sound, in decibels, at which a person may sustain hearing loss is 95 dB.

**e** For what period of time during the 4-hour concert will the decibel level be dangerous for patrons? 2 marks

Normal conversation is held between 60 dB and 65 dB.

**f** Give a reason as to why normal conversation cannot be heard during the 4-hour concert. 1 mark

**g** Find $\frac{d}{dt}(10\sin(\pi t) + 100)$. 1 mark

**h** Hence solve the equation $f'(t) = 0$. 2 marks

The loudest recommended exposure with hearing protection is 140 dB. Even short-term exposure can cause permanent hearing loss at this level.

**i** Using your information from part **h**, explain why there is no possibility of permanent hearing loss for the concert patrons. 1 mark

**Question 3** (6 marks)

The velocity of a particle is described by the function $v(t) = 3t^3 - t - 2$, where $v(t)$ is in m/s and $t$ is in seconds, $t \geq 0$.

**a** By finding the discriminant of a quadratic factor in $v(t)$, show that its graph has only one $t$-intercept. 2 marks

**b** Find when the particle momentarily stops in its journey. 1 mark

**c** What distance does the particle travel in the first 5 seconds? 3 marks

**Question 4** (16 marks)

The function of $f(x) = x^3 + ax^2 + bx + c$ has a stationary point at (1, 300).

**a** Find the values of $a$ and $b$ in terms of $c$. 3 marks

A company models its monthly profits using the function $f(x) = x^3 + ax^2 + bx + c$, where $f(x)$ is the monthly profit in \$ and $x$ is the day of the month, where, for example, $x = 1$ represents the 1st of the month and $x = 20$ represents the 20th day of the month. The mathematicians in the company find that profits are satisfactory when $a = b$.

**b** Find the value of $c$ for which $a = b$. 1 mark

**c** For the domain $x \geq 1$, and using your values of $a$, $b$ and $c$, find $f'(x)$, and hence show that the maximum or minimum point(s) occur at the end of the domain. 3 marks

**d** Find the minimum and maximum profits for a 30-day month, and the days of the month on which they occur. 3 marks

**e** Sketch the graph of $f: [1, 30] \to R, y = f(x)$, labelling the coordinates of the endpoints. 3 marks

**f** Find the value of $c$ for which $a = 2b$. 1 mark

**g** Using the values of $c$ found in part **f**, describe what this does to the profit of the company. 2 marks

**Question 5** (10 marks)

The pitch of a roof in the northern states of America is, by standard, a certain value so that the heavy snow in the area doesn't weigh too heavily on the roof. In general, as the pitch of the roof increases, the snow slides off more easily, and the load on the roof decreases. But a steeper roof is more expensive to build. To encourage snow to slide off, the roof should have a minimum pitch of 3 : 12, meaning that it drops more than 3 feet for every 12 feet of roof, but a pitch of 4 : 12 is better.

**a** What is the gradient of a 3 : 12 roof? 1 mark

**b** What is the gradient of a 4 : 12 roof? 1 mark

**c** If $\tan(45°) = 1$ and equals the ratio $\frac{12}{12}$, what approximate ratio in the form of $\frac{x}{12}$ does $\tan(14°)$ equal? 1 mark

**d** If $\tan(45°) = 1 =$ the ratio $\frac{12}{12}$, what approximate ratio in the form of $\frac{x}{12}$ does $\tan(18.5°)$ equal? 1 mark

**e** A straight line of roofing has a pitch of 4 : 12 and goes through the point (0, 0). What is the equation of the line? 1 mark

**f** A straight line of roofing has a pitch of 3 : 12 and goes through the point (2, 10). What is the equation of the line? 2 marks

Let $\tan(\theta) = \frac{x}{12}$ so that $\theta = \tan^{-1}\left(\frac{x}{12}\right)$, where $\theta°$ is the angle of the roof to the horizontal and $x$ is the vertical height of the roof, in inches.

**g** Find $\frac{d}{dx}(\theta)$ and hence find the value of $x$ for which the angle of the roof is at its maximum. 3 marks

**Question 6** (14 marks)

Let $f(x) = x + 2$ and $g(x) = 2x^2 + ax + 4$ for maximal domains, where $a$ is a real constant.

If $f(x) = g(x)$, find the value(s) of $a$ for which there is

**a** exactly one solution to the equation $f(x) = g(x)$. 2 marks

**b** more than one solution to the equation $f(x) = g(x)$. 2 marks

**c** no solution to the equation $f(x) = g(x)$. 1 mark

**d** For the case of exactly one solution to the equation $f(x) = g(x)$, where $a > 0$, sketch the graph of $h(x) = f(x) - g(x)$. 3 marks

**e** Find $h'(x)$, and hence find the coordinates of any stationary point in the graph of $h(x)$. 3 marks

**f** Find the simplified transformed equation for $h_T(x) = -h(2x - 3) + 7$. 1 mark

**g** Describe in words the step-by-step transformation to get to the image $h_T(x)$. 2 marks

**Question 7** (11 marks)

Consider the function $f(x) = (x - 1)(x^2 - k)$, where $k$ is a real constant.

**a** Find the rule for $g(x)$ if $g(x) = -f(2x - 1) + 3$. Give your answer in the form of $ax^3 + bx^2 + cx + d$. 2 marks

**b** Find the equation of the tangent line to the curve $g(x)$, when $x = 1$, in terms of $k$. 3 marks

**c** Find the equation of the line that is perpendicular to the tangent found in part **b**, also going through the point when $x = 1$, in terms of $k$. 3 marks

**d** Find the value(s) of $k$, if it is possible, such that the tangent found in part **b** is also a tangent to $f(x) = (x - 1)(x^2 - k)$ at $x = 1$. 3 marks

**Question 8** (13 marks)

Paddy inadvertently throws his cricket ball over the fence into the parkland behind his house. The ball follows the path of a parabola with the equation $h(x) = -x^2 + 10x + 1$, where $h(x)$ metres is the vertical height of the ball from the ground, at a horizontal distance, $x$ metres, from Paddy's hand.

**a** At what height from the ground does the ball leave Paddy's hand? 1 mark

**b** What is the maximum height that the ball will reach? 1 mark

**c** The fence into the parkland has a height of 2 metres and is at a horizontal distance of 2 metres from Paddy. Will his ball hit the fence? Why/why not? 2 marks

A jogging path has been built around the parkland and Imogen is jogging just as Paddy throws the ball. She sees the ball and stops jogging. Imogen is 1.5 metres tall and the ball hits her on the head.

**d** Where is Imogen standing, correct to two decimal places, when the ball hits her? 1 mark

Imogen picks up the ball, runs away with it and puts it on the ground at the coordinates (30, 0).

**e** What is the shortest distance from the ball on the ground to Paddy's hand, still at 1 metre vertically from the ground? 2 marks

**f** Paddy finds a way to crawl through the fence and runs to the ball. How far does Paddy run? 1 mark

**g** Find the equation of the tangent to the curve $h(x)$ at $x = 2$. 2 marks

**h** **i** Hence state an integral to find the area between the tangent at $x = 2$, the curve $h(x)$ and the $y$-axis. 2 marks

**ii** Evaluate this area. 1 mark

**Question 9** (13 marks)

The function $f$ is defined by $f: [0, 2\pi] \to R$, where $f(x) = 3e^{\frac{x}{10}}g(x)$, where $g(x) = \sin(2x)$.

**a** Show that the solutions to the equation $g(x) = 0$ for $x \in [0, 2\pi]$ are $x = 0, \frac{\pi}{2}, \pi, \frac{3\pi}{2}, 2\pi$. 2 marks

**b** Hence state the solutions to the equation $f(x) = 0$. 1 mark

**c** Using the product rule, find $f'(x)$ and hence find the coordinates, correct to two decimal places, of the stationary points of the graph of $f(x)$. 3 marks

**d** Sketch the graph of $f(x)$, labelling axial intercepts and endpoints. 3 marks

**e** Let the points $A(a, f(a))$ and $B(b, f(b))$ be the first two points of intersection between the graphs of $f(x)$ and $y = 3e^{\frac{x}{10}}$. The line that joins $A$ and $B$ has a gradient of $m$.

Show that $m = \frac{f(b) - f(a)}{\pi}$. 2 marks

**f** Hence, find an expression for the gradient, $n$, of points $C$ and $D$, where $C$ and $D$ are the first 2 points of intersection between the graphs of $f(x)$ and $y = -3e^{\frac{x}{10}}$. 2 marks

**Question 10** (8 marks)

A section of a rollercoaster is modelled by a function with the equation $p(x) = a(bx^2 - cx)^2$.

**a** Show that the $x$-intercepts of the function are at $x = 0$ and $x = \frac{c}{b}$. 2 marks

**b** Show that the stationary points of the function are at $x = 0$, $x = \frac{c}{2b}$ and $x = \frac{c}{b}$. 2 marks

This section of the rollercoaster has the domain $x \in [-1, 5]$, where $x$ and $y$ are respectively the horizontal and vertical displacement, in metres, from the point $(0, 0)$. The point $(1, 0)$ is where the safety officer sits.

**c** We know that $p'\left(\frac{10}{3}\right) = 0$. Show that $\frac{10}{3} = \frac{c}{b}$. 1 mark

**d** It is also true that $p(5) = 62.5$ and $p'(0) = 0$. Find the values of $a$, $b$ and $c$. 1 mark

**e** The safety officer looks directly above him and decides that the rollercoaster is inoperable if the gradient is greater than 5. Does he shut down the rollercoaster? 2 marks

**Question 11** (21 marks)

Consider the function $f: [-1, 3] \to R, f(x) = 3x^2 - x^3$.

**a** Find the coordinates of the stationary points of $f$. 2 marks

**b** Sketch the graph of $f$, labelling endpoints and stationary points with their coordinates. 3 marks

**c** What is the maximum value of $f$ and for what $x$ value(s) does the maximum exist? 2 marks

**d** What is the minimum value of $f$ and for what $x$ value(s) does the minimum exist? 2 marks

**e** For what $x$ values is the graph of $f$ strictly increasing? 1 mark

**f** A tangent is drawn to the curve at $x = \frac{7}{4}$. What is the equation of this tangent? 2 marks

**g** **i** An area is formed that is enclosed by the curve of $f$ and the tangent. Write down the integral for finding this area. 3 marks

**ii** Hence evaluate this area. 1 mark

A different area is formed when a line is drawn parallel to the $y$-axis, starting at the point $(2, 0)$ and ending at $(2, 4)$, as shown in the diagram.

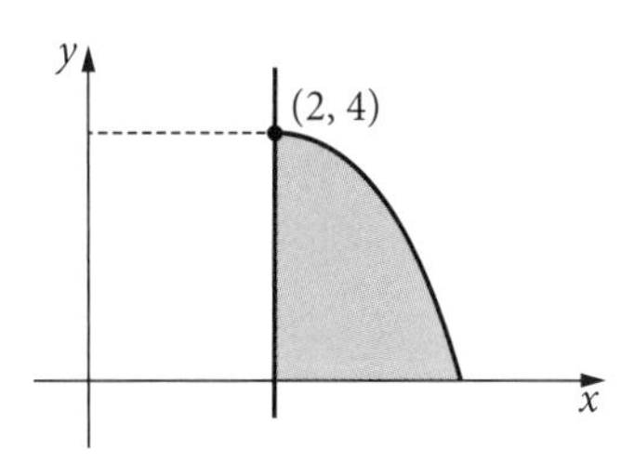


An approximation to the area between the line $x = 2$ and the curve to the right side of the line is found by a series of rectangles of width 1 unit, as shown.

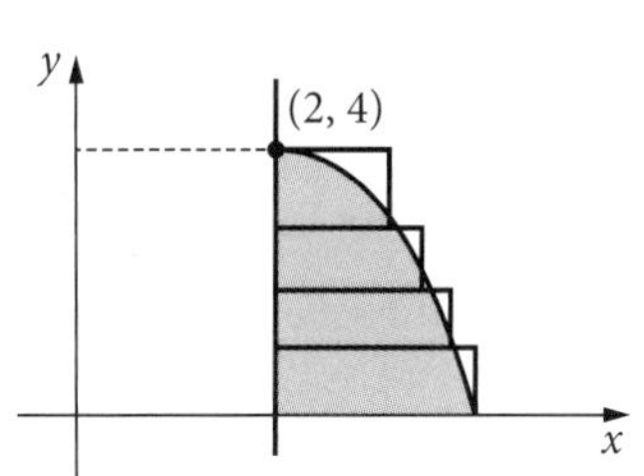


**h** Find an approximation to the area using the four rectangles. Give your answer correct to two decimal places. 3 marks

**i** Find the ratio between the shaded area found by integration and your approximated area from part **h**. 2 marks

**Question 12** (12 marks)

The population of the common loon in one particular lake in the Great Lakes region of North America varies according to the rule $p(t) = 10\,000 - 5000\cos\left(\frac{\pi t}{6}\right)$, where $p$ is the population of the common loon and $t$ is the number of months after 1 January 2014 with $t \in [0, 12]$.

Lucas is the conservation officer in charge of maintaining the population health of the loon birds.

**a** State the period and amplitude of the function $p$. 2 marks

**b** Find the maximum number of common loons in this lake and state when this maximum occurs. 2 marks

**c** Find the minimum number of common loons in this lake and state when this minimum occurs. 1 mark

In his first year of work, starting on 1 January 2014, Lucas reported that the loon population is healthy when the rate of population change is greater than 1000 birds per month.

**d** Find the fraction of time, over the first 12 months of his job, when Lucas decides that the population is healthy. Give your answer as an interval for $t$, correct to two decimal places. 2 marks

**e** Sketch the graph of $p(t)$, labelling axial intercepts and endpoints. 2 marks

Lucas moves to another lake for the second year of his job, starting on 1 January 2015.

This time, the population of the yellow-billed loon varies according to the rule $y(t) = 5000\sin\left(\frac{\pi t}{4}\right) + 80\,000$, where $y$ is the population of the yellow-billed loon and $t$ is the number of months after 1 January 2015. The standard for a healthy loon population stays the same.

**f** Find the ratio of time when the yellow-billed loon population is healthy compared to when the common loon population is healthy. 3 marks

9780170465366

**Question 13** (8 marks)

Toby Jones is hiking through the woods towards a river and is deciding whether he will swim or walk to his destination. He is able to walk at a rate of 6 metres per second and swim at a rate of $k$ metres per second, where $k$ is a constant. Toby has a 'mud map' that he is following, sketched below.

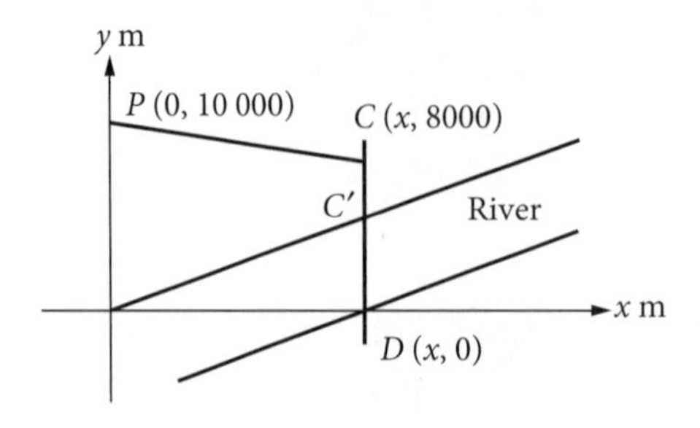


In the diagram above, $P$ is Toby's starting point, $C$ is the coffee shop and $D$ is his destination on the opposite bank of the river. $P$ has coordinates $(0, 10\,000)$. $D$ has coordinates $(x, 0)$. $C$ has coordinates $(x, 8000)$. $C'$ is where Toby hits the river. Let $x$ be the horizontal distance between the origin and $D$. Let $(1000 - \sqrt{x})$ metres be the distance between $C$ and $C'$.

**a** State an expression for $T(x)$, the time in seconds that Toby walks and swims for if he goes from $P$ to $C$ and then from $C$ to $D$. 2 marks

**b** Find $T'(x)$ and hence find the minimum time that Toby walks if $k = 10$. Give your answer to the nearest second. 2 marks

**c** What is the restriction on $x$ in this derivative? 1 mark

**d** Find the time taken, to the nearest second, if Toby decides to walk in a line directly from $P$ to the nearest bank of the river. Assume that this direct line is at right angles to the bank of the river. 3 marks

# UNIT 4

# Chapter 4
# Area of Study 4: Data analysis, probability and statistics

## Content summary notes

**Data analysis, probability and statistics**

### Random variables

- Random variable as a real function
- Parameters
- Properties
- Discrete random variable
- Continuous random variable
- Interpretation

### Visual representation

- Tree diagrams
- Tables
- Karnaugh maps
- Venn diagrams
- Probability functions

### Continuous random variables

- Probability density function
  Non-negative functions of a real variable
- Normal distribution $X \sim N(\mu, \sigma^2)$
  $\mu = \mathrm{E}(X) =$ mean
  $\sigma^2 =$ variance, $\sigma =$ standard deviation
- Standard normal distribution $X \sim Z(0, 1)$
- Variation of parameters of normal distribution
- Probability graph

### Measures of spread

- Range
- Variance
- Standard deviation
- Interpretation

### Basic probability

- Set notation
- Sample space
- Complement
- Addition rule
- Independent events
- Mutually exclusive events
- Conditional probability

### Discrete random variables

- Tables of data/sample space
- $\mu = \mathrm{E}(X) =$ mean
  $\mathrm{E}(X^2) - (\mathrm{E}(X))^2 =$ variance
- Binomial distribution $X \sim \mathrm{Bi}(n, p)$
  $n =$ number of independent events
  $p =$ probability of success
- Parameters of binomial distribution
- Probability graph

### Central measures

- Mean, median and mode
- Interpretation

### Statistical inference

- Sample proportions
- Definition of sample proportion
- Distribution of sample proportion
- Simulations
- Confidence intervals
- 1, 2 and 3 standard deviations

In this area of study students cover discrete and continuous random variables, their representation using tables, probability functions (specified by rule and defining parameters as appropriate); the calculation and interpretation of central measures and measures of spread; and statistical inference for sample proportions. The focus is on understanding the notion of a random variable, related parameters, properties and application and interpretation in context for a given probability distribution.

This area of study includes:

- random variables, including the concept of a random variable as a real function defined on a sample space and examples of discrete and continuous random variables
- discrete random variables:
  - specification of probability distributions for discrete random variables using graphs, tables and probability mass functions
  - calculation and interpretation of mean, $\mu$, variance, $\sigma^2$, and standard deviation of a discrete random variable and their use
  - Bernoulli trials and the binomial distribution, $\text{Bi}(n, p)$, as an example of a probability distribution for a discrete random variable
  - effect of variation in the value(s) of defining parameters on the graph of a given probability mass function for a discrete random variable
  - calculation of probabilities for specific values of a random variable and intervals defined in terms of a random variable, including conditional probability
- continuous random variables:
  - construction of probability density functions from non-negative functions of a real variable
  - specification of probability distributions for continuous random variables using probability density functions
  - calculation and interpretation of mean, $\mu$, variance, $\sigma^2$, and standard deviation of a continuous random variable and their use
  - standard normal distribution, $\text{N}(0, 1)$, and transformed normal distributions, $\text{N}(\mu, \sigma^2)$, as examples of a probability distribution for a continuous random variable
  - effect of variation in the value(s) of defining parameters on the graph of a given probability density function for a continuous random variable
  - calculation of probabilities for intervals defined in terms of a random variable, including conditional probability (the cumulative distribution function may be used but is not required)
- statistical inference, including definition and distribution of sample proportions, simulations and confidence intervals:
  - distinction between a population parameter and a sample statistic and the use of the sample statistic to estimate the population parameter
  - simulation of random sampling, for a variety of values of $p$ and a range of sample sizes, to illustrate the distribution of $\hat{P}$ and variations in confidence intervals between samples
  - concept of the sample proportion $\hat{P} = \dfrac{X}{n}$ as a random variable whose value varies between samples, where $X$ is a binomial random variable which is associated with the number of items that have a particular characteristic and $n$ is the sample size
  - approximate normality of the distribution of $\hat{P}$ for large samples and, for such a situation, the mean $p$ (the population proportion) and standard deviation, $\sqrt{\dfrac{p(1-p)}{n}}$

- determination and interpretation of, from a large sample, an approximate confidence interval $\left(\hat{p} - z\sqrt{\frac{\hat{p}(1-\hat{p})}{n}}, \hat{p} + z\sqrt{\frac{\hat{p}(1-\hat{p})}{n}}\right)$, for a population proportion where $z$ is the appropriate quantile for the standard normal distribution, in particular the 95% confidence interval as an example of such an interval where $z \approx 1.96$ (the term standard error may be used but is not required).

# 4.1 Review of probability (Units 1 and 2 concepts)

This section reviews the concepts of a sample space, visual displays of the sample space, an event and the probability of a favourable event. The focus is on understanding the notion of a random variable, related parameters, properties and application and interpretation in context for a given probability distribution.

VCE Mathematics Study Design 2023–2027 p. 100, © VCAA 2022

- $0 \le \Pr(A) \le 1$ for all events $A$ in the sample space
- $\Pr(A') = 1 - \Pr(A)$ or $\Pr(A) + \Pr(A') = 1$

  $A'$ is the **complement** of $A$
- **Addition rule**: the probability of $A$ or $B$ or both

  $\Pr(A \cup B) = \Pr(A) + \Pr(B) - \Pr(A \cap B)$
- **Independent events**

  $A$ and $B$ are independent if $\Pr(A \cap B) = \Pr(A) \times \Pr(B)$
- **Mutually exclusive events**

  $A$ and $B$ are mutually exclusive if $\Pr(A \cap B) = 0$
- **Conditional probability** This is the probability of event $A$ occurring, given that event $B$ has occurred.

  $\Pr(A \mid B) = \dfrac{\Pr(A \cap B)}{\Pr(B)}$

  Remember that any of the rules learned in Year 11 can be used in Year 12 examination questions.

**Hint**
Watch out especially for conditional probability questions. This sub-topic is often used in both Technology Free and Technology Active contexts.

## 4.1.1 Conditional probability

### Example 1

In a particular school, the probability that a student studies Maths Methods is 0.6, and the probability that a student studies both Maths Methods and Psychology is 0.4. What is the probability that a randomly selected student studies Psychology given we know they also study Maths Methods?

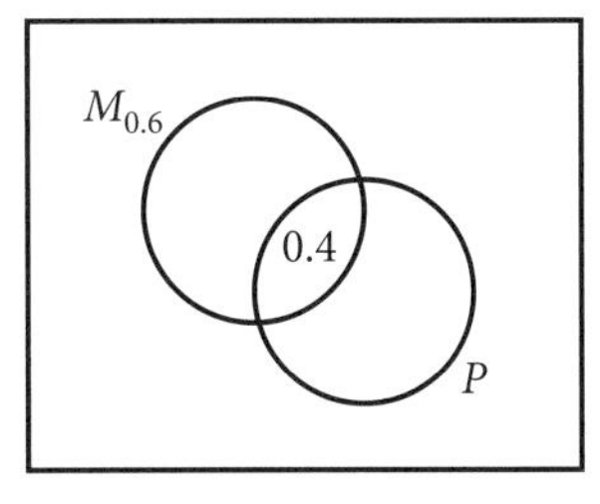


**Solution**

$$\Pr(P \mid M) = \frac{\Pr(P \cap M)}{\Pr(M)}$$

$$\Pr(P \mid M) = \frac{0.4}{0.6}$$

gives $$\Pr(P \mid M) = \frac{2}{3}$$

**Hint**
Don't leave your answer as decimals within a fraction and make sure you completely simplify your fraction answer.

## 4.1.2 The language of probability

Probability is the mathematics of chance. Every day we are involved with events that may or may not occur with different degrees of certainty.

Will it be fine today?
Will I feel well today?
Will my sports team win?

Probability tells us the likelihood of an event occurring. We can calculate the probability of an event occurring by assigning a number between 0 and 1 to describe the likelihood of the event.

- An impossible event will have a probability of zero.
- A certain event will have a probability of one.
- All other events between these two extremes will have a probability between 0 and 1.

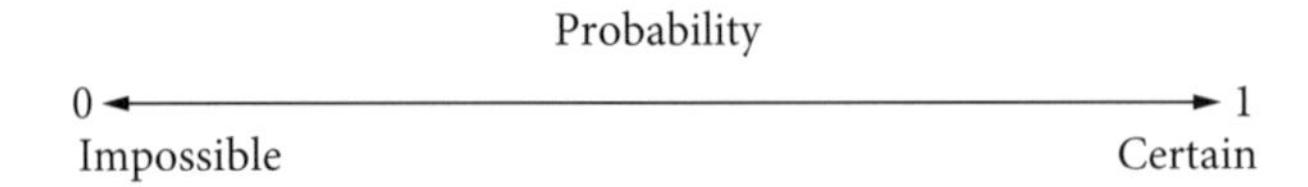


The **sample space** is the full range of possibilities that occur when we carry out an experiment. For example, when we throw a die, the sample space of the number shown uppermost is written as $\{1, 2, 3, 4, 5, 6\}$.

---

**Example 2**

If the probability of event $A$ occurring is 0.3, of event $B$ occurring is 0.5, and the probability of them both happening is 0.1, what is the probability of either event $A$ or $B$ occurring or both?

**Solution**

Using the addition formula $\quad \Pr(A \cup B) = \Pr(A) + \Pr(B) - \Pr(A \cap B)$

$\Pr(A \cup B) = 0.3 + 0.5 - 0.1$

gives $\quad \Pr(A \cup B) = 0.7$

## 4.1.3 Visual displays

Select any of the following visual displays to help solve problems.

- **tree diagrams**

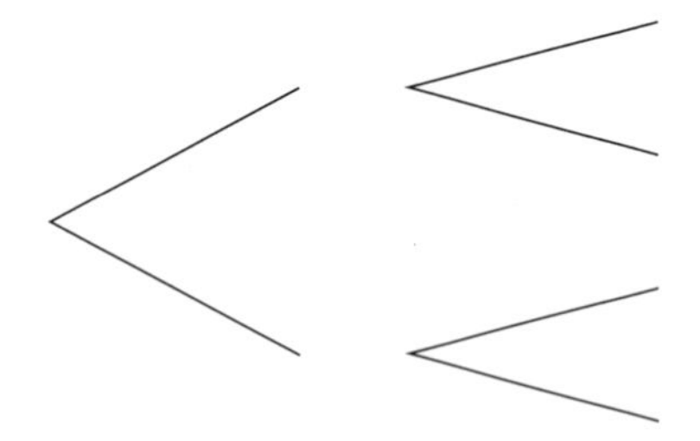

- **Venn diagrams**

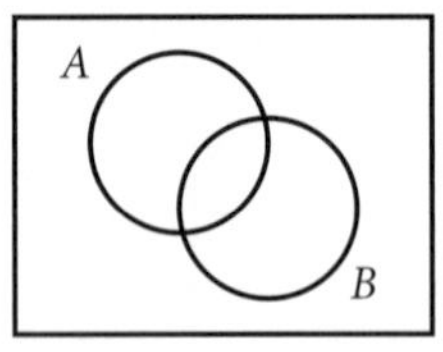


- **Karnaugh maps/probability tables**

| | A | A′ | |
|---|---|---|---|
| B | | | |
| B′ | | | |
| | | | 1 |

- **grids or arrays**

| 1, 1 | 1, 2 | 1, 3 | 1, 4 | 1, 5 | 1, 6 |
|---|---|---|---|---|---|
| 2, 1 | 2, 2 | 2, 3 | 2, 4 | | |
| 3, 1 | | | | | |
| 4, 1 | | | | | |
| 5, 1 | | | | | |
| 6, 1 | | | | | |

or in an alternative form with 'edging/labelling'

| | | Die | | | | | |
|---|---|---|---|---|---|---|---|
| | | 1 | 2 | 3 | 4 | 5 | 6 |
| Spinner | 1 | | | | | | |
| | 2 | | | | | | |
| | 3 | | | | | | |
| | 4 | | | | | | |

## 4.2 Random variables

This section introduces

- random variables, including the concept of a random variable as a real function defined on a sample space and examples of discrete and continuous random variables.

VCE Mathematics Study Design 2023–2027 p. 100, © VCAA 2022

A **random variable** is one whose value is determined by the outcome of a random experiment. There are two different types of random variables.

- **Discrete random variables**: can only take particular (discrete) values, for example, the number of pets in a household, the number of children in a class.

  The outcomes that can occur in any event can be listed in discrete random variables.

- **Continuous random variables**: can take any value in a given interval, for example, the height of students in my class, the BMI in a list of patients at a surgery clinic.

  The number of outcomes that can occur in any event cannot be listed in continuous random variables. For example, the weight of a parcel to be posted overseas makes sense measured as 8 kg, or 8.1 kg, or 8.01 kg; all may be possible and valid.

### 4.2.1 Discrete random variables

The experiment of tossing two coins produces the sample space or list of outcomes $\{HH, HT, TH, TT\}$.

These are all the possible outcomes that can be produced.

Let $X$ stand for the number of heads in two tosses of a coin, then $X$ can take the values from the set of outcomes $\{0, 1, 2\}$.

$X$ is said to be **discrete** because it can only take a **countable** number of values; the number of outcomes can actually be **counted**.

The experiment displayed as a tree diagram:

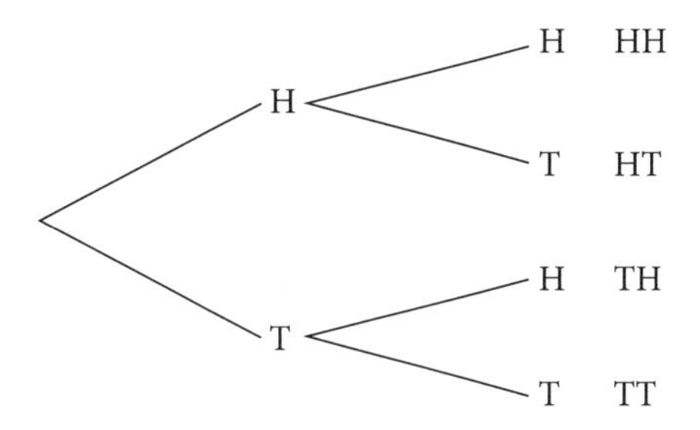


The values do not have to be whole numbers or integer values. An example of this is shoe size, which is discrete because although a size of 8 or $8\frac{1}{2}$ or 9 can be counted, a shoe size of $8\frac{1}{7}$ does not exist.

### 4.2.2 Continuous random variables

Finding the height of students in a class produces outcomes that are continuous. It makes sense for a student to have a height of 120 cm or 120.01 cm, or 120.002 cm, depending on the accuracy of the measuring equipment. Here the number of outcomes that can occur in any event cannot be listed.

If $X$ is a random variable which takes its values as 'height in metres' of a person, then $X$ is said to be a continuous random variable, and $X$ may be any non-negative real number.

In this case, $X$ is measured rather than counted.

## 4.3 Discrete random variables

This section further develops discrete random variables, specifying probability distributions for discrete random variables using probability formulas and concepts supported by visual displays.

- specification of probability distributions for discrete random variables using graphs, tables and probability mass functions
- calculation and interpretation of mean, $\mu$, variance, $\sigma^2$, and standard deviation of a discrete random variable and their use
- Bernoulli trials and the binomial distribution, $\text{Bi}(n, p)$, as an example of a probability distribution for a discrete random variable
- effect of variation in the value(s) of defining parameters on the graph of a given probability mass function for a discrete random variable
- calculation of probabilities for specific values of a random variable and intervals defined in terms of a random variable, including conditional probability.



Discrete random variable distributions give all possible values that $X$ can take, matched with the probability for each value.

- The sum of all the probabilities must total 1.
- The formula for the sum of probabilities is $\Sigma \Pr(X = x) = 1$.

#### Example 3

The number of pets in a household in a country town, and the probability of them occurring, are listed in the table below.

| $x$ | 0 | 1 | 2 | 3 | 4 |
|---|---|---|---|---|---|
| $\Pr(X = x)$ | 0.1 | 0.2 | 0.4 | 0.1 | 0.2 |

**a** Show that the formula $\Sigma \Pr(X = x) = 1$ is true for this discrete distribution table.

**b** Find $\Pr(X > 1)$.

**c** Calculate $\Pr(X > 1 \mid X < 4)$.

#### Solution

**a** $0.1 + 0.2 + 0.4 + 0.1 + 0.2 = 1$
so $\Sigma \Pr(X = x) = 1$.

**b** $\Pr(X > 1) = 0.4 + 0.1 + 0.2 = 0.7$

**c** Using $\Pr(A \mid B) = \dfrac{\Pr(A \cap B)}{\Pr(B)}$,

we get $\Pr(X > 1 \mid X < 4) = \dfrac{\Pr(X > 1 \cap X < 4)}{\Pr(X < 4)}$

$$= \frac{\Pr(X = 2) + \Pr(X = 3)}{\Pr(X = 0) + \Pr(X = 1) + \Pr(X = 2) + \Pr(X = 3)}$$

$$= \frac{0.4 + 0.1}{0.1 + 0.2 + 0.4 + 0.1} = \frac{0.5}{0.8} = \frac{5}{8}$$

**Hint**

Remember not to leave decimals inside fractions.

A useful way to 'turnaround' a question to make calculations easier is shown.

$$\Pr(X > 1 \mid X < 4) = \frac{\Pr(X = 2) + \Pr(X = 3)}{1 - \Pr(X = 4)}$$
$$= \frac{0.4 + 0.1}{1 - 0.2}$$
$$= \frac{0.5}{0.8}$$
$$= \frac{5}{8}$$

The information about the number of pets in households can be expressed as a probability graph, also called a **probability mass function**.

**Hint**

Note that in contrast to an algebraic function graph, zero is not placed at an origin.

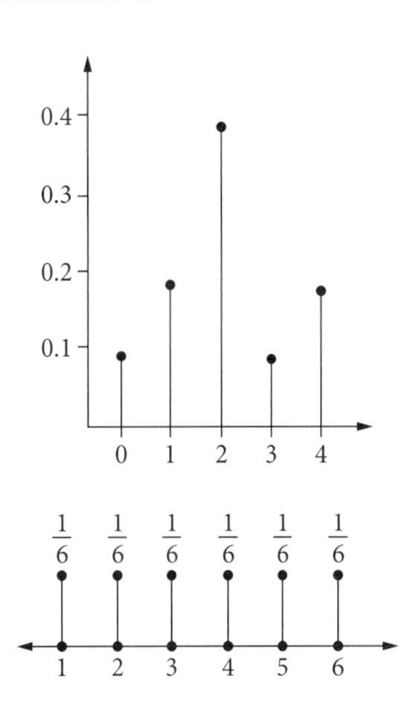


A common probability mass function used in teaching is the distribution of the result of rolling a fair die. The graph to the right shows the probability mass function of a fair die. Each of the numbers on the die have an equal chance of appearing uppermost when the die stops rolling.

## 4.4 Mean, variance and standard deviation of discrete random variables

This section continues with discrete random variables: calculating, interpreting and using the mean, $\mu$, variance, $\sigma^2$, and standard deviation, $\sigma$, of a discrete random variable.

### 4.4.1 Mean or expected value of a discrete probability distribution

The **mean** is denoted by $\mu$ or $\text{E}(X)$.

$\text{E}(X)$ is also known as the **expected value**.

The formula for the mean is $\mu = \text{E}(X) = \Sigma x \Pr(X = x)$.

This is the long-run average value of $X$. We can interpret this as what we could expect in the future if this pattern continues, hence the words 'expected value'.

The mean is calculated as the sum of each value of $x \times$ its matching probability.

So, the formula $\Sigma x \Pr(X = x)$ asks you to multiply each $x$ value by its matching probability and then add them up.

---

**Example 4**

Using the data about the number of household pets again:

| $x$ | 0 | 1 | 2 | 3 | 4 |
|---|---|---|---|---|---|
| $\Pr(X = x)$ | 0.1 | 0.2 | 0.4 | 0.1 | 0.2 |

The expected number of pets in a household = $\mu = E(X) = \Sigma x \Pr(X = x)$

$$\begin{aligned} E(X) &= (0 \times 0.1) + (1 \times 0.2) + (2 \times 0.4) + (3 \times 0.1) + (4 \times 0.2) \\ &= 0.2 + 0.8 + 0.3 + 0.8 \\ &= 2.1 \end{aligned}$$

We leave the decimal in the **expected number** of pets.

We don't round this number but leave it as it is. This means, on average, we expect 2.1 pets per household.

$$\text{mean} = 2.1$$

The mean is a **measure of central tendency**.

#### 4.4.1.1 Interpretation of mean: useful formulas

$$E(aX + b) = aE(X) + b$$

$$E(X + Y) = E(X) + E(Y)$$

---

**Example 5**

Using the household pet data from Examples 3 and 4, find the value of $E(2X + 1)$.

**Solution**

Use the formula

$E(aX + b) = aE(X) + b$

$$\begin{aligned} E(2X + 1) &= 2E(X) + 1 \\ &= 2 \times 2.1 + 1 \\ &= 5.2 \end{aligned}$$

If we did not use this formula, we can still reach the same answer.

$$\begin{aligned} &E(2X + 1) \\ &= ((2 \times 0 + 1) \times 0.1) + ((2 \times 1 + 1) \times 0.2) + ((2 \times 2 + 1) \times 0.4) \\ &\quad + ((2 \times 3 + 1) \times 0.1) + ((2 \times 4 + 1) \times 0.2) \\ &= 0.1 + 0.6 + 2 + 0.7 + 1.8 \\ &= 5.2 \text{ as before} \end{aligned}$$

We can see that using the formula is simpler.

## 4.4.3 Variance and standard deviation of a discrete probability distribution

We now look at **measures of spread**.

- The **variance** is represented by $\sigma^2$ or Var $(X)$.
- The formula for the variance is $\sigma^2 = \text{Var}(X) = \Sigma(x - \mu)^2 \Pr(X = x)$.

  The alternative formula for variance, which is easier to use, is

$$\sigma^2 = \text{Var}(X) = \text{E}(X^2) - [\text{E}(X)]^2$$

  or

$$\sigma^2 = \text{Var}(X) = \text{E}(X^2) - \mu^2$$

- The **standard deviation** is given by $\sigma$ or SD$(X)$ or $\sqrt{\text{Var}(X)}$.

  This gives another formula for standard deviation, $\sigma = \text{SD}(X) = \sqrt{\text{E}(X^2) - \mu^2}$.

The measure of spread, the variance, is important as through it we find the standard deviation. The standard deviation describes how far sections of the distribution are from the measure of central tendency, the mean.

Imagine, after a topic test in a Year 12 class, the students know that the mean of their scores is 70%. This is only one piece of information. The second significant piece of information is how far the class scores are spread above and below that score of 70%. Two quite different pictures are given: if all the results cluster around 70% or if individual scores are spread from 30% to 90%.

---

**Example 6**

Using the household pet data, find the variance and standard deviation.

| $x$ | 0 | 1 | 2 | 3 | 4 |
|---|---|---|---|---|---|
| $\Pr(X = x)$ | 0.1 | 0.2 | 0.4 | 0.1 | 0.2 |

**Solution**

Begin with finding $\text{E}(X^2)$.

$$\text{E}(X^2) = (0^2 \times 0.1) + (1^2 \times 0.2) + (2^2 \times 0.4) + (3^2 \times 0.1) + (4^2 \times 0.2)$$
$$= 0.2 + 1.6 + 0.9 + 3.2 = 5.9$$

Use the formula $\sigma^2 = \text{Var}(X) = \text{E}(X^2) - \mu^2$

$$\text{Var}(X) = \text{E}(X^2) - \mu^2 = 5.9 - 2.1^2 = 1.49$$

Use the formula $\sigma = \text{SD}(X) = \sqrt{\text{Var}(X)}$

$$\text{SD}(X) = \sqrt{\text{Var}(X)} = \sqrt{1.49} = 1.2206\ldots$$
$$= 1.221, \text{ correct to three decimal places}$$

#### 4.4.3.1 Interpretation of Variance: useful formula

$$\text{Var}(aX + b) = a^2\,\text{Var}(X)$$

---

**Example 7**

Using the household pet data above, find $\text{Var}(2X + 1)$.

**Solution**

Use the formula

$\text{Var}(aX + b) = a^2\,\text{Var}(X)$

$$\text{Var}(2X + 1) = 2^2\,\text{Var}(X) = 4\,\text{Var}(X)$$
$$= 4 \times 1.49 = 5.96$$

This is easier than going through the process of finding $(2X + 1)$ and then finding the variance of that expression.

> **Hint**
> Contrast $E(2X + 1)$ and $\text{Var}(2X + 1)$, where the '+ 1' seems to matter to the $E(2X + 1)$ calculations and not the $\text{Var}(2X + 1)$ calculations. Variance is a measure of spread and it is not relevant what its translation is.

## 4.5 Binomial distribution

This section introduces the most significant example of a discrete probability distribution, called the **binomial distribution**. A **Bernoulli trial** is a random experiment with exactly two possible outcomes, 'success' and 'failure'. A series of Bernoulli trials, in which the probability of success is the same every time the experiment is conducted and we count the total number of successes, produces a binomial distribution.

We define the binomial distribution $X \sim \text{Bi}(n,p)$, where

$n$ = the number of independent trials

$p$ = the probability of success.

'$q$' is often used in this context as the other case in the binomial context of 'not $p$', where $q = 1 - p$.

A binomial experiment consists of the following features:

- There are $n$ identical trials.
- The trials are independent.
- Each trial has only two possible outcomes (success or failure).
- The probability of success is $p$ and is constant for each trial.
- The probability of failure $q$ is $1 - p$.
- It can involve the idea of sampling with replacement.

If $x$ is the number of successes in $n$ trials, with the probability of each success equal to $p$, then the formula for the probability is given by

$$\Pr(X = x) = {}^nC_x\, q^{n-x} p^x \quad \text{or} \quad \Pr(X = x) = {}^nC_x (1 - p)^{n-x} p^x$$

where $q = 1 - p,\ x = 0, 1, 2 \ldots n$ and ${}^nC_x = \dfrac{n!}{x!(n - x)!}$.

You will have met the formulas for combinations, ${}^nC_x = \dfrac{n!}{x!(n - x)!}$ and permutations, ${}^nP_x = \dfrac{n!}{(n - x)!}$ in the topic of Counting methods or Combinatorics in Units 1 and 2. The binomial theorem uses the combinations formula as this models **sampling with replacement**.

The constants $n$ and $p$ are called the **parameters** of the binomial distribution.

### Example 8

An experiment involves independently rolling a fair die 3 times and success on each of these trials is deemed 'getting a 6'.

**a** Identify $n$ and $p$.

**b** Use the formula $\Pr(X = x) = {}^nC_x(1-p)^{n-x}p^x$ to find the probability of 'getting a 6' only twice.

### Solution

**a** Number of independent trials $= n = 3$

Probability of success $= p = \frac{1}{6}$

**b** Use $X \sim \text{Bi}(n,p)$ with $X \sim \text{Bi}\left(3,\frac{1}{6}\right)$

$$\text{So } \Pr(X = 2) = {}^3C_2\left(1-\frac{1}{6}\right)^{3-2}\left(\frac{1}{6}\right)^2$$

$$= {}^3C_2\left(\frac{5}{6}\right)^{3-2}\left(\frac{1}{6}\right)^2$$

$$= \frac{3!}{2!(3-2)!}\left(\frac{5}{6}\right)\left(\frac{1}{6}\right)^2$$

$$= \frac{15}{216}$$

$$= \frac{5}{72}$$

Checking with CAS distribution

Using binomial PDf

$x = 2, n = 3, p = \frac{1}{6}$

```
binomialPDf(2, 3, 1/6)
                    0.06944444444
15/216
                    0.06944444444
```

## 4.5.1 Expected value (mean) of the binomial distribution

If a coin is tossed 100 times, we can expect a head to turn up 50 times.

That is, the expected value of $x = E(X) = \mu = 100 \times \frac{1}{2} = 50$.

- The formula for the mean of the binomial distribution is $\mu = np$.

Note that this only applies when it has been determined that the distribution is binomial.

### 4.5.1.1 Variance of the binomial distribution

$$\text{Var}(X) = \sigma^2 = npq = np(1-p), \text{ where } q = 1 - p$$

### 4.5.1.2 Standard deviation of the binomial distribution

$$\text{SD}(X) = \sigma = \sqrt{npq} = \sqrt{np(1-p)}, \text{ where } q = 1 - p$$

### Example 9

The probability of scoring a goal in a netball game is 0.8. The probability of scoring a goal is **independent** for each scoring shot. For a particular player, Sharon, who shoots for goal 10 times, what is the probability that she scores 8 goals?

**Hint**

Sharon's scoring shots are considered independent because each time she goes for a new shot at goal, the result (score/not score) is considered unaffected by the previous or next shot at goal.

**Solution**

This question describes a binomial experiment because

- there are 10 identical trials
- the trials are independent
- there are only two possible outcomes (goal or not goal)
- the probability of success is 0.8 and is constant for each trial.

Use $X \sim \text{Bi}(n,p)=\text{Bi}(10,0.8)$

where $n$ = the number of independent trials and
$p$ = the probability of success

Using $X \sim \text{Bi}(10,0.8)$

$$\begin{aligned}\Pr(X=8) &= {}^{10}C_8(1-0.8)^2 0.8^8 \\ &= {}^{10}C_8 0.2^2 0.8^8 \\ &= \frac{10!}{8!2!}0.2^2 0.8^8 \\ &= \frac{10\times 9}{2\times 1}0.2^2 0.8^8 \\ &= 0.30199\end{aligned}$$

Checking with CAS distribution

Using binomial PDf

$x = 8$, $n = 10$, $p = 0.8$

```
binomialPDf(8,10,0.8)
                0.301989888
```

## 4.6 Parameters of binomial distribution

This section explores varying the parameters of the binomial distribution and the difference this variation makes to the probability graph. It also looks at questions that ask for a specific value in binomial distributions or intervals defined in terms of the random variable, including conditional probability.

### 4.6.1 The graph of the binomial probability distribution

By sketching a few graphs with varying $p$, consider the effect of $p$ on the shape of a probability graph for the binomial distribution.

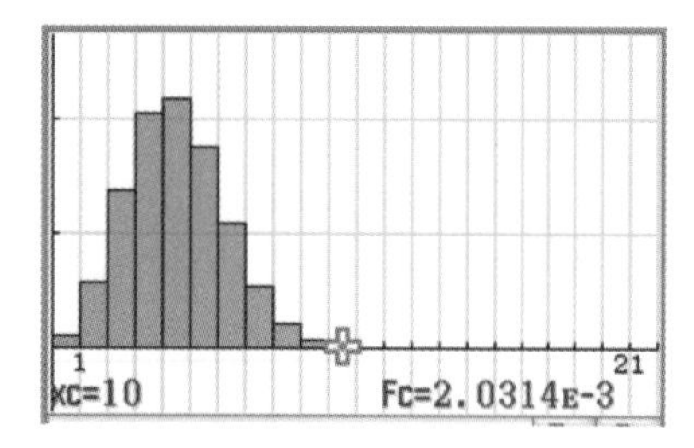


The above graph shows $p = 0.2$, $n = 20$ (highlighting the score at $x = 10$).

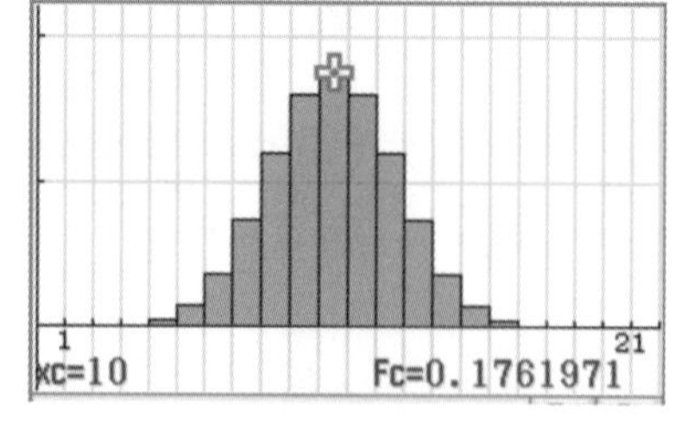


The above graph shows $p = 0.5$, $n = 20$ (highlighting the score at $x = 10$).

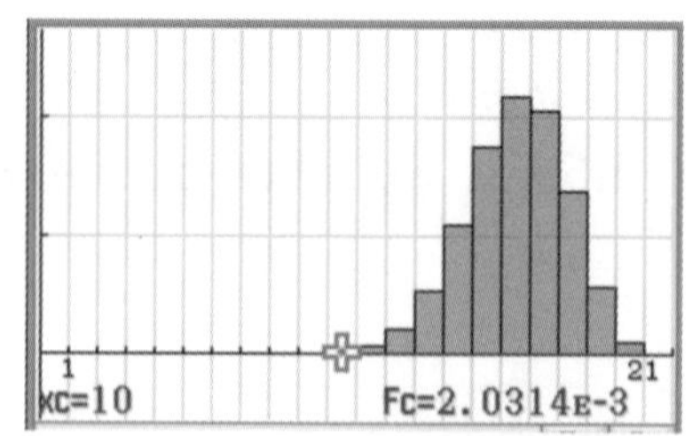


The above graph shows $p = 0.8$, $n = 20$ (highlighting the score at $x = 10$).

- $p = 0.5$ gives a symmetrical distribution.
- $p = 0.2$ gives a distribution that is skewed to the right, i.e., positively skewed.
- $p = 0.8$ gives a distribution that is skewed to the left, i.e., negatively skewed.

### Example 10

Emma stays at home, unwell with asthma, on **20%** of her school days. It seems for Emma that it is not connected whether she has asthma one day or another. In the next **10 days**, what is the probability that she will stay at home unwell with asthma for

**a** exactly two days?

**b** at least one day?

**c** two days given that Emma stays at home for at least one day?

### Solution

Let $X$ = number of times Emma stays home unwell with asthma. Identify that this is a binomial distribution with 10 independent trials and probability of 'success' = 0.2.

> **Hint**
> The 'success' is not what you perceive as better, it is just the outcome asked for in the question.

This question describes a binomial experiment because

- there are 10 identical trials
- the trials are independent
- there are only two possible outcomes (stays home unwell or doesn't)
- the probability of success is 0.2 and is constant for each trial.

Using $X \sim \text{Bi}(n, p)$

$n$ = the number of independent trials
$p$ = the probability of success

so $X \sim \text{Bi}(10, 0.2)$.

**a** exactly two days

$$\Pr(X = 2) = {}^{10}C_2 0.8^8 0.2^2$$
$$= \frac{10!}{8!\,2!} 0.8^8 0.2^2$$
$$= 0.3020$$

```
binomialPDf(2,10,0.2)
                0.301989888
```

**b** at least one day

$$\Pr(X \geq 1) = \Pr(X = 1) + \Pr(X = 2) + \Pr(X = 3) \ldots \Pr(X = 10)$$
$$= 1 - \Pr(X = 0)$$
$$= 1 - {}^{10}C_0\, 0.8^{10} 0.2^0$$
$$= 1 - 0.1074$$
$$= 0.8926$$

> **Note**
> 1 – Pr($X$ = 0) uses the 'turnaround type' referred to earlier.

Checking with CAS distribution:

Using binomial PDf, (single value of $x$)

1 minus ($x = 0$, $n = 10$, $p = 0.2$)

or

Using binomial CDf (range of values of $x$)

lower = 1, upper = 10,

$n = 10$, $p = 0.2$

```
1-binomialPDf(0,10,0.2)
                0.8926258176
binomialCDf(1,10,10,0.2)
                0.8926258176
```

**c** two days given that Emma stays at home for at least one day

$$\Pr(X = 2 | X \geq 1) = \frac{\Pr(X = 2)}{\Pr(X \geq 1)}$$
$$= \frac{\Pr(X = 2)}{1 - \Pr(X = 0)}$$
$$= \frac{0.3020}{0.8926}$$
$$= 0.3383$$

```
binomialPDf(2,10,0.2)
binomialCDf(1,10,10,0.2)
                    0.338316327
```

# 4.7 Continuous random variables, probability density function

This section develops continuous random variables:

- construction of probability density functions from non-negative functions of a real variable
- specification of probability distributions for continuous random variables using probability density functions.

VCE Mathematics Study Design 2023–2027 p. 100, © VCAA 2022

Continuous random variables can take any value in a given interval, so the **probability density function** (called hereafter a PDF) of a continuous random variable is defined as

$$\Pr(a < X < b) = \int_a^b f(x)\,dx$$
$$= F(b) - F(a), \quad \text{where } F'(x) = f(x).$$

Note that $\Pr(a < X < b)$ for continuous random variables is treated the same as $\Pr(a \leq X \leq b)$, as the edges of the interval can be included or not, making little difference to the continuous data.

A probability density function has the following properties:

1 $f(x) \geq 0$ (the function must be on or above the $x$-axis)

2 $\int_a^b f(x)\,dx = 1$ (the area equals 1)

3 The probability of an outcome where $a \leq x \leq b$ can be found by calculating the area under the curve from $x = a$ to $x = b$.

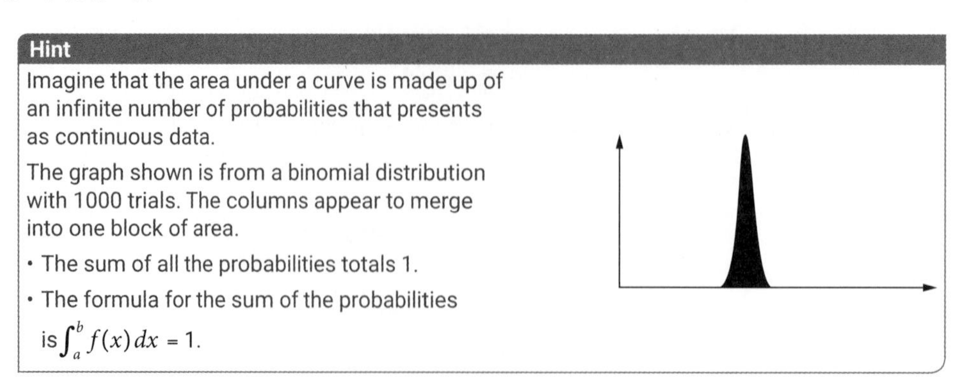

**Hint**

Imagine that the area under a curve is made up of an infinite number of probabilities that presents as continuous data.

The graph shown is from a binomial distribution with 1000 trials. The columns appear to merge into one block of area.

- The sum of all the probabilities totals 1.
- The formula for the sum of the probabilities is $\int_a^b f(x)\,dx = 1$.

### Example 11

Consider a random variable $X$ having its probability density function given by

$$f(x) = \begin{cases} 12x^2(1-x), & 0 \le x \le 1 \\ 0 & \text{otherwise} \end{cases}$$

**a** Sketch the graph of $f(x)$.

**b** Find $\Pr(X < 0.6)$.

### Solution

**a**

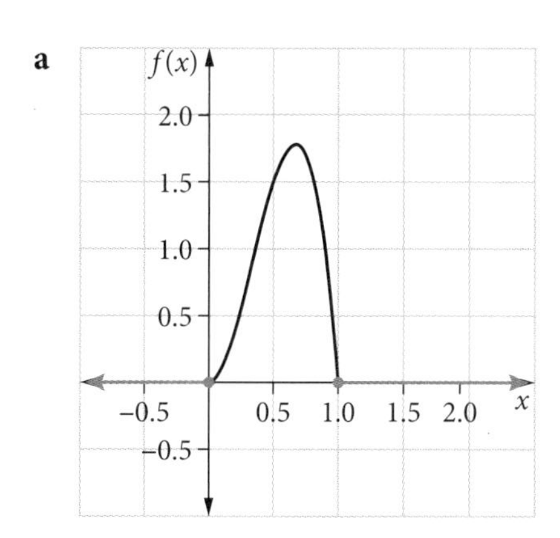


Note the horizontal lines for '0 elsewhere'

**b**

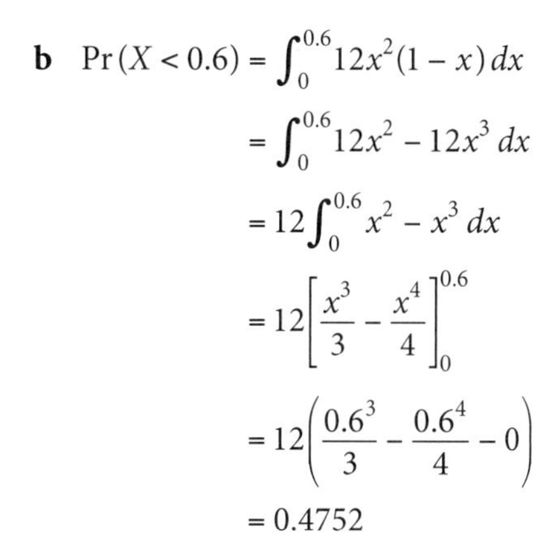

$$\begin{aligned} \Pr(X < 0.6) &= \int_0^{0.6} 12x^2(1-x)\,dx \\ &= \int_0^{0.6} 12x^2 - 12x^3\,dx \\ &= 12\int_0^{0.6} x^2 - x^3\,dx \\ &= 12\left[\frac{x^3}{3} - \frac{x^4}{4}\right]_0^{0.6} \\ &= 12\left(\frac{0.6^3}{3} - \frac{0.6^4}{4} - 0\right) \\ &= 0.4752 \end{aligned}$$

The shading of the integral from 0 to 0.6 shows a total area of 0.4752.

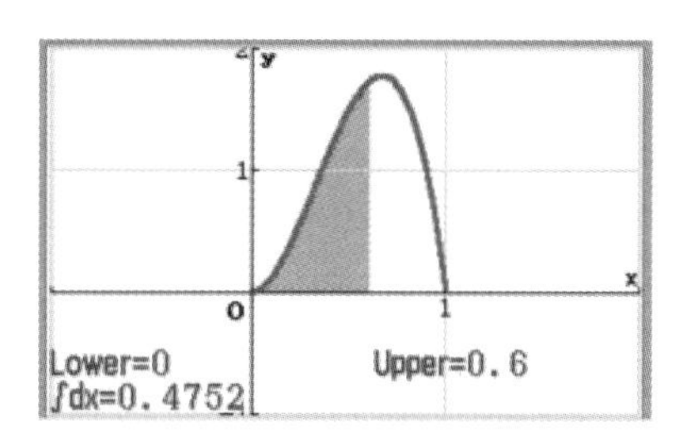


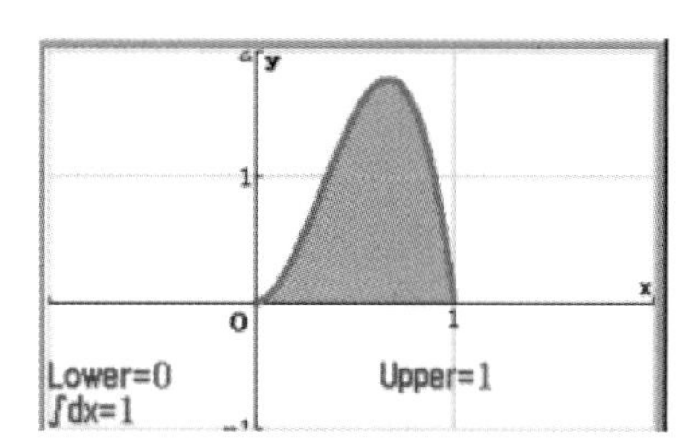


The shading and the integral from 0 to 1 shows a total area of 1.

# 4.8 Mean, variance and standard deviation of PDFs

This section continues to develop probability density functions:

- calculation and interpretation of mean, $\mu$, variance, $\sigma^2$, and standard deviation of a continuous random variable and their use.



## 4.8.1 Mean or expected value of a continuous probability density function

The mean is denoted by $\mu$ or E($X$).

The formula for the mean is $\mu = \text{E}(X) = \int_a^b x\, f(x)\, dx$.

Note the similarity between the mean of a continuous probability density function and the mean of a discrete distribution.

The mean for a **discrete random variable** is given by

$$\mu = \text{E}(X) = \Sigma x \Pr(X = x)$$

**Note**

Compare $\int_a^b x\, f(x)\, dx$ with $\Sigma x \Pr(X = x)$ both evaluating $x$ times its probability.

---

**Example 12**

Let $X$ be a probability density function such that

$$f(x) = \begin{cases} \frac{1}{2}x & 0 \le x \le 2 \\ 0 & \text{elsewhere} \end{cases}$$

Find the value of E($X$) using the formula $\text{E}(X) = \int_a^b x\, f(x)\, dx$.

**Solution**

$$\begin{aligned} \text{E}(X) &= \int_0^2 x \times \frac{1}{2}x\, dx \\ &= \int_0^2 \frac{1}{2}x^2\, dx \\ &= \left[\frac{1}{6}x^3\right]_0^2 \\ &= \frac{1}{6}(2^3 - 0^3) \\ &= \frac{8}{6} \\ &= \frac{4}{3} \end{aligned}$$

## 4.8.2 Variance of a continuous probability density function

The formula for variance is

$$\sigma^2 = \text{Var}(X) = \int_a^b x^2 f(x)\,dx - \mu^2$$

Note the similarity between the variance of a continuous probability density function and the variance of a discrete distribution.

The variance for **discrete random variables** is given by

$$\sigma^2 = \text{Var}(X) = \Sigma x^2 \Pr(X = x) - \mu^2$$



---

### Example 13

Let $X$ be a probability density function such that

$$f(x) = \begin{cases} \frac{1}{2}x & 0 \le x \le 2 \\ 0 & \text{elsewhere} \end{cases}$$

Find Var $(X)$ and SD$(X)$ using the formulas Var $(X) = \int_a^b x^2 f(x)\,dx - \mu^2$ and SD$(X) = \sqrt{\int_a^b x^2 f(x)\,dx - \mu^2}$.

### Solution

$$\begin{aligned}\text{Var}(X) &= \int_0^2 x^2 \times \frac{1}{2}x\,dx - \mu^2 \\ &= \int_0^2 \frac{1}{2}x^3\,dx - \mu^2 \\ &= \left[\frac{1}{8}x^4\right]_0^2 - \mu^2 \qquad \text{where } \mu = \frac{4}{3} \text{ (from Example 12)} \\ &= \frac{1}{8}(2^4 - 0^4) - \left(\frac{4}{3}\right)^2 \\ &= 2 - \frac{16}{9} \\ &= \frac{2}{9}\end{aligned}$$

and SD$(X) = \sqrt{\frac{2}{9}} = \frac{\sqrt{2}}{3}$.

9780170465366

# 4.9 Normal distribution

This section will cover:

- standard normal distribution, $N(0, 1)$, and transformed normal distributions, $N(\mu, \sigma^2)$, as examples of a probability distribution for a continuous random variable
- effect of variation in the value(s) of defining parameters on the graph of a given probability density function for a continuous random variable
- calculation of probabilities for intervals defined in terms of a random variable, including conditional probability (the cumulative distribution function may be used but is not required).



The **normal distribution** is a special case of a probability density function (PDF).

Continuous random variables can take any value in a given interval. The normal random variable may take any value within a given domain; these are **not** distinct, or countable values.

**Hint**
Conditional probability will appear in a range of probability examples.

Note that $\Pr(a < X < b)$ for continuous random variables is treated the same as $\Pr(a \le X \le b)$.

The equation of the normal distribution curve, $N(\mu, \sigma^2)$, helps us to understand the exponential effects in the normal distribution formula, hence the asymptotic nature of the curve.

$$f(x) = \frac{1}{\sigma\sqrt{2\pi}}e^{-\frac{1}{2}\left(\frac{x-\mu}{\sigma}\right)^2}$$

The normal distribution curve has the following shape:

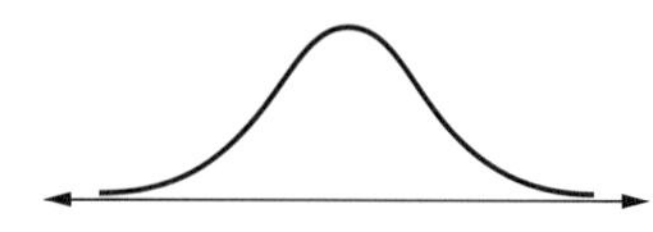

To calculate $\Pr(a < X < b)$, the formula is

$$\int_a^b f(x)\,dx = \int_a^b \frac{1}{\sigma\sqrt{2\pi}}e^{-\frac{1}{2}\left(\frac{x-\mu}{\sigma}\right)^2}dx$$

CAS is used to find the area using either integration as above, or the CAS normal distribution option.

---

**Example 14**

For the normal distribution $N(20, 5^2)$, ($\mu = 20$, $\sigma = 5$), calculate $\Pr(0 < X < 21)$ using the formula

$$\int_a^b f(x)\,dx = \int_a^b \frac{1}{\sigma\sqrt{2\pi}}e^{-\frac{1}{2}\left(\frac{x-\mu}{\sigma}\right)^2}dx$$

and the CAS normal distribution option.

**Solution**

$$\int_0^{21} f(x)\,dx$$

$$= \int_0^{21} \frac{1}{5\sqrt{2\pi}}e^{-\frac{1}{2}\left(\frac{x-20}{5}\right)^2}dx$$

$= 0.5792$

$$\int_0^{21} \frac{1}{5\cdot\sqrt{2\cdot\pi}}\cdot e^{-\frac{1}{2}\cdot\left(\frac{x-20}{5}\right)^2}dx$$

0.5792280382

and using the distribution normCDf.

```
normCDf(0,21,5,20)
                0.5792280382
```

**Hint**
We would expect the answer for the probability between 0 and 21, with a mean of 20 to be just over 0.5.

## 4.9.1 Standard normal distribution

The $x$-coordinates on the **standardised** normal distribution curve, where $\mu = 0$ and $\sigma = 1$, are called **z-scores**.

Substituting the values of $\mu = 0$ and $\sigma = 1$, we get the equation for the **standard normal distribution**

$$g(x) = \frac{1}{\sqrt{2\pi}} e^{-\frac{1}{2}x^2}$$

giving the following shape:

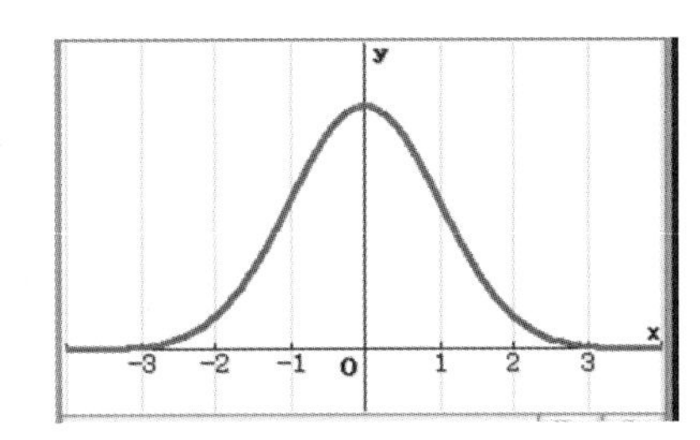

We can see that the area under the standard normal curve approximates 1, either by finding the area under the graph, or using the integral from $-\infty$ to $\infty$, as expected for a PDF.

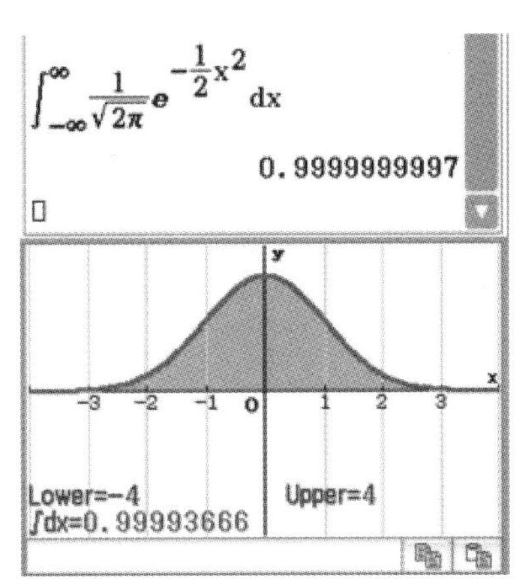


By further exploring, we find that the area under the curve between $-1$ and 1 is approximately 0.68

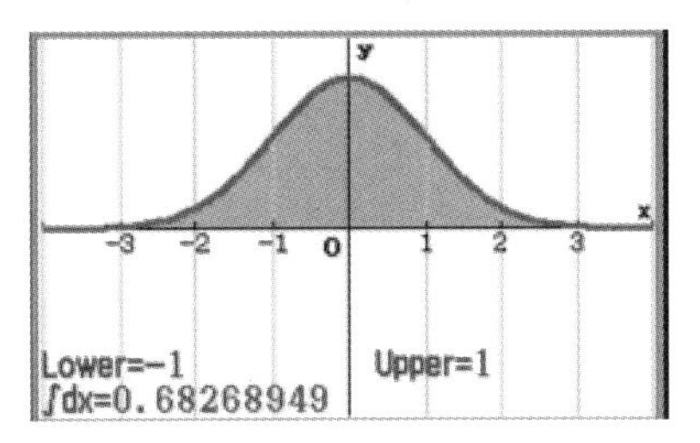


and between $-2$ and 2 is 0.95

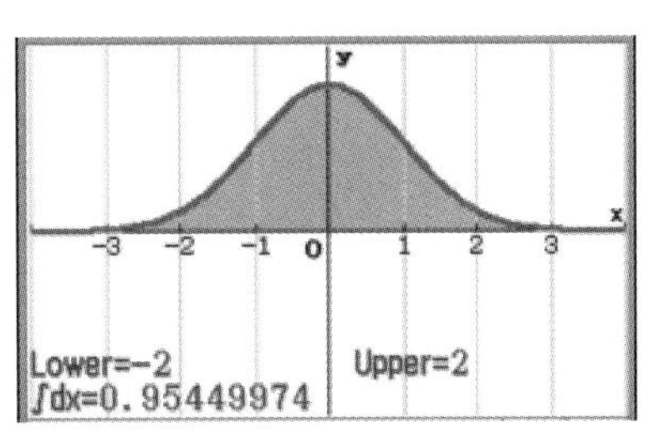


and between $-3$ and 3 is 0.997.

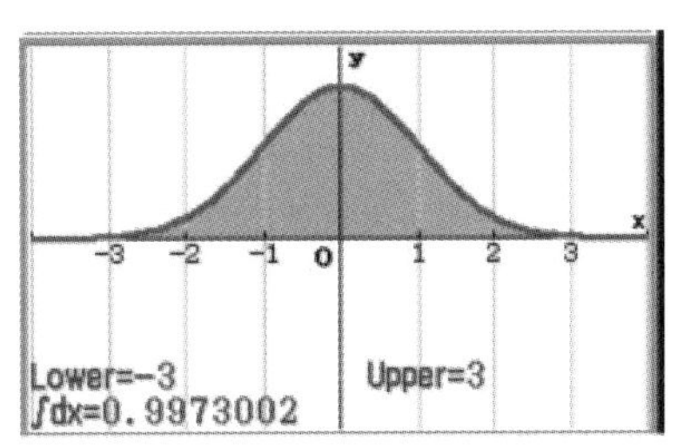


These results demonstrate the proportion/percentage of results within 1, 2 and 3 standard deviations of the mean.

## 4.9.2 Confidence intervals

The normal probability density function has two parameters: $\mu$ (mean) and $\sigma$ (standard deviation).

For many random variables (discrete and continuous), approximately 95% of their probability distribution lies within two standard deviations of the mean. This result is taken from the area under the graph of the normal distribution, as shown on the previous page.

- $\Pr(\mu - \sigma \le X \le \mu + \sigma) \approx 0.68$
- $\Pr(\mu - 2\sigma \le X \le \mu + 2\sigma) \approx 0.95$
- $\Pr(\mu - 3\sigma \le X \le \mu + 3\sigma) \approx 0.997$

**Note**
The Study Design mentions, particularly, the 95% confidence interval.

These are called **confidence intervals**.

The properties above are true for the normal distribution. They are used as a measure of spread for other distributions.

If $\sigma$ is small, then the values of the distribution are close to the mean; if $\sigma$ is large, then the distribution is widely spread.

Here again we consider the concept of **measures of central tendency** (as $\mu$ moves) and **measures of spread** (as $\sigma$ changes).

The normal curve is **bell-shaped,** with the mean, median and mode all occurring at the centre.

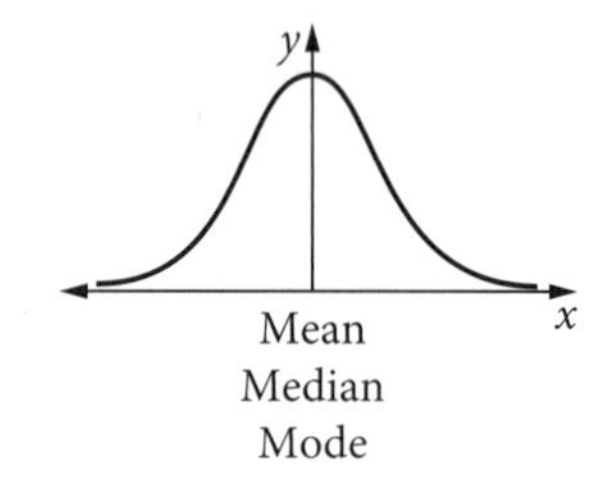


### 4.9.2.1 68%, 95% and 99.7% confidence intervals (CI)

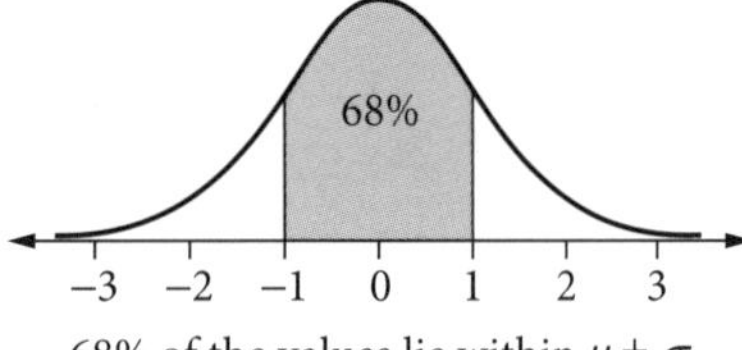

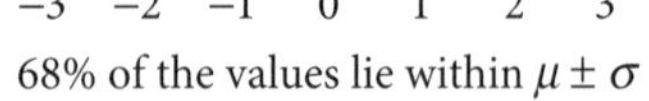
68% of the values lie within $\mu \pm \sigma$

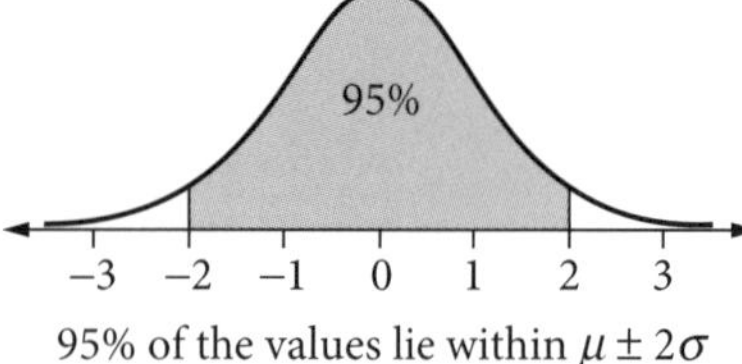

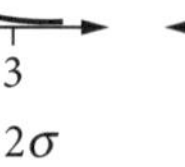
95% of the values lie within $\mu \pm 2\sigma$

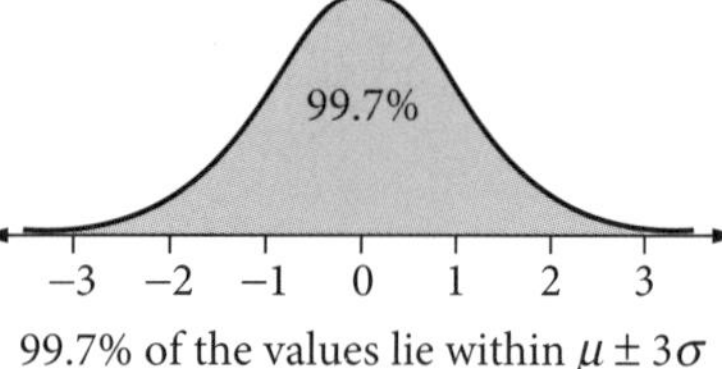

99.7% of the values lie within $\mu \pm 3\sigma$

## 4.9.3 Standardising $x$ values to $z$ values

Remember that standard $z$ values have $\mu = 0$, $\sigma = 1$.

In the formula for the normal distribution, we can see that it includes the fraction $\frac{x - \mu}{\sigma}$.

This gives the relationship $z = \frac{x - \mu}{\sigma}$ to transfer between the $x$ values and $z$ values.

---

**Example 15**

Find the $z$ value, where the $X$ values have $\mu = 100$, $\sigma = 10$, and $\Pr(X > 105) = \Pr(Z > z)$.

**Solution**

Using the formula $z = \frac{x - \mu}{\sigma}$ $\qquad z = \frac{x - \mu}{\sigma}$

Hence $\qquad z = \frac{105 - 100}{10}$

$\Pr(X > 105) = \Pr\left(Z > \frac{1}{2}\right)$ $\qquad z = \frac{5}{10} = \frac{1}{2}$

This is extremely useful when comparing an $X$ curve to a $Z$ curve.

**Note**
In this case, we would expect the answer for the $z$ score of $\frac{1}{2}$ to be just to the right of the mean of the standard normal distribution, which is 0, because 105 is slightly higher than the mean for $X$.

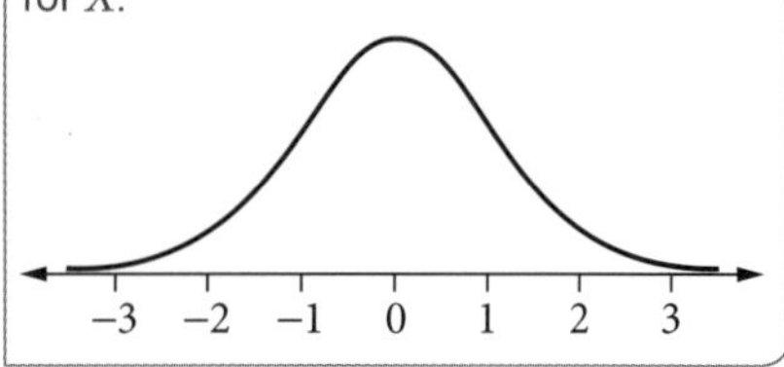

## 4.9.4 How to determine probabilities using the area under the normal curve

When solving normal distribution problems, it is always useful to draw a simple diagram and shade the area required. Here we explore several examples.

In this first case, we can see from the diagram that $\Pr(Z < 0) = 0.5$.

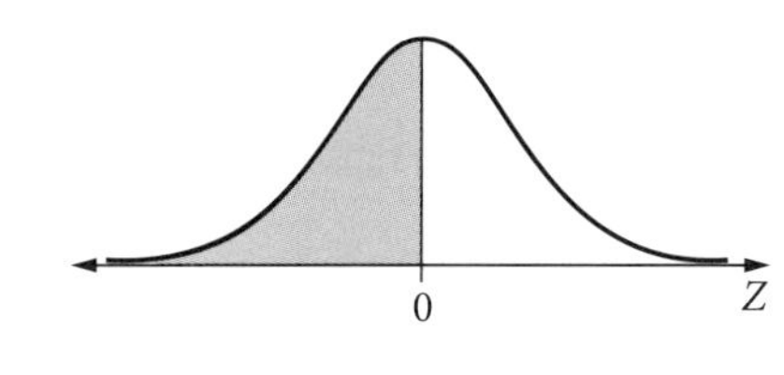


On CAS:

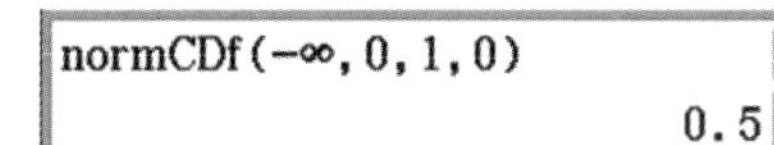


> **Hint**
> Note that in this screenshot the lower limit is entered as −∞, due to the horizontal asymptotic nature of the formula for the normal curve.

Examining a different case, $\Pr(Z < 1) = 0.8413$

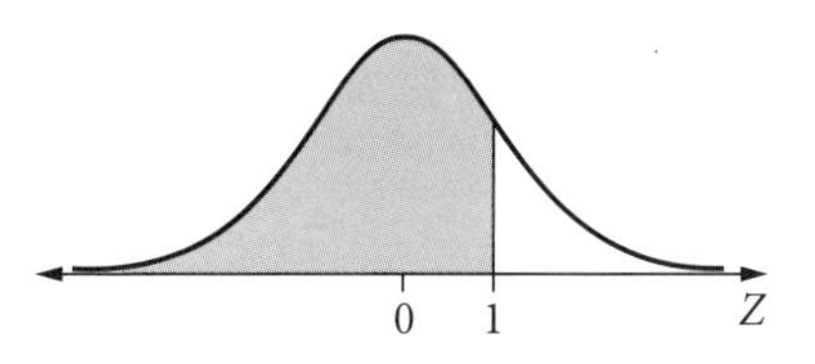


On CAS:

```
normCDf(−∞, 1, 1, 0)
                0.8413447461
```

We can then use the symmetry of the curve to rearrange as below:

$$\begin{aligned}\Pr(Z > 1) &= 1 - \Pr(Z < 1)\\ &= 1 - 0.841345\\ &= 0.1589\end{aligned}$$

On CAS:

```
normCDf(1, ∞, 1, 0)
                0.1586552539
```

$$\begin{aligned}\Pr(Z < -2.3) &= \Pr(Z > 2.3)\\ &= 0.0107\end{aligned}$$

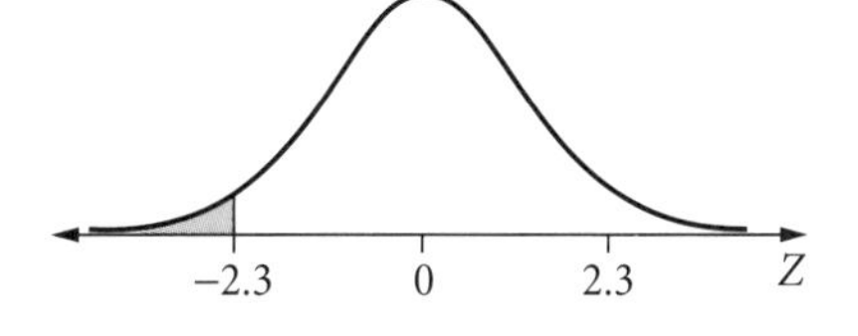


On CAS, showing both sides of the symmetry:

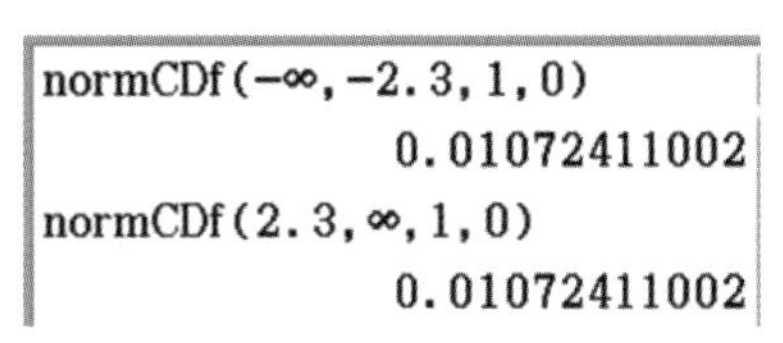


Now between two values:

$$\begin{aligned}\Pr(-1.246 < Z < 2.368) &= \Pr(Z < 2.368) - \Pr(Z < -1.246)\\ &= 0.8847\end{aligned}$$

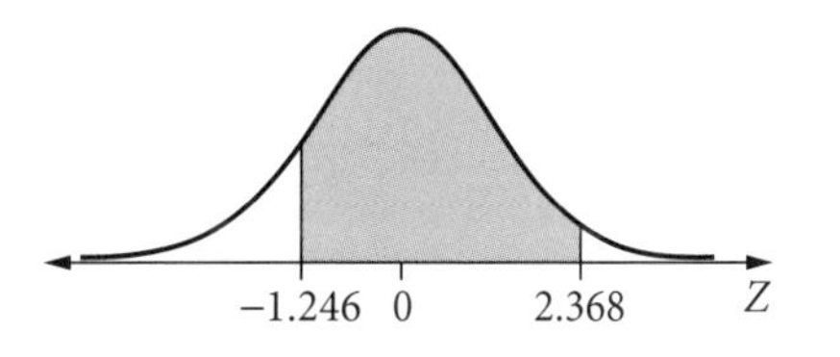


On CAS:

```
normCDf(−1.246, 2.368, 1, 0)
                0.8846755345
```

### 4.9.4.1 Probabilities other than the standard normal distribution

In worded problems, you will be given normal distributions that are not in terms of the standard normal distribution. In these cases, use the values of $\mu$ and $\sigma$ that are given in the problem to transform the $x$ values into $z$ values.

---

**Example 16**

If $X$ is a normal random variable with $\mu = 100$, $\sigma = 5$, find $\Pr(X < 109)$.

**Solution**

Use the formula $z = \dfrac{x - \mu}{\sigma}$.

$$\Pr(X < 109) = \Pr\left(z < \frac{109 - 100}{5}\right)$$

$$\text{So } \Pr(X < 109) = \Pr\left(z < \frac{9}{5}\right) = \Pr(z < 1.8) = 0.9641$$

**Note**
The CAS screen shows the answer of 0.9641 using, in the 1st case, the $z$-score with $\mu = 0$ and $\sigma = 1$ and, in the 2nd case, the $x$-score with $\mu = 100$ and $\sigma = 5$.

```
normCDf(-∞, 1.8, 1, 0)
                    0.9640696809
normCDf(-∞, 109, 5, 100)
                    0.9640696809
```

The shaded area shows $\Pr(X < 109) = 0.9641$ and $\Pr(z < 1.8) = 0.9641$.

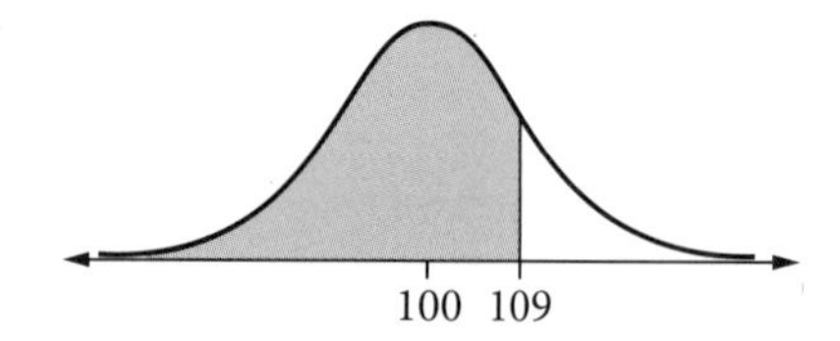


Using technology, it would be easier to directly find $\Pr(X < 109)$ with $\mu = 100$, $\sigma = 5$, rather than going via the $z$-score.

There are often real-life worded problems in normal distribution questions. Refer to Examples 18 and 19.

## 4.9.5 Inverse normal questions

By using **Inverse Normal** on CAS, we can find the $z$- or $x$-score that gives a certain probability.

---

**Example 17**

Find $b$, where $\Pr(Z < b) = 0.9$.

**Solution**

Use the **Inverse Normal** command on CAS.

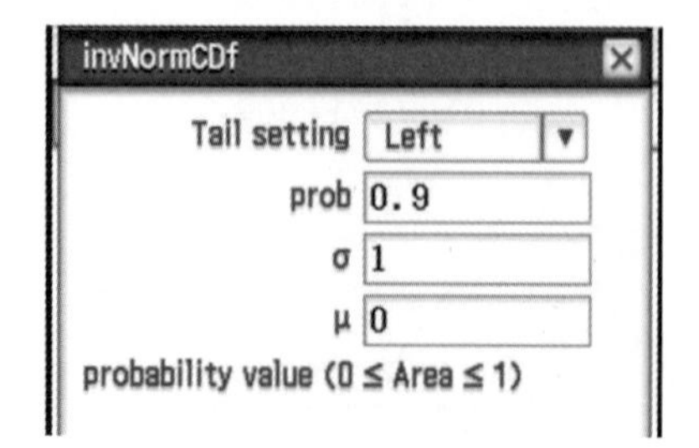


The Tail setting 'Left' here means < or ≤ as the distribution is continuous.

```
invNormCDf("L", 0.9, 1, 0)
                    1.281551566
```

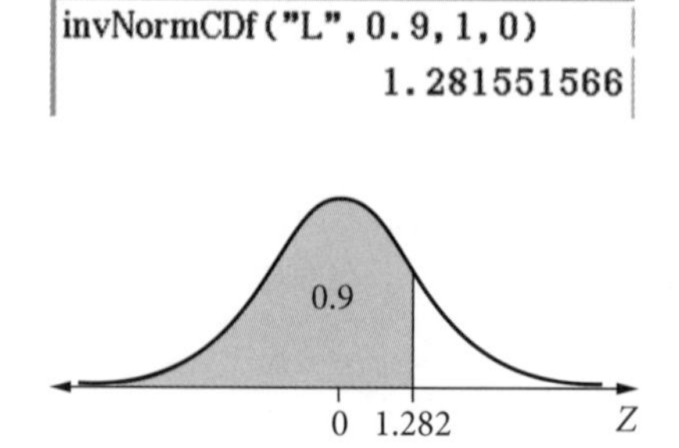


So $b = 1.282$.

There are often worded problems in normal distribution exam questions using **Inverse Normal**.

### Example 18 ©VCAA 2020 2BQ3ab

A transport company has detailed records of all deliveries.

The number of minutes a delivery is made before or after its scheduled delivery time can be modelled as a normally distributed random variable, $T$, with a mean of zero and a standard deviation of four minutes. A graph of the probability distribution of $T$ is shown below.

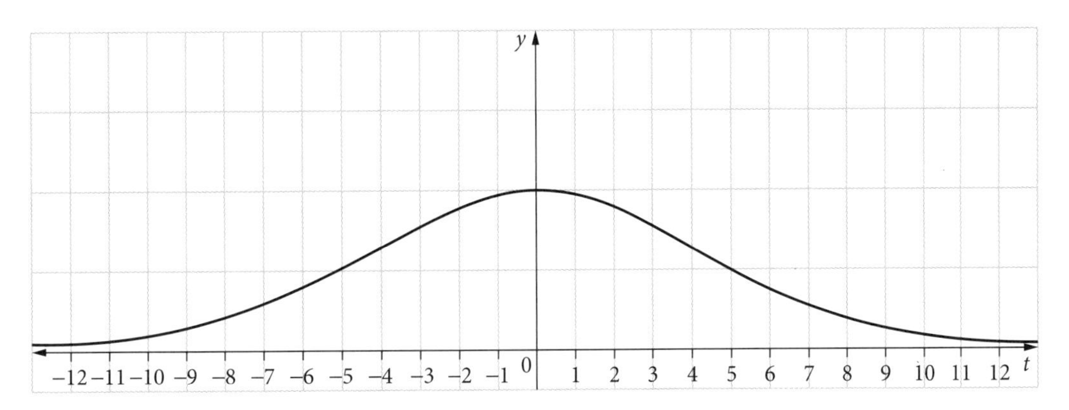


**a** If $\Pr(T \le a) = 0.6$, find $a$ to the nearest minute.

**b** Find the probability, correct to three decimal places, of a delivery being no later than three minutes after its scheduled delivery time, given that it arrived after its scheduled delivery time.

### Solution

$T \sim N(0, 42)$

**a** Given $\Pr(T \le a) = 0.6$

Using Inverse Normal $\Pr(T \le a) = 0.6$

gives $a = 1.0134$.

To the nearest minute, $a = 1$.

```
invNormCDf("L",0.6,4,0)
                1.013388413
```

| invNormCDf | |
|---|---|
| Tail setting | Left |
| prob | 0.6 |
| σ | 4 |
| μ | 0 |

**b** $\Pr(T < 3 \mid T > 0) = \dfrac{\Pr(0 < T < 3)}{\Pr(T > 0)} = 0.547$

```
normCDf(0,3,4,0)
normCDf(0,∞,4,0)
                0.5467452952
```

### Example 19

In a class of maths students, the average height is 150 cm and the standard deviation is 5 cm. Find the height of a student when the probability of being greater than that height equals 0.35.

### Solution

This means: Find $a$ where $\Pr(X > a) = 0.35$.

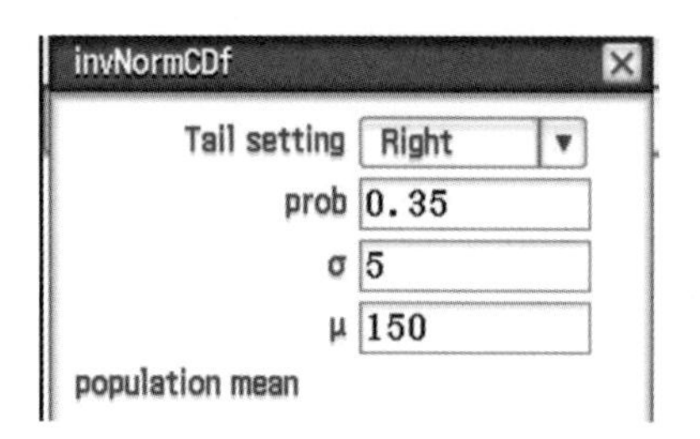

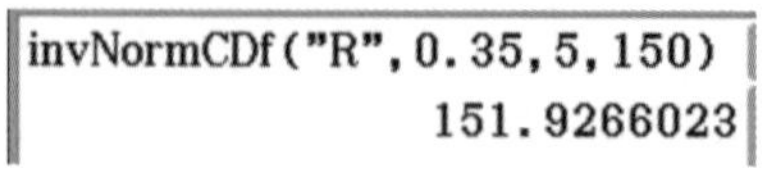

```
invNormCDf("R",0.35,5,150)
                 151.9266023
```

The answer is 151.93 cm.

## 4.10 Parameters of normal distribution

In this section we consider the effect of variation in the value(s) of defining parameters on the graph of a given probability for a normal distribution, where we can calculate the probabilities for intervals defined in terms of the random variable. We observe the difference this variation makes to the probability graph.

The normal density function has two parameters: $\mu$ (mean) and $\sigma$ (standard deviation).

If two normal distributions have the same standard deviation, $\sigma$, but different means, they will be exactly of the same size. If $\sigma_1 > \sigma_2$, then the graph with standard deviation $\sigma_1$ will be more spread out than the graph with $\sigma_2$.

The location of the curve on the horizontal axes is determined by the mean, but the steepness and spread is determined by $\sigma$.

Regardless of the values of $\mu$ and $\sigma$, the percentages of values that lie within a given number of standard deviations from the mean is always the same.

The following figures illustrate the effect of $\mu$ and $\sigma$ on the shape of a probability graph for the normal distribution.

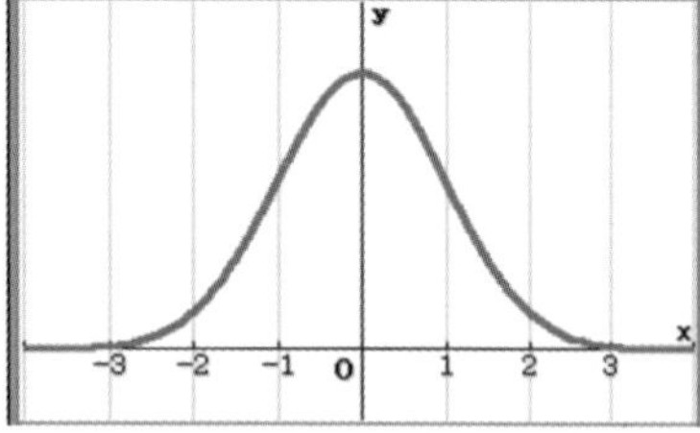


This graph shows $\mu = 0$ and $\sigma = 1$.

Note the different $y$-axis scales.

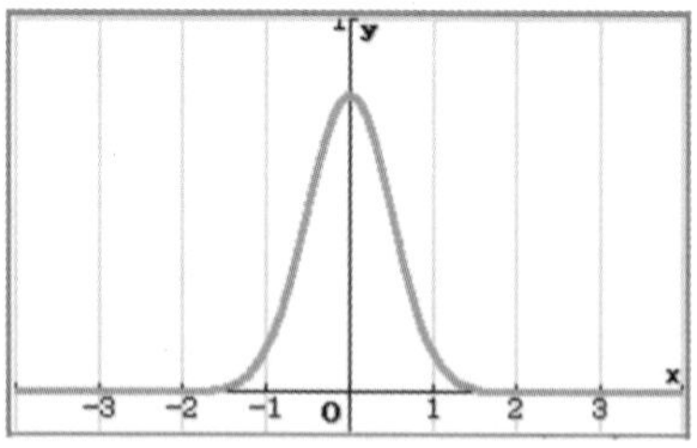

This graphs shows $\mu = 0$ and $\sigma = 0.5$.

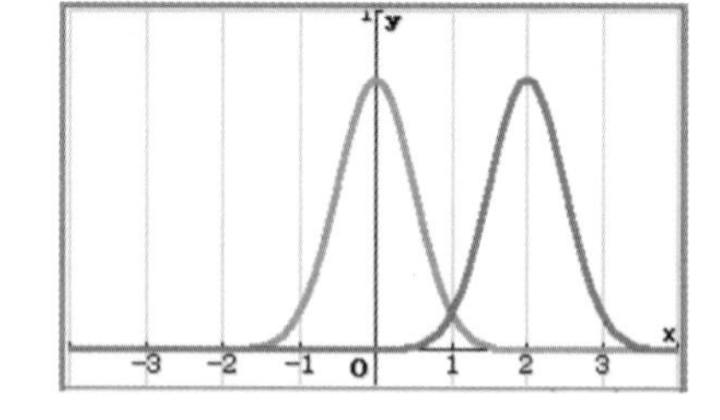


This graph shows $\mu = 0$ and $\sigma = 0.5$, as well as $\mu = 2$ and $\sigma = 0.5$. Here we can see the effect of a translation to the right by 2 units, given by the mean, but the same steepness given by the standard deviation of 0.5.

Some multiple-choice questions ask you to recognise the effect of the parameters $\mu$ and $\sigma$ on the normal distribution curve.

# 4.11 Statistical inference

This section will cover:

- statistical inference, including definition and distribution of sample proportions, simulations and confidence intervals:
  - distinction between a population parameter and a sample statistic and the use of the sample statistic to estimate the population parameter
  - simulation of random sampling, for a variety of values of $p$ and a range of sample sizes, to illustrate the distribution of $\hat{P}$ and variations in confidence intervals between samples
  - concept of the sample proportion $\hat{P} = \frac{X}{n}$ as a random variable whose value varies between samples, where $X$ is a binomial random variable which is associated with the number of items that have a particular characteristic and $n$ is the sample size
  - approximate normality of the distribution of $\hat{P}$ for large samples and, for such a situation, the mean $p$ (the population proportion) and standard deviation, $\sqrt{\frac{p(1-p)}{n}}$
  - determination and interpretation of, from a large sample, an approximate confidence interval $\left(\hat{p} - z\sqrt{\frac{\hat{p}(1-\hat{p})}{n}}, \hat{p} + z\sqrt{\frac{\hat{p}(1-\hat{p})}{n}}\right)$, for a population proportion where $z$ is the appropriate quantile for the standard normal distribution, in particular the 95% confidence interval as an example of such an interval where $z \approx 1.96$ (the term standard error may be used but is not required).



- The **population** is the whole group from which data could be collected.
- A **sample** is part of the population.
- A **parameter** is a characteristic of a particular population.
- A **statistic** is an estimate of a parameter obtained using a sample.

For any variable or group, the population (or population of interest) is the whole group from which data could be collected. A **population parameter (pp)** is a characteristic value of the population. We have come across parameters previously in the normal distribution where the parameters are $\mu$ = mean and $\sigma$ = standard deviation: $X \sim N(\mu, \sigma^2)$. So the standard normal distribution is defined by $N(0, 1^2)$. We can say that the population parameters for the standard normal distribution are $\mu = 0$, $\sigma = 1$.

If we are interested in the IQ of Year 12 students across Australia, then 'Year 12 students in Australia' becomes the population, and mean and standard deviation for IQ become population parameters of interest.

Dealing with a population is not often practical, economical, possible or relevant, because of reasons such as the population being too large. The solution is to consider data from a smaller part of the population, that is, a **sample**, in the hope that what we find out about the sample is also true about the population it comes from. Dealing with a sample is generally quicker and cheaper than dealing with the whole population, and a well-chosen sample will give much useful information about this population.

A **sample statistic (ss)** is an estimate of the parameter obtained using a sample.

What is significant here is how well we select a sample. The polling for an election cannot fully be relied on because of all the biases, conscious or unconscious, used when selecting a sample.

How do we sample the Year 12 students in Australia?

How the sample is selected then becomes a very important issue.

In order to make valid conclusions about a population from a sample, we would like the sample chosen to be representative of the population as a whole. This means that all the different subgroups present in the population must appear in similar proportions in the sample as they do in the population.

To select a sample of Year 12 students across Australia, we should select students who represent each subgroup of all Year 12 students across Australia. If a single teacher only selected her own Year 12 class to measure IQ, it would not be a representative sample.

---

**Example 20**

A sample of 20 people waiting in an ATM queue at 7:30 am was asked how much cash they intended to withdraw. The smallest amount was \$20, the average amount was \$78 and the greatest was \$500.

Identify

**a** the population: the whole group that could be asked about the amount they intend to withdraw from an ATM.

**b** parameters: clearly defined values from the whole population. (You don't need to know the value to define it clearly.)

**c** statistics: values attained from the sample. The number of people (20) is not a statistic because it is not an estimate of a parameter.

**Solution**

**a** The population is all the people who use ATMs.

**b** There are three parameters: the minimum withdrawal, the average amount withdrawn and the maximum withdrawal.

**c** There are three statistics: the minimum withdrawal (\$20), the average withdrawal (\$78) and the maximum withdrawal (\$500).

### 4.11.1.1 Simulation

One very useful method of drawing random samples is to generate random numbers, using a calculator or computer.

The 1st example is selecting 1 number randomly between 1 and 100.

The 2nd example is selecting 1 number randomly between 16 and 100.

The 3rd example is selecting 5 numbers randomly between 10 and 20.

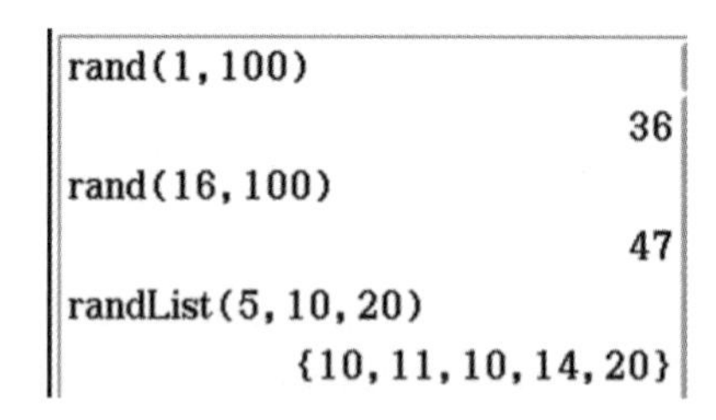


If we select again, different numbers will be generated.

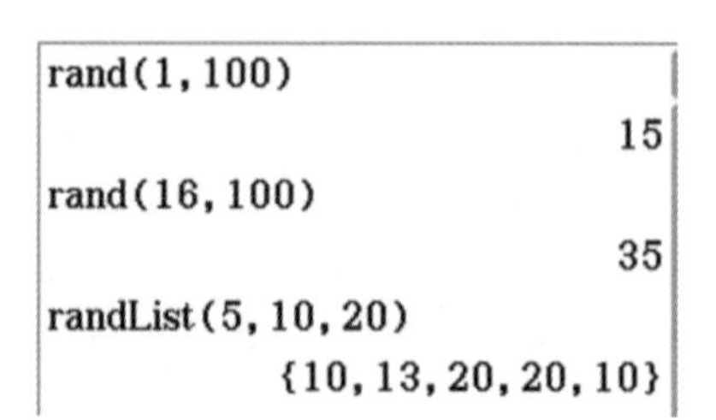


To summarise, the study of statistics revolves around the study of data sets. Two important types of data sets are populations and samples.

9780170465366

### 4.11.1.2 Population vs sample

- A **population** includes all of the elements from the group of interest.
- A **sample** consists of one or more members of the population.
- Depending on the sampling method, a sample can have fewer observations than the population, the same number of observations, or more observations (under sampling methods where the same member is observed more than once). More than one sample can be derived from the same population.

## 4.11.2 Sample proportion

- A **sample proportion** is the ratio of the number of times a property (or characteristic) occurs in a sample, divided by the number in the sample.
- The sample proportion is denoted by $\hat{P}$ (read as $p$-hat).
- The occurrence of the property is normally called a success, so for $x$ successes in a sample of $n$, the sample proportion is given by $\hat{P} = \frac{X}{n}$, where $X$ is a binomial random variable which is associated with the number of items that have a particular characteristic.

---

**Example 21**

If you were checking whether Vitamin C supplements decrease the chances of catching a cold, you might ask one sample of people to take no Vitamin C, another to take one capsule per day and a third sample to take 2 capsules per day.

For each sample, the variable would be a random binomial variable with success being 'catching a cold'. You would then look at the frequencies of people in each group who caught colds. The actual frequencies would not be as important as the ratio of the numbers of people who caught colds to the numbers in the samples.

Each ratio is an example of a sample proportion $\hat{P} = \frac{X}{n}$.

From 20 people who took no Vitamin C, 8 got colds during one winter. What is the sample proportion of catching a cold?

**Solution**

Success is catching a cold.

Use the formula $\hat{P} = \frac{X}{n} = \frac{8}{20} = 0.4$.

The sample proportion for those who caught colds is 0.4.

**Hint**

The sample proportion $\hat{P}$ for the occurrence of a property (characteristic) in a population is an estimator of the probability, $p$, of the occurrence in the population.

For large samples, we use the normal distribution to approximate such a situation giving the sample distribution $\hat{P}$, where mean = $\hat{p} = p$ (the population proportion) and standard deviation, $\sqrt{\frac{p(1-p)}{n}}$.

## 4.11.3 The normal approximation to the binomial

A rule of thumb to apply for the normal approximation to the binomial is that both $np$ and $n(1 - p)$, that is, $np$ and $nq$, should be > 5.

The distribution of sample proportions gets closer to the normal distribution as $n \to \infty$ and $p \to 0.5$.

---

### Example 22

Given that about 15% of Australians are left-handed, what is the probability that in a sample of 200 Australians, 20 to 30 of them are left-handed?

### Solution

To solve this, we verify that we can use the normal approximation, then find the standard values of $\hat{p}$ using 19.5 and 30.5 (to include both 20 and 30). Then use normCDf to find the probability.

Find the mean and standard deviation. $\overline{X} = p = 0.15$

$$\text{SD} = \sqrt{\frac{p(1-p)}{n}} = \sqrt{\frac{pq}{n}} = \sqrt{\frac{0.15 \times 0.85}{200}}$$
$$\approx 0.0252$$

Check $np$ and $nq$.

$$np = 200 \times 0.15 = 30$$

$$nq = 200 \times 0.85 = 170$$

Given that $np > 5$ and $nq > 5$, the normal distribution can be used.

Standard values of $\hat{p}$:

$$\frac{19.5}{200} \text{ gives } \frac{0.0975 - 0.15}{0.025} \approx -2.1$$

$$\frac{30.5}{200} \text{ gives } \frac{0.1525 - 0.15}{0.025} \approx 0.1$$

$$\Pr(20 \le X \le 30) = \Pr(-2.1 \le \hat{p} \le 0.1) = 0.52$$

```
normCDf(-2.1,0.1,1,0)
                0.5219634167
```

The probability that 20 to 30 Australians from a sample of 200 are left-handed is about 0.52.

### 4.11.3.1 Finding the proportion of the standard normal distribution

To understand what happens to a statistic as sample size increases, it is useful to examine the distribution of the statistics for multiple samples. The distribution of a statistic for many samples is called a **sampling distribution**.

Remember, sample proportion $\hat{P} = \frac{X}{n}$.

Collection of real data for multiple samples is very time-consuming so we use a simulation. Consider samples of Australian high school students and the property of having blue eyes. For Australian students, the probability of having blue eyes is 0.32. The occurrence of blue eyes is a Bernoulli random variable with a probability of success of 0.32. A sample of 20 students corresponds to 20 trials. The number of successes is a value $x$ of the random binomial variable $X$ and the sample proportion is $\frac{x}{20}$.

By standardising the distribution of a random variable, $\frac{\hat{p} - p}{\sigma}$ is approximated by standard normal random variable $Z$.

We know that $\Pr(-1.96 < z < 1.96) = 0.95$

```
normCDf(-1.96,1.96,1,0)
                0.9500042097
```

Or we can use **Inverse Normal** to create the 1.96.

```
invNormCDf("C",0.95,1,0)
                -1.959963985
```

So
$$\Pr\left(-1.96 < \frac{\hat{p} - p}{\sigma} < 1.96\right) = 0.95$$

Multiplying by the standard deviation,
$$\Pr(-1.96\,\sigma < \hat{p} < 1.96\,\sigma) = 0.95$$

so
$$\Pr(\hat{p} - 1.96\,\sigma < p < \hat{p} + 1.96\,\sigma) = 0.95$$

This is the final result that we use which becomes
$$\left(\hat{p} - z\sqrt{\frac{\hat{p}(1-\hat{p})}{n}}, \hat{p} + z\sqrt{\frac{\hat{p}(1-\hat{p})}{n}}\right)$$

Unless otherwise stated, confidence intervals are taken
to be symmetric about the mean. For such confidence intervals, the **margin of error** is half the width of the interval. It is the distance of the mean from the edge of the interval.

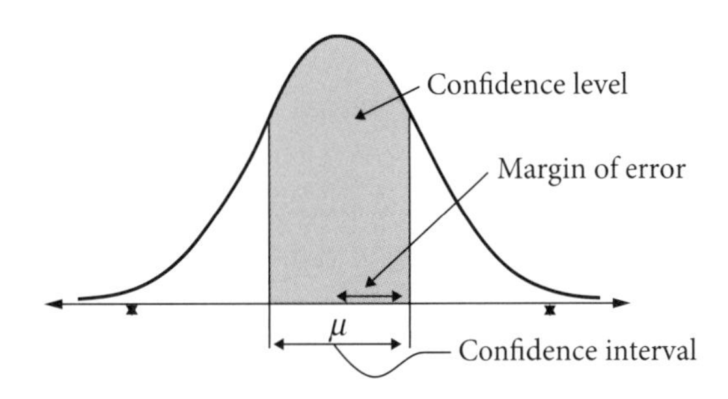

### Example 23

Find the approximate 95% confidence interval (CI) for the proportion $p$ of primary school children having brown eyes, using a random sample of 200 children when it is found that the sample proportion $\hat{p}$ is 0.7.

### Solution

Since $\hat{p} = 0.7$ and $n = 200$, $\sqrt{\frac{\hat{p}(1-\hat{p})}{n}} = 0.0324$.

## 4.11.3.2 Using a 95% confidence interval

$$95\%\ \text{CI} = \left(\hat{p} - z\sqrt{\frac{\hat{p}(1-\hat{p})}{n}}, \hat{p} + z\sqrt{\frac{\hat{p}(1-\hat{p})}{n}}\right)$$

gives $\left(0.7 - 1.96\sqrt{\frac{0.7(1-0.7)}{200}}, 0.7 + 1.96\sqrt{\frac{0.7(1-0.7)}{200}}\right)$

$= (0.7 - 1.96 \times 0.0324, 0.7 + 1.96 \times 0.0324)$

$= (0.636, 0.764)$

## 4.11.3.3 CAS statistics using confidence intervals

Agreeing with the CI above = (0.636, 0.764).

Lower 0.6364899
Upper 0.7635101
p̂ 0.7
n 200
<< Back
Help
OnePropZInt

# Glossary

**addition rule for probability**
$\Pr(A \cup B) = \Pr(A) + \Pr(B) - \Pr(A \cap B)$

**Bernoulli trial** A random experiment with two possible outcomes, 'success' or 'failure'.

**biased sample** An unfair sample that is not representative of the population, which favours or neglects a particular part of the population.

**binomial distribution** A discrete probability distribution of the results of a series of Bernoulli trials.

**census** A collection of data from the entire population for statistical study.

**central tendency** The tendency for the values of a random variable to cluster round its mean, mode or median.

**complement of a set** The set containing all the elements that are not in the original set. On a Venn diagram, $A'$ is everything except set $A$.

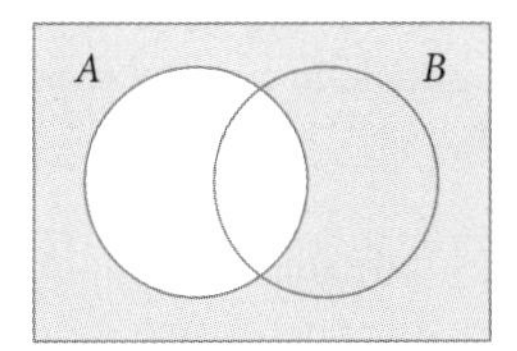


**compound event** Two or more simple events being considered together in probability, such as a bus running late and it being a rainy day.

**conditional probability** The probability that an event occurs given that another event also occurs, for example, the probability that a soccer team will win a match given that it is a rainy day. $\Pr(A|B) = \frac{\Pr(A \cap B)}{\Pr(B)}$, where $\Pr(A|B)$ means 'the probability of $A$ occurring, given $B$ occurs'.

**confidence interval** An interval centred on the mean of a distribution that contains a specific proportion of the values of the distribution. For example, a 60% confidence interval contains 60% of the values of the distribution.

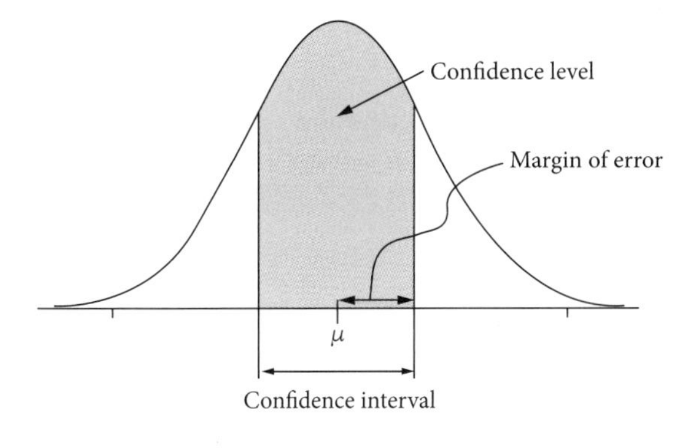

**A+ DIGITAL FLASHCARDS**
Revise this topic's key terms and concepts by scanning the QR code or typing the URL into your browser.

https://get.ga/aplus-vce-maths-methods

**confidence level** The proportion of values contained in a confidence interval, for example, a 95% confidence interval has a confidence level of 0.95.

**continuous random variable** A random variable that can take any value in a given interval, for example, the heights of students.

**discrete random variable** A random variable that can only take a countable number of values, for example, the number of pets in a household.

**expected value, E(X)** The mean of a probability distribution. For discrete probability distributions, $E(X) = \Sigma x \cdot p(x)$. For continuous probability distributions, $E(X) = \int_{-\infty}^{\infty} x f(x)\,dx$.

**fair sample** A sample that accurately represents the population. *See also* **simple random sample**.

**independent events** Events whose outcomes (and probabilities) do not depend on each other, for example, the number rolled on a second die does not depend on the number rolled on the first die. For two independent events $A$ and $B$, $\Pr(A \cap B) = \Pr(A) \times \Pr(B)$.

**intersection of sets** A set containing elements common to two or more sets. On a Venn diagram, $A \cap B$ is where sets $A$ and $B$ overlap.

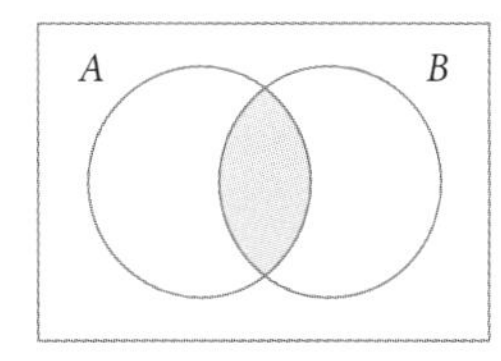


**inverse normal distribution** The distribution used to find the value of $X$ from a probability or proportion in a normal distribution.

**Karnaugh map** A 2 × 2 probability table.

**margin of error** In a confidence interval, the difference between the mean and the extremes. For the confidence interval $(\mu - E, \mu + E)$, the margin of error is E.

**mean** *See* **expected value**.

**measure of central tendency** A central or typical value for a probability distribution. The three measures of centre are mean, mode and median. They provide a measure of an expected or likely outcome for the distribution.

**measure of spread** A measure to describe how similar or varied the set of observed values are for a particular variable (data item). Measures of spread include the range, quartiles and the interquartile range, variance and standard deviation.

**mode** The most frequent value in a distribution.

**mutually exclusive events** Events or categories that have no items in common. Sets $A$ and $B$ are mutually exclusive if $A \cap B$ has no elements.

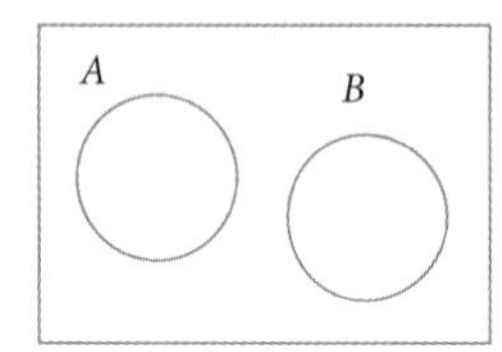


**normal distribution** A continuous probability distribution whose graph is a bell-shaped curve. People's heights or reaction times approximate a normal distribution.

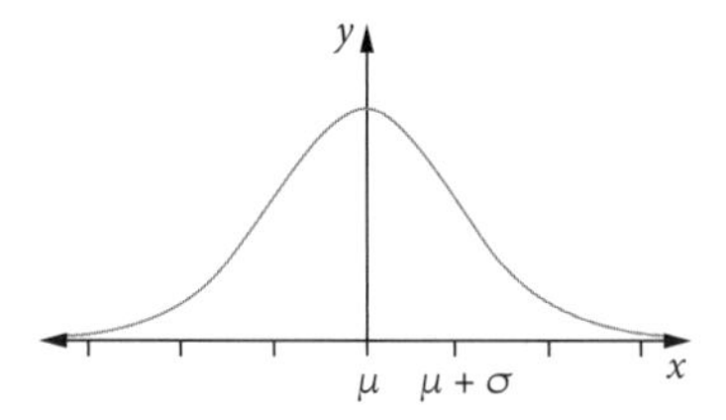


**parameter** A quantity that influences the behaviour of a mathematical function. In a binomial distribution, the number of trials, $n$, and the probability of success, $p$, are both parameters.

**population** The whole group from which data could be collected for statistical study, for example, all high school students if high school students were being surveyed.

**population parameter (pp)** A numerical value describing a population of interest, for example, the average height of all high school students.

**probability density function (PDF)** The function that describes a continuous probability distribution. It describes the relative probability of a random variable taking a value within a given range. The area under the graph of a PDF gives the probability of the random variable taking on that value within a given range, and its value is found by integration: $\Pr(a < X < b) = \int_a^b f(x)\,dx$

**probability distribution** A function that gives the probabilities of all the possible values of a random variable. The probability distribution for a discrete random variable is a probability mass function and the distribution for a continuous random variable is a probability density function.

**probability mass function** A function that assigns a probability to every possible value of a discrete random variable.

**quantile ($t_\alpha$)** In a statistical distribution, the value below which the proportion $\alpha$ of the distribution lie. For example, 35% of the distribution lie below $t_{0.35}$.

**random sample** *See* **simple random sample**.

**random variable** A variable, $X$, whose value is determined by the outcome of a random experiment. *See also* **discrete random variable** and **continuous random variable**.

**sample** A part of a population selected for statistical study.

**sample proportion** The proportion of a sample that has a particular property, expressed as a decimal. For example, if 6 people in a sample of 10 have blue eyes, then $\hat{p} = 0.6$.

**sample space** A full range of possibilities that occur when we carry out an experiment. For example, when we throw a die, the sample space of the number shown uppermost is written as $\{1, 2, 3, 4, 5, 6\}$.

**sample statistic (ss)** An estimate of a population parameter found from a sample.

**sampling with replacement** Selecting several items, replacing each before selecting the next. *See* **independent events**.

**sampling without replacement** Selecting several items, **not** replacing each before selecting the next.

**simple random sample** A type of fair sample in which every member of the population has an equal chance of being selected.

**standard deviation** A measure of spread of a distribution, equal to the square root of the variance. $\sigma = \text{SD}(X) = \sqrt{\text{Var}(X)}$.

**standard normal distribution** A normal distribution with a mean of 0 and a standard deviation of 1. The letter $Z$ denotes a standard normal variable where $Z \sim \text{N}(0, 1)$.

**statistic** A numerical value describing observations on a group, for example average height. This can refer either to a sample statistic or to a population parameter. *See also* **sample statistic, population parameter**.

**tree diagram** A diagram of branches for listing all of the possible outcomes of a multi-step chance experiment.

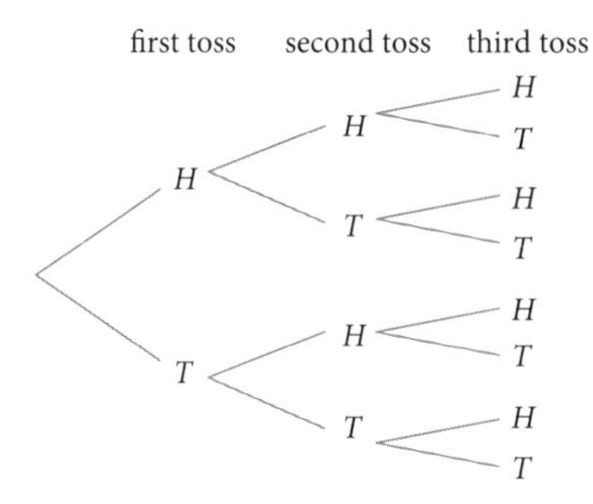


**union of sets** A set combining the elements of two or more sets. On a Venn diagram, $A \cup B$ are sets $A$ and $B$ combined.

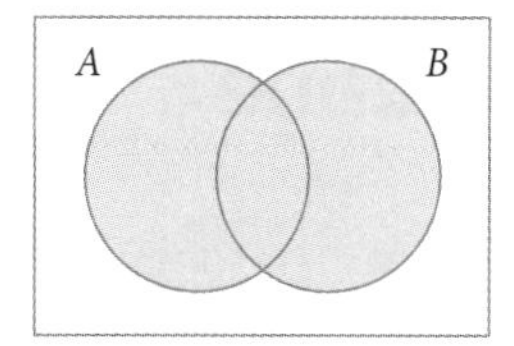


**variance** A measure of the spread of a distribution, equal to the square of the standard deviation.

$\sigma = \text{Var}(X) = E(X^2) - \mu^2$, where $\mu = E(X)$ and

$E(X^2) = \Sigma x^2 \cdot p(x)$ for a discrete probability distribution,

$E(X^2) = \int_{-\infty}^{\infty} x^2 f(x)dx$ for a continuous probability distribution.

**Venn diagram** A diagram of circles (usually overlapping) showing how multiple sets are related.

**z-score** A statistical measurement that describes a value's relationship to the mean of a group of values, measured in terms of standard deviations from the mean. If a z-score is 0, it indicates that the data point's score is identical to the mean score.

$$z = \frac{x - \mu}{\sigma}$$

# Exam practice

## Short-answer questions

### Technology free: 20 questions

Solutions to this section start on page 239.

**Question 1** (2 marks)

IQ scores are normally distributed with a mean of 100 and a standard deviation of 15. It is known that $\Pr(Z > 1.5) = 0.0668$. What is the score, $x$, that a randomly selected student will score for the IQ test if $\Pr(X < x) = 0.0668$?

**Question 2** (2 marks)

If $X$ is a normal variable with $\mu = 10$ and $\sigma = 5$, when $X = 18$, what does the standard normal variable $z$ equal?

**Question 3** (2 marks)

For the following discrete random variable $Z$, calculate the mean.

| $z$ | 1 | 3 | 5 |
|---|---|---|---|
| $\Pr(Z = z)$ | 0.5 | 0.3 | 0.2 |

**Question 4** (2 marks)

The following table represents a discrete probability distribution:

| $y$ | −2 | −1 | 0 | 1 |
|---|---|---|---|---|
| $\Pr(Y = y)$ | 0.4 | 0.3 | 0.1 | 0.2 |

**a** Calculate the mean. 1 mark

**b** Find $\Pr(Y \le 0)$. 1 mark

**Question 5** (1 mark)

If $E(X) = 2$, state the value for $E(3X - 2)$.

**Question 6** (3 marks)

The following table represents a discrete probability distribution with a mean of 1. Find the values for $a$ and $b$.

| $x$ | 0 | 1 | 2 | 3 |
|---|---|---|---|---|
| $\Pr(X = x)$ | $a$ | $b$ | 0.1 | 0.1 |

**Question 7** (1 mark)

The probability of catching the common cold is 0.2. At an insurance company there are 500 employees. How many of these would you expect to catch the common cold?

**Question 8** (4 marks)

A binomial random variable has a mean of 12 and a probability of success of 0.2.

**a** How many trials were conducted? 1 mark

**b** What is the variance? 1 mark

**c** For this binomial random variable, write an expression for $\Pr(X = 1)$. 2 marks

**Question 9** (4 marks)

Let $X$ = number of boys in a two-child family.

Assume that the probability of a boy for each birth is $\frac{1}{2}$.

**a** Complete the table for the probability distribution. 3 marks

| $x$ | 0 | 1 | 2 |
|---|---|---|---|
| $\Pr(X = x)$ | | | |

**b** Calculate $\Pr(X \leq 1)$. 1 mark

**Question 10** (3 marks)

The probability that Sam ($S$) catches a plane on time is 0.8, but if it is raining ($R$) on the way to the airport, it is known that the probability decreases to 0.5. If it rains 20% of the time, calculate the probability that it rains if Sam catches a plane on time.

**Question 11** (4 marks)

A multiple-choice test has four questions, with five alternative solutions for each question, with only one alternative being correct for each question. If a student guesses the answers to all four questions, calculate the probability that

**a** exactly four questions are correct 1 mark

**b** up to one question is correct 1 mark

**c** at least three questions are correct 2 marks

**Question 12** (1 mark)

A normally distributed variable has $\mu = 18$ and $\sigma = 2.7$. Find the range between which approximately 95% of the values lie.

**Question 13** (3 marks)

A particular clothing store has mean Saturday sales of 80 dresses, normally distributed with a standard deviation of 5, and mean Sunday sales of 50 dresses, normally distributed with a standard deviation of 6. The store sells well on a particular Saturday, with 90 dresses sold, and on the following Sunday it sells 60 dresses. On which of these two days did it do better, relative to usual sales?

**Question 14** (5 marks)

For a random variable $X$ with probability density function defined by

$$f(x) = \begin{cases} 12x^2(1 - x) & 0 \leq x \leq 1 \\ 0 & \text{elsewhere} \end{cases}$$

**a** sketch the graph of $f(x)$, shading on the graph the area that corresponds to $\Pr\left(X < \frac{1}{2}\right)$ 3 marks

**b** calculate $\Pr\left(X < \frac{1}{2}\right)$. 2 marks

**Question 15** (4 marks)

If

$$f(x) = \begin{cases} kx & 0 \le x \le 1 \\ 2 - x & 1 < x \le 2 \\ 0 & \text{elsewhere} \end{cases}$$

**a** find $k$ such that $f(x)$ is a PDF — 2 marks

**b** find the mean of the PDF. — 2 marks

**Question 16** (3 marks)

Consider a random variable $X$ with a probability density defined by

$$f(x) = \begin{cases} 2k\sin(\pi x) & 0 \le x \le 1 \\ 0 & \text{elsewhere} \end{cases}$$

where $k$ is a real constant.

Find $k$ such that $f(x)$ is a PDF.

**Question 17** (3 marks)

A sample of 20 people waiting in a queue at the MCG on a particular Saturday afternoon were asked how much money they intended to pay in total for tickets. The smallest amount was \$10, the average amount was \$50 and the greatest was \$100.

Identify

**a** the population — 1 mark

**b** three population parameters — 1 mark

**c** three sample statistics. — 1 mark

**Question 18** (4 marks)

It is given that $\hat{p} = \frac{X}{4}$, where $\Pr(X = x) = {}^4C_x\, 0.7^x\, 0.3^{4-x}$ and $X$ is binomial.

**a** Complete the discrete random variable probability table for a sample of 4. — 3 marks

| $x$ | 0 | 1 | 2 | 3 | 4 |
|---|---|---|---|---|---|
| $\hat{p} = \frac{x}{4}$ | | | | $\frac{3}{4}$ | |
| $\Pr(X = x)$ | | ${}^4C_1\, 0.7^1 0.3^3$ | | | |

**b** Show that $E(\hat{p}) = p$. — 1 mark

**Question 19** (2 marks)

If you were checking whether taking the bus to work increases the chances of catching a cold from other passengers, you might ask one sample of people to not catch the bus for 5 days, another to catch one bus for the week and a third sample to take buses every day.

For each sample, the variable would be a random binomial variable with success being 'catching a cold'. You would then look at the frequencies of people in each of the three sample groups.

From 100 people who didn't catch the bus, 20 caught colds during one week. What is the sample proportion of colds?

**Question 20** (3 marks)

Find the approximate 95% CI for the proportion $p$ of students, using a random sample of 100 students, when it is found that the sample proportion $\hat{p}$ is 0.5.

## Multiple-choice questions

### Technology active: 50 questions

Solutions to this section start on page 242.

**Question 1**

A fair coin is tossed twice and the outcomes are recorded. The sample space of the outcomes is

**A** $\{H, T\}$ **B** $\{H, HT, T\}$ **C** $\{HH, HT, TT\}$

**D** $\{HH, HT, TH, TT\}$ **E** $\{T, TH, H\}$

**Question 2**

A fair die is thrown and the outcomes are recorded. The sample space of the outcomes is

**A** $\{11, 22, 33, 44, 55, 66\}$ **B** $\{1, 2, 3, 4, 5, 6\}$ **C** $\{1, 3, 5\}$

**D** $\{2, 4, 6\}$ **E** $\{H, T\}$

**Question 3**

Which one of the following random variables is **not** discrete?

**A** The number of people in your hall for a school assembly.

**B** The shoe size of students in your class.

**C** The number of people ahead of you in the queue at your school canteen.

**D** The height of footballers in a team.

**E** The number of apples you ate for lunch this week.

**Question 4**

Which one of the following random variables is discrete?

**A** The weight of students in your class.

**B** The shoe size of students in your class.

**C** The height of people ahead of you in the queue at your school canteen.

**D** The height of footballers in a team.

**E** The volume of apples you ate for lunch this week.

**Question 5**

For a discrete probability distribution, $X$, if $\text{Var}(X) = 2$, then $\text{Var}(2X + 1)$ equals

**A** $\sqrt{2}$ **B** 3 **C** 4 **D** 5 **E** 8

**Question 6**

For a discrete probability distribution, $X$, if $\text{Var}(X) = 4$, then $\text{Var}(-X)$ equals

**A** $-4$ **B** $-1$ **C** 2 **D** 4 **E** 9

**Question 7**

For a discrete probability distribution, $X$, if $E(X) = 1.1$, then $E(-X - 2)$ equals

**A** −3.1 **B** −1.1 **C** −0.9 **D** −0.79 **E** 0.9

**Question 8**

Four coins are tossed. The probability of getting four tails is

**A** $\frac{1}{16}$ **B** $\frac{1}{8}$ **C** $\frac{1}{4}$ **D** $\frac{1}{2}$ **E** $\frac{15}{16}$

**Question 9**

For the graph of the binomial distribution shown, which of the following is an **incorrect** statement?

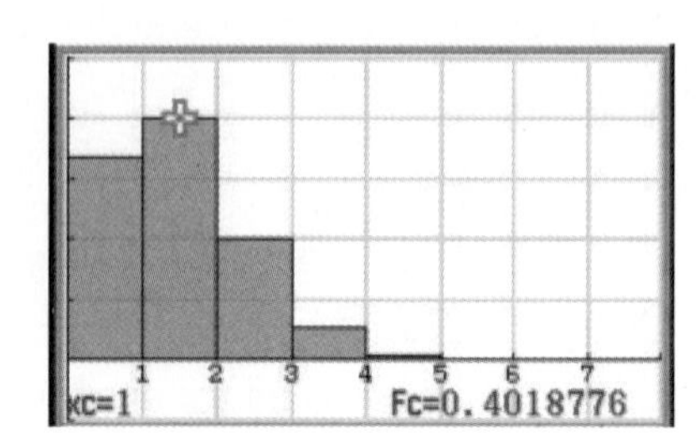


**A** The probability of success, $p$, is less than 0.5.

**B** The graph is skewed to the right.

**C** The expected value is approximately 1.

**D** The graph is skewed negatively.

**E** The graph is not symmetrical.

**Question 10**

A fair die is rolled 60 times. The expected number of sixes is

**A** 5 **B** 10 **C** 15 **D** 20 **E** 30

**Question 11**

For a binomial experiment, it is **not** true to say

**A** there are $n$ independent trials

**B** the probability of success, $p$, is the same for each trial

**C** there are only two possible outcomes for each trial

**D** binomial trials form a continuous probability distribution

**E** the sum of the probabilities of all the possible outcomes is 1.

**Question 12**

If a binomial variable $X$ has the probability function $\Pr(X = x) = {}^5C_x(0.8)^{5-x}(0.2)^x$, where $x = 0, 1 \ldots 5$, then the probability of success is

**A** 0.2 **B** 0.8 **C** $5 - x$ **D** 1 **E** $x$

**Question 13**

Consider the following table of a discrete probability distribution.

| $x$ | −1 | 0 | 1 | 2 | 3 |
|---|---|---|---|---|---|
| $\Pr(X = x)$ | 0.3 | 0.1 | 0.3 | 0.2 | 0.1 |

The mean of $X$ is

**A** 0 **B** 0.7 **C** 1 **D** 1.81 **E** 2.3

**Question 14**

Consider the following table of a discrete probability distribution.

| $x$ | −1 | 0 | 1 | 2 | 3 |
|---|---|---|---|---|---|
| $\Pr(X = x)$ | 0.3 | 0.1 | 0.3 | 0.2 | 0.1 |

The expected value of $X^2$ is

**A** 0 **B** 0.7 **C** 1 **D** 1.81 **E** 2.3

**Question 15**

Consider the following table of a discrete probability distribution.

| $x$ | −1 | 0 | 1 | 2 | 3 |
|---|---|---|---|---|---|
| $\Pr(X = x)$ | 0.3 | 0.1 | 0.3 | 0.2 | 0.1 |

The variance of $X$ is

**A** 0 **B** 0.7 **C** 1 **D** 1.81 **E** 2.3

**Question 16**

Consider the following table of a discrete probability distribution.

| $x$ | −1 | 0 | 1 | 2 | 3 |
|---|---|---|---|---|---|
| $\Pr(X = x)$ | 0.3 | 0.1 | 0.3 | 0.2 | 0.1 |

E($X$) equals

**A** 0.1 **B** 0.3 **C** 0.6 **D** 0.7 **E** 0.8

**Question 17**

A discrete random variable $X$ is given below.

| $x$ | −1 | 0 | 1 |
|---|---|---|---|
| $\Pr(X = x)$ | $2a$ | $0.5a$ | $a$ |

The value of $a$ and the mean of the distribution respectively equal

**A** $a = \frac{2}{7}, \frac{2}{7}$ **B** $a = -\frac{2}{7}, \frac{2}{7}$ **C** $a = 0, 0$ **D** $a = \frac{2}{7}, -\frac{2}{7}$ **E** $a = -\frac{2}{7}, \frac{2}{7}$

**Question 18**

The graph shown has $E(X^2) =$

**A** 1

**B** 2

**C** 2.5

**D** 3

**E** 5.4

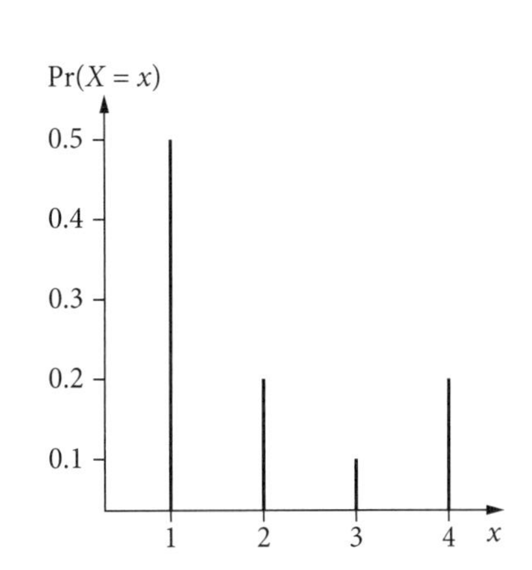

**Question 19**

For the graph shown in question 18, the mean of $X$ is

**A** 0.5 **B** 1 **C** 2 **D** 3 **E** 4

**Question 20**

Which one of the following is a binomial trial?

**A** Drawing 6 marbles from a bag of marbles without replacement and recording the number of yellow marbles.

**B** Rolling a die 10 times and recording each number.

**C** Selecting 4 students from a group of 10 students to form a committee.

**D** Drawing 6 cards from a deck of cards, with replacement, and recording the number of picture cards.

**E** Selecting 4 jellybeans from a jar containing 2 different-colour jelly beans and eating them after each selection.

**Question 21**

If $X$ is a binomial random variable with $n = 40, p = \frac{3}{4}$, then the mean and standard deviation, respectively, are:

**A** $15, \frac{15}{4}$ **B** $5, \frac{\sqrt{15}}{2}$ **C** $15, \frac{\sqrt{15}}{2}$ **D** $30, \sqrt{\frac{15}{2}}$ **E** $40, \frac{\sqrt{30}}{2}$

**Question 22**

If $X$ has a binomial distribution and $\Pr(X = x) = {}^4C_x(0.3)^x(0.7)^{4-x}$, then $\Pr(X \geq 3)$ equals

**A** $4(0.3)(0.7)^3 + (0.7)^4$

**B** $4(0.3)^3(0.7) + (0.3)^4$

**C** $4(0.3)^3(0.7)$

**D** $1 - (0.7)^4 - 4(0.3)(0.7)^3$

**E** $1 - 4(0.3)(0.7)^3 - 6(0.3)^2(0.7)^2$

**Question 23**

$X$ and $Y$ are discrete random variables. If $\text{Var}(X) = 4$ and $Y = 2X + 3$, then $\text{Var}(Y)$ equals

**A** 2 **B** 4 **C** 11 **D** 16 **E** 19

**Question 24**

$X$ and $Y$ are discrete random variables. If $\text{E}(X) = 4$ and $Y = 2X + 3$, then $\text{E}(Y)$ equals

**A** 2 **B** 4 **C** 11 **D** 16 **E** 19

**Question 25**

A random variable $X$ has the following probability distribution:

| $x$ | 0 | 2 | 3 | 5 |
|---|---|---|---|---|
| $\Pr(X = x)$ | 0.2 | 0.1 | 0.5 | 0.2 |

$\text{E}((X - 1)^2)$ is equal to

**A** 1.7 **B** 2.89 **C** 5.5 **D** 6.29 **E** 7.3

**Question 26**

Two dice are rolled. Let $X$ be the sum of the numbers on the uppermost faces. If a discrete distribution table is set up with all the outcomes and their probabilities, it would look like

**A**

| $x$ | 2 | 3 | 4 | 5 | 6 | 7 | 8 | 9 | 10 | 11 | 12 |
|---|---|---|---|---|---|---|---|---|---|---|---|
| $\Pr(X = x)$ | $\frac{1}{36}$ | $\frac{2}{36}$ | $\frac{3}{36}$ | $\frac{4}{36}$ | $\frac{5}{36}$ | $\frac{6}{36}$ | $\frac{5}{36}$ | $\frac{4}{36}$ | $\frac{3}{36}$ | $\frac{2}{36}$ | $\frac{1}{36}$ |

**B**

| $x$ | 1 | 2 | 3 | 4 | 5 | 6 | 7 | 8 | 9 | 10 | 11 | 12 |
|---|---|---|---|---|---|---|---|---|---|---|---|---|
| $\Pr(X = x)$ | $\frac{1}{36}$ | $\frac{1}{36}$ | $\frac{2}{36}$ | $\frac{3}{36}$ | $\frac{4}{36}$ | $\frac{5}{36}$ | $\frac{5}{36}$ | $\frac{5}{36}$ | $\frac{4}{36}$ | $\frac{3}{36}$ | $\frac{2}{36}$ | $\frac{1}{36}$ |

**C**

| $x$ | 2 | 4 | 6 | 8 | 10 | 12 |
|---|---|---|---|---|---|---|
| $\Pr(X = x)$ | $\frac{1}{36}$ | $\frac{3}{36}$ | $\frac{5}{36}$ | $\frac{5}{36}$ | $\frac{3}{36}$ | $\frac{1}{36}$ |

**D**

| $x$ | 2 | 3 | 4 | 5 | 6 | 7 | 8 | 9 | 10 | 11 | 12 |
|---|---|---|---|---|---|---|---|---|---|---|---|
| $\Pr(X = x)$ | $\frac{1}{6}$ | $\frac{2}{6}$ | $\frac{3}{6}$ | $\frac{4}{6}$ | $\frac{5}{6}$ | $\frac{6}{6}$ | $\frac{5}{6}$ | $\frac{4}{6}$ | $\frac{3}{6}$ | $\frac{2}{6}$ | $\frac{1}{6}$ |

**E**

| $x$ | 1 | 2 | 3 | 4 | 5 | 6 | 7 | 8 | 9 | 10 | 11 | 12 |
|---|---|---|---|---|---|---|---|---|---|---|---|---|
| $\Pr(X = x)$ | $\frac{1}{6}$ | $\frac{1}{6}$ | $\frac{2}{6}$ | $\frac{3}{6}$ | $\frac{4}{6}$ | $\frac{5}{6}$ | $\frac{6}{6}$ | $\frac{5}{6}$ | $\frac{4}{6}$ | $\frac{3}{6}$ | $\frac{2}{6}$ | $\frac{1}{6}$ |

**Question 27**

A random variable $X$ is defined by the probability density function

$$f(x) = \begin{cases} k(x-1)^2 & 1 \le x \le 3 \\ 0 & \text{otherwise} \end{cases}$$

The value of $k$ is

**A** $\frac{3}{8}$ **B** $\frac{\sqrt{6}}{4}$ **C** 1 **D** $\frac{8}{3}$ **E** 3

**Question 28**

$X$ is a discrete random variable, and if $\text{Var}(X) = 6$ and $\text{E}(X) = 2$, then $\text{E}(X^2)$ equals

**A** 1 **B** 4 **C** 7 **D** 9 **E** 10

**Question 29**

The probability that Amy ($A$) drives to work is 0.7. Independently of Amy, the probability that her sister Claire ($C$) walks to work is 0.5. Which of the following is **incorrect**?

**A** $\Pr(A \cap C) = 0.35$

**B** $A$ and $C$ are independent events.

**C** $\Pr(A \cup C) = 0.85$

**D** $A$ and $C$ are mutually exclusive events

**E** $\Pr(A | C) = 0.7$

9780170465366

### Question 30

A binomial probability distribution has a mean of 20 and a variance of 10. The probability of success, $p$, and the number of trials, $n$, are

**A** $p = \frac{1}{2}, n = 36$
**B** $p = \frac{1}{2}, n = 40$
**C** $p = \frac{1}{6}, n = 40$
**D** $p = \frac{1}{6}, n = 180$
**E** $p = \frac{1}{2}, n = 4$

### Question 31

A bag of marbles has 5 red and 5 blue marbles in it. I reach into the bag and randomly select a marble and set it aside. I reach into the bag again and select another marble. The probability of selecting one of each colour is

**A** $\frac{25}{90}$
**B** $\frac{5}{18}$
**C** $\frac{4}{9}$
**D** $\frac{1}{2}$
**E** $\frac{5}{9}$

### Question 32

The probability that Sue, a cricketer, will score one run each time she faces a bowler is 0.7. If Sue has 6 shots at facing a bowler in every over, and she never takes more than one run at a time, the probability that she will score at least 2 runs in an over can be expressed as

**A** $1 - (0.3)^6 - 6(0.3)^5(0.7)^1$
**B** $1 - (0.3)^6 - 6(0.3)^5(0.7)^1 - 15(0.3)^4(0.7)^2$
**C** $1 + (0.3)^6 + 6(0.3)^5(0.7)^1$
**D** $(0.3)^6 + 6(0.3)^5(0.7)^1$
**E** $(0.3)^6$

### Question 33

A binomial variable, $X$, has the probability function $\Pr(X = x) = {}^3C_x(0.2)^{3-x}(0.8)^x$ where $x = 0, 1, 2, 3$.

The probability distribution is

**A**

| $x$ | 0 | 1 | 2 | 3 |
|---|---|---|---|---|
| $\Pr(X = x)$ | $(0.8)^3$ | $3(0.2)^2(0.8)$ | $3(0.2)^2(0.8)$ | $(0.2)^3$ |

**B**

| $x$ | 0 | 1 | 2 | 3 |
|---|---|---|---|---|
| $\Pr(X = x)$ | $(0.2)^3$ | $3(0.2)^2(0.8)$ | $3(0.2)(0.8)^2$ | $(0.8)^3$ |

**C**

| $x$ | 0 | 1 | 2 | 3 |
|---|---|---|---|---|
| $\Pr(X = x)$ | 0.008 | 0.009 | 0.081 | 0.512 |

**D**

| $x$ | 0 | 1 | 2 | 3 |
|---|---|---|---|---|
| $\Pr(X = x)$ | 0.729 | 0.2 | 0.07 | 0.001 |

**E**

| $x$ | 0 | 1 | 2 | 3 |
|---|---|---|---|---|
| $\Pr(X = x)$ | 0.001 | 0.12 | 0.15 | 0.729 |

**Question 34**

If the mean of a binomial distribution is 3 and the standard deviation is 1.5, then the number of trials, $n$, and the probability of success, $p$, are respectively

**A** $n = 3, p = \frac{3}{2}$ **B** $n = 10, p = 0.9$ **C** $n = 15, p = 0.6$

**D** $n = 12, p = \frac{1}{4}$ **E** $n = 12, p = \frac{3}{4}$

**Question 35**

Anna takes a short maths test every week. There are always 20 questions on the test and Anna has found that on average she usually answers 15 of them correctly. The probability that she correctly answers exactly 18 questions on a particular test is closest to

**A** 0.000 003 **B** 0.000 182 **C** 0.0669 **D** 0.07 **E** 0.202

**Question 36**

If $f(x) = 1.5(1 - x^2)$, $0 \le x \le 1$ represents a probability density function, then $\Pr(X \le 0.3)$ equals

**A** $\frac{1127}{2000}$ **B** $\frac{1703}{2000}$ **C** $\frac{5157}{80\,000}$ **D** $\frac{873}{200}$ **E** $\frac{873}{2000}$

**Question 37**

If a random variable $X$ has a probability density function $f(x) = 1.5(x - 1)^2$, $0 \le x \le 2$, and 0 elsewhere, then the mean of $X$ is

**A** 0 **B** 1 **C** $\frac{8}{5}$ **D** 0 and 2 **E** 2

**Question 38**

If $f(x)$ is a probability density function such that

$$f(x) = \begin{cases} x & 0 \le x \le 1 \\ 2 - x & 1 < x \le 2 \\ 0 & \text{elsewhere} \end{cases}$$

then the mean of $X$ is

**A** $\frac{1}{2}$ **B** 1 **C** $\frac{7}{6}$ **D** $\frac{5}{4}$ **E** 2

**Question 39**

If $f(x)$ is a probability density function such that

$$f(x) = \begin{cases} x & 0 \le x \le 1 \\ 2 - x & 1 < x \le 2 \\ 0 & \text{elsewhere} \end{cases}$$

then variance of $X$ is

**A** $\frac{1}{6}$ **B** 1 **C** $\frac{7}{6}$ **D** $\frac{5}{4}$ **E** 2

**Question 40**

If $X$ is a normal random variable with $\mu = 10$ and $\sigma = \frac{1}{3}$, then the graph of this distribution is most likely to be

**A**

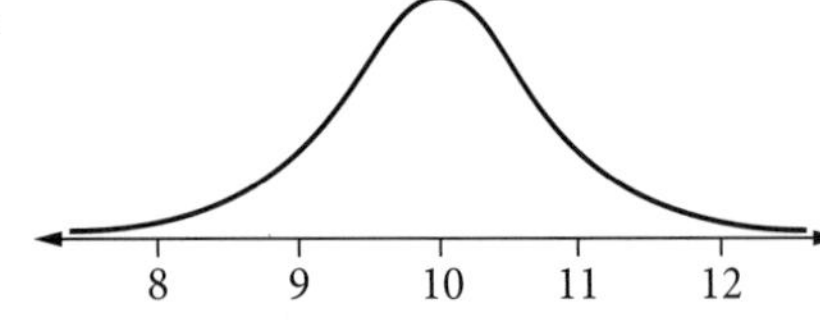


**B**

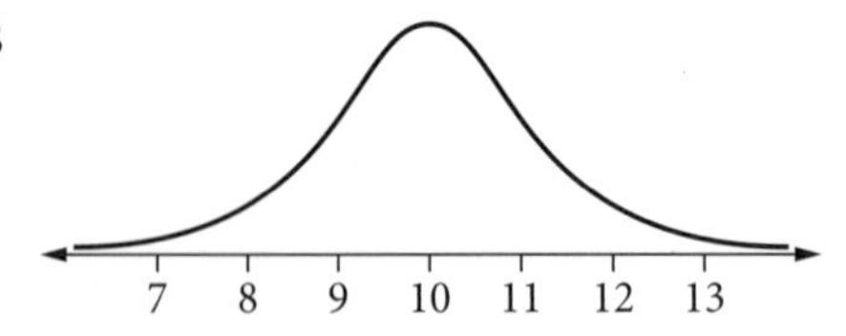


**C**

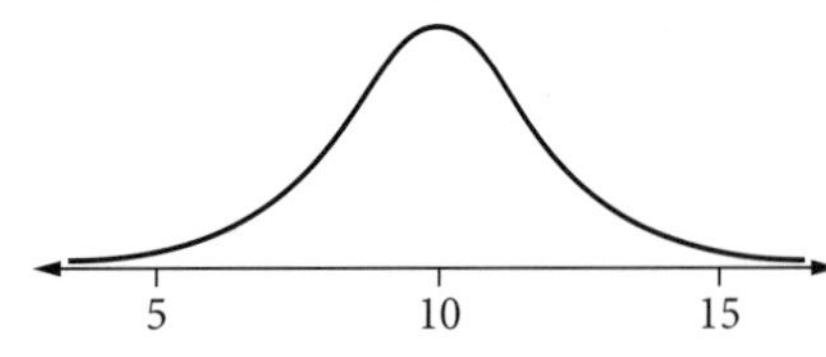


**D**

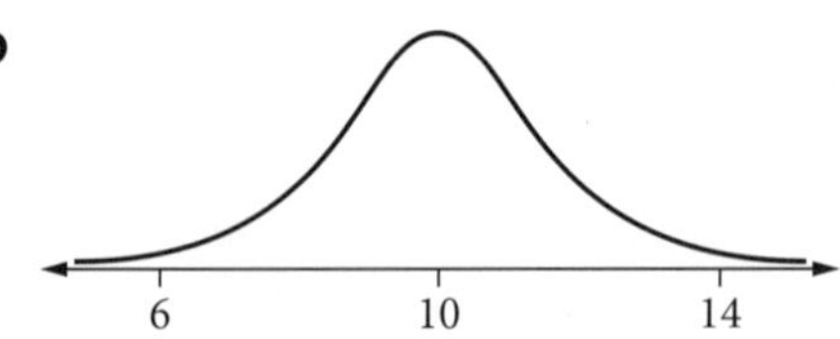


**E**

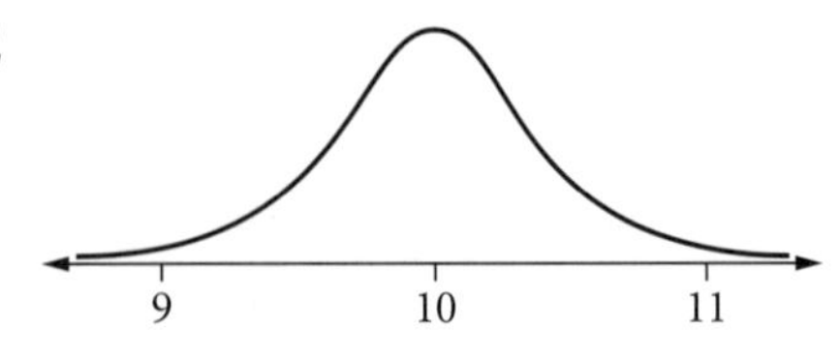


**Question 41**

Heights of students, in cm, in a particular classroom are normally distributed with a mean of 150 and a standard deviation of 8. The probability that a randomly selected student is taller than 166 cm is approximately

**A** 0.425 **B** 0.475 **C** 0.84 **D** 0.925 **E** 0.975

**Question 42**

If $Z$ is the standard normal variable, $\Pr(Z < 1.2)$ is closest to

**A** 0.03 **B** 0.05 **C** 0.11 **D** 0.86 **E** 0.88

**Question 43**

If $\Pr(Z \leq k) = 0.8$ and $Z$ is a standard normal variable, then $k$ is closest to

**A** −1.282 **B** −0.842 **C** 0.524 **D** 0.726 **E** 0. 842

**Question 44**

If $X$ is normally distributed with a mean of 15 and a standard deviation of 2, and $\Pr(X < m) = 0.48$, then $m$ is closest to

**A** 12.1 **B** 14.8 **C** 14.9 **D** 16.8 **E** 17.9

**Question 45**

IQ scores are currently normally distributed with a mean of 100 and a standard deviation of 15. When a randomly selected student sits the IQ test, the probability that he scores higher than 102 is closest to

**A** 0.102 **B** 0.288 **C** 0.356 **D** 0.447 **E** 0.553

**Question 46**

The weight of bags of lollies is normally distributed. The bags of lollies have a mean weight of 100 g. If a bag weighs 95 g or less, it is unacceptable. Tests show that 3% of bags were unacceptable. The standard deviation of the weight (in grams) of the bags is

**A** 2.054 **B** 2.435 **C** 2.658 **D** 5.941 **E** 10.03

**Question 47**

In a binomial distribution it is found that 70% of 17-year-olds have their L plates. For a sample size of 20, the probability that the sample proportion is equal to the population proportion (0.7) is closest to

**A** 0.0002 **B** 0.006 **C** 0.1916 **D** 0.9196 **E** 14

**Question 48**

In a binomial distribution it is found that 70% of 17-year-olds have their L plates.

For a sample size of 20, the probability that the percentage of 17-year-olds who have their L plates lies within one standard deviation of the population proportion is closest to

**A** 0.7795 **B** 0.7796 **C** 0.9165 **D** 0.9166 **E** 0.9752

**Question 49**

The approximate 95% CI for the proportion $p$ of primary school children, using a random sample of 2000 children when it is found that the sample proportion $\hat{p}$ is 0.6, is

**A** $(0.042, 0.042)$ **B** $(0.579, 0.621)$ **C** $(0.670, 0.730)$

**D** $(0.6, 0.95)$ **E** $(0.690, 0.710)$

**Question 50**

It is known that 60% of families in a district of over 20 000 families own a piano. If you choose a sample of 1000 families, the probability that the proportion of families in the sample is between 58% and 62% is closest to

**A** 0% **B** 1.65% **C** 58% **D** 60% **E** 80.31%

## Extended-answer questions

### Technology active: 18 questions

Solutions to this section start on page 248.

**Question 1** (24 marks)

A binomial random variable distribution has probability $p$ for success.

**a** **i** For three independent trials state an expression, in terms of $p$, that would find $\Pr(X > 2)$. 2 marks

**ii** For three independent trials state an expression, in terms of $p$, that would find $\Pr(X = 0)$. 2 marks

**b** **i** For what value of $p$ is $\Pr(X > 2) = \Pr(X = 0)$? 1 mark

**ii** Explain why you would expect this answer for $p$ in this situation. 1 mark

For a 3-child family, the distribution table is shown below, where $X$ = number of boys in a family and the probability of having a boy for each birth is $p = 0.5$.

**c** Complete the discrete distribution table below in terms of $p$. 4 marks

| $x$ | 0 | 1 | 2 | 3 |
|---|---|---|---|---|
| $\Pr(X = x)$ | | | | |

**d** What is the mean number of boys in a family for this situation? 3 marks

The situation is changed and for a particular genetic community, a 3-child family, where $Y$ = number of boys in a family, the probability of having a boy is 0.55.

**e** Complete the discrete distribution table below, correct to four decimal places. 2 marks

| $y$ | 0 | 1 | 2 | 3 |
|---|---|---|---|---|
| $\Pr(Y = y)$ | | | | |

**f** **i** What is the mean and variance of the number of boys in a family for this situation? Give your answers correct to three decimal places. 3 marks

**ii** Find the probability that the outcome falls within one standard deviation of the mean. 2 marks

**iii** In a family where the probability of a boy is 0.55, what is the probability, correct to three decimal places, that there will be at least 2 boys in the family? 1 mark

**g** Counting each family of 3 children as a trial, and with the probability of a boy for each birth being 0.55, how many trials will be needed for the probability of at least 2 boys in at least 2 families to be at least 0.6? 3 marks

**Question 2** (16 marks) ©VCAA 2018 2BQ4 ●●●

Doctors are studying the resting heart rate of adults in two neighbouring towns: Mathsland and Statsville. Resting heart rate is measured in beats per minute (bpm).

The resting heart rate of adults in Mathsland is known to be normally distributed with a mean of 68 bpm and a standard deviation of 8 bpm.

**a** 87% Find the probability that a randomly selected Mathsland adult has a resting heart rate between 60 bpm and 90 bpm. Give your answer correct to three decimal places. 1 mark

The doctors consider a person to have a slow heart rate if the person's resting heart rate is less than 60 bpm. The probability that a randomly chosen Mathsland adult has a slow heart rate is 0.1587. It is known that 29% of Mathsland adults play sport regularly. It is also known that 9% of Mathsland adults play sport regularly and have a slow heart rate.

Let S be the event that a randomly selected Mathsland adult plays sport regularly and let $H$ be the event that a randomly selected Mathsland adult has a slow heart rate.

**b** **i** 57% Find $\Pr(H|S)$, correct to three decimal places. 1 mark

**ii** 44% Are the events $H$ and $S$ independent? Justify your answer. 1 mark

**c** **i** 71% Find the probability that a random sample of 16 Mathsland adults will contain exactly one person with a slow heart rate. Give your answer correct to three decimal places. 2 marks

**ii** 41% For random samples of 16 Mathsland adults, $\hat{P}$ is the random variable that represents the proportion of people who have a slow heart rate.

Find the probability that $\hat{P}$ is greater than 10%, correct to three decimal places. 2 marks

**iii** 11% For random samples of $n$ Mathsland adults, $\hat{P}_n$ is the random variable that represents the proportion of people who have a slow heart rate.

Find the least value of $n$ for which $\Pr\left(\hat{P}_n > \frac{1}{n}\right) > 0.99$. 2 marks

The doctors took a large random sample of adults from the population of Statsville and calculated an approximate 95% confidence interval for the proportion of Statsville adults who have a slow heart rate. The confidence interval they obtained was $(0.102, 0.145)$.

**d** **i** 45% Determine the sample proportion used in the calculation of this confidence interval. 1 mark

**ii** 11% Explain why this confidence interval suggests that the proportion of adults with a slow heart rate in Statsville could be different from the proportion in Mathsland. 1 mark

Every year at Mathsland Secondary College, students hike to the top of a hill that rises behind the school. The time taken by a randomly selected student to reach the top of the hill has the probability density function M with the rule

$$M(t) = \begin{cases} \dfrac{3}{50}\left(\dfrac{t}{50}\right)^2 e^{-\left(\frac{t}{50}\right)^3} & t \geq 0 \\ 0 & t < 0 \end{cases}$$

where $t$ is given in minutes.

**e** 58% Find the expected time, in minutes, for a randomly selected student from Mathsland Secondary College to reach the top of the hill. Give your answer correct to one decimal place. 2 marks

Students who take less than 15 minutes to get to the top of the hill are categorised as 'elite'.

**f** 56% Find the probability that a randomly selected student from Mathsland Secondary College is categorised as elite. Give your answer correct to four decimal places. 1 mark

**g** 8% The Year 12 students at Mathsland Secondary College make up $\frac{1}{7}$ of the total number of students at the school. Of the Year 12 students at Mathsland Secondary College, 5% are categorised as elite. Find the probability that a randomly selected non-Year 12 student at Mathsland Secondary College is categorised as elite. Give your answer correct to four decimal places. 2 marks

**Question 3** (7 marks) ©VCAA 2013 2BQ2ab

FullyFit is an international company that owns and operates many fitness centres (gyms) in several countries. At every one of FullyFit's gyms, each member agrees to have his or her fitness assessed every month by undertaking a set of exercises called S. There is a five-minute time limit on any attempt to complete S and if someone completes S in less than three minutes, they are considered fit.

At FullyFit's Melbourne gym, it has been found that the probability that any member will complete S in less than three minutes is $\frac{5}{8}$. This is independent of any other member.

In a particular week, 20 members of this gym attempt S.

**a** **i** 77% Find the probability, correct to four decimal places, that at least 10 of these 20 members will complete S in less than three minutes. 2 marks

**ii** 58% Given that at least 10 of these 20 members complete S in less than three minutes, what is the probability, correct to three decimal places, that more than 15 of them complete S in less than three minutes? 3 marks

**b** 62% Paula is a member of FullyFit's gym in San Francisco. She completes **S** every month as required, but otherwise does not attend regularly and so her fitness level varies over many months. Paula finds that if she is fit one month, the probability that she is fit the next month is $\frac{3}{4}$, and if she is not fit one month, the probability that she is not fit the next month is $\frac{1}{2}$.

If Paula is not fit in one particular month, what is the probability that she is fit in exactly two of the next three months? 2 marks

**Question 4** (7 marks) ©VCAA 2016 2BQ3abh MODIFIED ●●●

A school has a class set of 22 new laptops kept in a recharging trolley. Provided each laptop is correctly plugged into the trolley after use, its battery recharges.

On a particular day, a class of 22 students uses the laptops. All laptop batteries are fully charged at the start of the lesson. Each student uses and returns exactly one laptop. The probability that a student does **not** correctly plug their laptop into the trolley at the end of the lesson is 10%. The correctness of any student's plugging-in is independent of any other student's correctness.

**a** 80% Determine the probability that at least one of the laptops is **not** correctly plugged into the trolley at the end of the lesson. Give your answer correct to four decimal places. 2 marks

**b** 51% A teacher observes that at least one of the returned laptops is not correctly plugged into the trolley. Given this, find the probability that fewer than five laptops are **not** correctly plugged in. Give your answer correct to four decimal places. 2 marks

**c** The laptop supplier also provides laptops to businesses. The probability density function for battery life, $x$ (in minutes), of a laptop after six months of use in a business is

$$f(x) = \begin{cases} \dfrac{(210 - x)e^{\frac{x-210}{20}}}{400} & 0 \le x \le 210 \\ 0 & \text{elsewhere} \end{cases}$$

**i** 67% Find the **mean** battery life, in minutes, of a laptop with six months of business use, correct to two decimal places. 1 mark

**ii** 66% Find the value of $m$, correct to the nearest integer, for which $\Pr(X < m) = 0.5$. 2 marks

**Question 5** (11 marks) ©VCAA 2015 2BQ3 ●●●

Mani is a fruit grower. After his oranges have been picked, they are sorted by a machine according to size. Oranges classified as **medium** are sold to fruit shops and the remainder are made into orange juice.

The distribution of the diameter, in centimetres, of medium oranges is modelled by a continuous random variable, $X$, with probability density function.

$$f(x) = \begin{cases} \dfrac{3}{4}(x - 6)^2(8 - x) & 6 \le x \le 8 \\ 0 & \text{elsewhere} \end{cases}$$

**a** **i** 82% Find the probability that a randomly selected medium orange has a diameter greater than 7 cm. 2 marks

**ii** 54% Mani randomly selects three medium oranges. Find the probability that exactly one of the oranges has a diameter greater than 7 cm. Express the answer in the form $\frac{a}{b}$, where $a$ and $b$ are positive integers. 2 marks

**b** 73% Find the mean diameter of medium oranges, in centimetres. 1 mark

For oranges classified as large, the quantity of juice obtained from each orange is a normally distributed random variable with a mean of 74 mL and a standard deviation of 9 mL.

**c** 57% What is the probability, correct to three decimal places, that a randomly selected large orange produces less than 85 mL of juice, given that it produces more than 74 mL of juice? 2 marks

Mani also grows lemons, which are sold to a food factory. When a truckload of lemons arrives at the food factory, the manager randomly selects and weighs four lemons from the load. If one or more of these lemons is underweight, the load is rejected. Otherwise it is accepted.

It is known that 3% of Mani's lemons are underweight.

**d** **i** 56% Find the probability that a particular load of lemons will be rejected. Express the answer correct to four decimal places. 2 marks

**ii** 42% Suppose that instead of selecting only four lemons, $n$ lemons are selected at random from a particular load.

Find the smallest integer value of $n$ such that the probability of at least one lemon being underweight exceeds 0.5 2 marks

**Question 6** (13 marks) ©VCAA 2012 2BQ3abicd

Steve, Katerina and Jess are three students who have agreed to take part in a psychology experiment. Each student is to answer several sets of multiple-choice questions. Each set has the same number of questions, $n$, where n is a number greater than 20. For each question there are four possible options (A, B, C or D), of which only one is correct.

**a** Steve decides to guess the answer to every question, so that for each question he chooses A, B, C or D at random. Let the random variable $X$ be the number of questions that Steve answers correctly in a particular set.

**i** 70% What is the probability that Steve will answer the first three questions of this set correctly? 1 mark

**ii** 60% Find, to four decimal places, the probability that Steve will answer at least 10 of the first 20 questions of this set correctly. 2 marks

**iii** 55% Use the fact that the variance of $X$ is $\frac{75}{16}$ to show that the value of $n$ is 25. 1 mark

If Katerina answers a question correctly, the probability that she will answer the next question correctly is $\frac{3}{4}$. If she answers a question incorrectly, the probability that she will answer the next question incorrectly is $\frac{2}{3}$.

In a particular set, Katerina answers Question 1 incorrectly.

**b** 44% Calculate the probability that Katerina will answer questions 3, 4 and 5 correctly. 3 marks

**c** 19% The probability that Jess will answer any question correctly, independently of her answer to any other question, is $p$ $(p > 0)$. Let the random variable $Y$ be the number of questions that Jess answers correctly in any set of 25.

If $\Pr(Y > 23) = 6\Pr(Y = 25)$, show that the value of $p$ is $\frac{5}{6}$. 2 marks

**d** 19% From these sets of 25 questions being completed by many students, it has been found that the time, in minutes, that any student takes to answer each set of 25 questions is another random variable, $W$, which is **normally distributed** with mean $a$ and standard deviation $b$.

It turns out that, for Jess, $\Pr(Y \geq 18) = \Pr(W \geq 20)$ and also $\Pr(Y \geq 22) = \Pr(W \geq 25)$.

Calculate the values of $a$ and $b$, correct to three decimal places. 4 marks

**Question 7** (12 marks) ©VCAA 2020 2BQ3 ●●●

A transport company has detailed records of all its deliveries.

The number of minutes a delivery is made before or after its scheduled delivery time can be modelled as a normally distributed random variable, $T$, with a mean of zero and a standard deviation of four minutes. A graph of the probability distribution of $T$ is shown below.

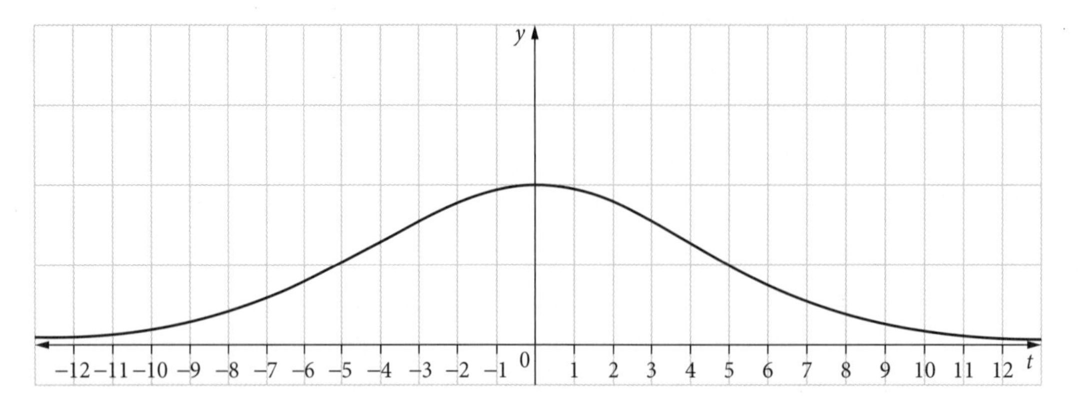


**a** 68% If $\Pr(T \le a) = 0.6$, find $a$ to the nearest minute. 1 mark

**b** 49% Find the probability, correct to three decimal places, of a delivery being no later than three minutes after its scheduled delivery time, given that it arrives after its scheduled delivery time. 2 marks

**c** 24% Using the model described, the transport company can make 46.48% of its deliveries over the interval $-3 \le t \le 2$. It has an improved delivery model with a mean of $k$ and a standard deviation of four minutes. Find the values of $k$, correct to one decimal place, so that 46.48% of the transport company's deliveries can be made over the interval $-4.5 \le t \le 0.5$. 3 marks

A rival transport company claims that there is a 0.85 probability that each delivery it makes will arrive on time or earlier. Assume that whether each delivery is on time or earlier is independent of other deliveries.

**d** 51% Assuming that the rival company's claim is true, find the probability that on a day in which the rival company makes eight deliveries, fewer than half of them arrive on time or earlier. Give your answer correct to three decimal places. 2 marks

**e** Assuming that the rival company's claim is true, consider a day in which it makes $n$ deliveries.

**i** 24% Express, in terms of $n$, the probability that one or more deliveries will **not** arrive on time or earlier. 1 mark

**ii** 23% Hence, or otherwise, find the minimum value of $n$ such that there is at least a 0.95 probability that one or more deliveries will **not** arrive on time or earlier. 1 mark

**f** 4% An analyst from a government department believes the rival transport company's claim is only true for deliveries made before 4 pm. For deliveries made after 4 pm, the analyst believes the probability of a delivery arriving on time or earlier is $x$, where $0.3 \le x \le 0.7$. After observing a large number of the rival transport company's deliveries, the analyst believes that the overall probability that a delivery arrives on time or earlier is actually 0.75. Let the probability that a delivery is made after 4 pm be $y$. Assuming that the analyst's beliefs are true, find the minimum and maximum values of $y$. 2 marks

**Question 8** (8 marks) ©VCAA 2009 2BQ3abcid

The Bouncy Ball Company (BBC) makes tennis balls whose diameters are normally distributed with mean 67 mm and standard deviation 1 mm. The tennis balls are packed and sold in cylindrical tins that each hold four balls. A tennis ball fits into such a tin if the diameter of the ball is less than 68.5 mm.

**a** 77% What is the probability, correct to four decimal places, that a randomly selected tennis ball produced by BBC fits into a tin? 2 marks

BBC management would like each ball produced to have diameter between 65.6 m and 68.4 mm.

**b** 77% What is the probability, correct to four decimal places, that the diameter of a randomly selected tennis ball made by BBC is in this range? 2 marks

**c** 42% What is the probability, correct to four decimal places, that the diameter of a tennis ball which fits into a tin is between 65.6 mm and 68.4 mm? 1 mark

BBC management wants engineers to change the manufacturing process so that 99% of all balls produced have a diameter between 65.6 mm and 68.4 mm. The mean is to stay at 67 mm, but the standard deviation is to be changed.

**d** 35% What should the new standard deviation be (correct to two decimal places)? 3 marks

**Question 9** (12 marks) ©VCAA 2014 2BQ4a–f

Patricia is a gardener and she owns a garden nursery. She grows and sells basil plants and coriander plants. The heights, in centimetres, of the basil plants that Patricia is selling are distributed normally with a mean of 14 cm and a standard deviation of 4 cm. There are 2000 basil plants in the nursery.

**a** 43% Patricia classifies the tallest 10 percent of her basil plants as super. What is the minimum height of a super basil plant, correct to the nearest millimetre? 1 mark

Patricia decides that some of her basil plants are not growing quickly enough, so she plans to move them to a special greenhouse. She will move the basil plants that are less than 9 cm in height.

**b** 54% How many basil plants will Patricia move to the greenhouse, correct to the nearest whole number? 2 marks

The heights of the coriander plants, $x$ centimetres, follow the probability density function $h(x)$, where

$$h(x) = \begin{cases} \frac{\pi}{100}\sin\left(\frac{\pi x}{50}\right) & 0 \le x \le 50 \\ 0 & \text{otherwise} \end{cases}$$

**c** 73% State the mean height of the coriander plants. 1 mark

Patricia thinks that the smallest 15% of her coriander plants should be given a new type of plant food.

**d** 33% Find the maximum height, correct to the nearest millimetre, of a coriander plant if it is to be given the new type of plant food. 2 marks

Patricia also grows and sells tomato plants that she classifies as either tall or regular. She finds that 20% of her tomato plants are tall. A customer, Jack, selects $n$ tomato plants at random.

**e** 30% Let $q$ be the probability that at least one of Jack's $n$ tomato plants is tall. Find the minimum value of $n$ so that $q$ is greater than 0.95. 2 marks

In another section of the nursery, a craftsman makes plant pots. The pots are classified as smooth or rough. The craftsman finishes each pot before starting on the next. Over a period of time, it is found that if one plant pot is smooth, the probability that the next one is smooth is 0.7, while if one plant pot is rough, the probability that the next one is rough is $p$, where $0 < p < 1$. The value of $p$ stays fixed for a week at a time, but can vary from week to week.

The first pot made each week is always a smooth pot.

**f** **i** 54% Find, in terms of $p$, the probability that the third pot made in a given week is smooth. 2 marks

**ii** 57% In one particular week, the probability that the third pot made is smooth is 0.61. Calculate the value of $p$ in this week. 2 marks

**Question 10** (6 marks) ©VCAA 2018N 2BQ2bgh

Rebecca's Robotics manufactures three types of components for robots: sensors, motors and controllers. The manufacturing processes for each type of component are independent.

It is known that 8% of all of the sensors manufactured are defective.

A random sample of 50 sensors is selected and it is found that the proportion of defective sensors in this sample is 0.08.

**a** Determine an approximate 90% confidence interval for the proportion of defective sensors, correct to four decimal places. 2 marks

The weight, $w$, in grams, of controllers is modelled by the following probability density function.

$$C(w) = \begin{cases} \dfrac{3}{640\,000}(330 - w)^2(w - 290) & 290 \le w \le 330 \\ 0 & \text{elsewhere} \end{cases}$$

**b** Determine the mean weight, in grams, of the controllers. 2 marks

**c** Determine the probability that a randomly selected controller weighs less than the mean weight of the controllers. Give your answer correct to four decimal places. 2 marks

**Question 11** (15 marks) ©VCAA 2011 2BQ2

In a chocolate factory, the material for making each chocolate is sent to one of two machines, machine $A$ or machine $B$.

The time, $X$ seconds, taken to produce a chocolate by machine $A$, is normally distributed with mean 3 and standard deviation 0.8.

The time, $Y$ seconds, taken to produce a chocolate by machine $B$, has the following probability density function.

$$f(y) = \begin{cases} 0 & y < 0 \\ \dfrac{y}{16} & 0 \le y \le 4 \\ 0.25e^{-0.5(y-4)} & y > 4 \end{cases}$$

**a** Find correct to four decimal places

**i** 77% $\Pr(3 \le X \le 5)$ 1 mark

**ii** 66% $\Pr(3 \le Y \le 5)$ 3 marks

**b** 52% Find the mean of $Y$, correct to three decimal places. 3 marks

**c** **i** 53% Find where the upper 50% of $Y$ exists 1 mark

**ii** 32% Find the value of $a$, correct to two decimal places, such that $\Pr(Y \le a) = 0.7$. 2 marks

**d** 57% It can be shown that $\Pr(Y \leq 3) = \frac{9}{32}$. A random sample of 10 chocolates **produced by machine *B*** is chosen. Find the probability, correct to four decimal places, that exactly 4 of these 10 chocolates took 3 or less seconds to produce. 2 marks

All of the chocolates produced by machine *A* and machine *B* are stored in a large bin. There is an equal number of chocolates from each machine in the bin.

It is found that if a chocolate, produced by either machine, takes longer than 3 seconds to produce then it can easily be identified by its darker colour.

**e** 19% A chocolate is selected at random from the bin. It is found to have taken longer than 3 seconds to produce. Find, correct to four decimal places, the probability that it was produced by machine *A*. 3 marks



**Question 12** (14 marks) ©VCAA 2007 2BQ5

In the Great Fun amusement park there is a small train called Puffing Berty which does a circuit of the park. The continuous random variable *T*, the time in minutes for a circuit to be completed, has a probability density function *f* with rule

$$f(t) = \begin{cases} \frac{1}{100}(t-10) & \text{if } 10 \leq t < 20 \\ \frac{1}{100}(30-t) & \text{if } 20 \leq t \leq 30 \\ 0 & \text{otherwise} \end{cases}$$

**a** 45% Sketch the graph of $y = f(t)$. 2 marks

**b** 59% Find the probability that the time taken by Puffing Berty to complete a full circuit is less than 25 minutes. Give the exact value. 2 marks

**c** 49% Find $\Pr(T \leq 15 \mid T \leq 25)$. Give the exact value. 2 marks

The train must complete six circuits between 9.00 am and noon. The management prefers Puffing Berty to complete a circuit in less than 25 minutes.

**d** 48% Find the probability, correct to four decimal places, that of the 6 circuits completed, at least 4 of them take less than 25 minutes each. 2 marks

For scheduling reasons the management wants to know the time, *b* minutes, for which the probability of exactly 3 or 4 out of the 6 circuits completed each taking less than *b* minutes, is maximised.

Let $\Pr(T < b) = p$

Let *Q* be the probability that exactly 3 or 4 circuits completed each take less than *b* minutes.

**e** 28% Show that $Q = 5p^3(1-p)^2(4-p)$. 2 marks

**f** **i** 19% Find the maximum value of *Q* and the value of *p* for which this occurs. (Give the exact value.) 2 marks

**ii** 11% Find, correct to one decimal place, the value of *b* for which this maximum occurs. 2 marks

**Question 13** (6 marks) ©VCAA 2008 2BQ1c

Sharelle is the goal shooter for her netball team. The time in hours that Sharelle spends training each day is a continuous random variable with probability density function given by

$$f(x) = \begin{cases} \frac{1}{64}(6 - x)(x - 2)(x + 2) & \text{if } 2 \le x \le 6 \\ 0 & \text{elsewhere} \end{cases}$$

**a** 61% Sketch the probability density function, and label the local maximum with its coordinates, correct to two decimal places. 2 marks

**b** 73% What is the probability, correct to four decimal places, that Sharelle spends less than 3 hours training on a particular day? 2 marks

**c** 58% What is the mean time (in hours), correct to four decimal places, that she spends training each day? 2 marks

**Question 14** (19 marks) ©VCAA 2017 2BQ3

The time Jennifer spends on her homework each day varies, but she does some homework every day. The continuous random variable $T$, which models the time, $t$, in minutes, that Jennifer spends each day on her homework, has a probability density function $f$, where

$$f(t) = \begin{cases} \frac{1}{625}(t - 20) & 20 \le t < 45 \\ \frac{1}{625}(70 - t) & 45 \le t \le 70 \\ 0 & \text{elsewhere} \end{cases}$$

**a** 62% Sketch the graph of $f$. 3 marks

**b** 73% Find $\Pr(25 \le T \le 55)$. 2 marks

**c** 64% Find $\Pr(T \le 25 \mid T \le 55)$. 2 marks

**d** 36% Find $a$ such that $\Pr(T \ge a) = 0.7$, correct to four decimal places. 2 marks

**e** The probability that Jennifer spends more than 50 minutes on her homework on any given day is $\frac{8}{25}$. Assume that the amount of time spent on her homework on any day is independent of the time spent on her homework on any other day.

- **i** 66% Find the probability that Jennifer spends more than 50 minutes on her homework on more than three of seven randomly chosen days, correct to four decimal places. 2 marks
- **ii** 65% Find the probability that Jennifer spends more than 50 minutes on her homework on at least two of seven randomly chosen days, given that she spends more than 50 minutes on her homework on at least one of those days, correct to four decimal places. 2 marks

Let $p$ be the probability that on any given day Jennifer spends more than $d$ minutes on her homework.

Let $q$ be the probability that on two or three days out of seven randomly chosen days she spends more than $d$ minutes on her homework.

**f** 36% Express $q$ as a polynomial in terms of $p$. 2 marks

**g**
- **i** 30% Find the maximum value of $q$, correct to four decimal places, and the value of $p$ for which this maximum occurs, correct to four decimal places. 2 marks
- **ii** 9% Find the value of $d$ for which the maximum found in part **g i** occurs, correct to the nearest minute. 2 marks

**Question 15** (14 marks) ©VCAA 2019N 2BQ3 ●●●

Concerts at the Mathsland Concert Hall begin $L$ minutes after the scheduled starting time. $L$ is a random variable that is normally distributed with a mean of 10 minutes and a standard deviation of four minutes.

**a** What proportion of concerts begin before the scheduled starting time, correct to four decimal places? 1 mark

**b** Find the probability that a concert begins more than 15 minutes after the scheduled starting time, correct to four decimal places. 1 mark

If a concert begins more than 15 minutes after the scheduled starting time, the cleaner is given an extra payment of $200. If a concert begins up to 15 minutes after the scheduled starting time, the cleaner is given an extra payment of $100. If a concert begins at or before the scheduled starting time, there is no extra payment for the cleaner.

Let $C$ be the random variable that represents the extra payment for the cleaner, in dollars.

**c** **i** Using your responses from part **a** and part **b**, copy and complete the following table, correct to three decimal places.

| $c$ | 0 | 100 | 200 |
|---|---|---|---|
| $\Pr(C = c)$ | | | |

1 mark

**ii** Calculate the expected value of the extra payment for the cleaner, to the nearest dollar. 1 mark

**iii** Calculate the standard deviation of $C$, correct to the nearest dollar. 1 mark

**d** The owners of the Mathsland Concert Hall decide to review their operation. They study information from 1000 concerts at other similar venues, collected as a simple random sample. The sample value for the number of concerts that start more than 15 minutes after the scheduled starting time is 43.

**i** Find the 95% confidence interval for the proportion of concerts that begin more than 15 minutes after the scheduled starting time. Give values correct to three decimal places. 1 mark

**ii** Explain why this confidence interval suggests that the proportion of concerts that begin more than 15 minutes after the scheduled starting time at the Mathsland Concert Hall is different from the proportion at the venues in the sample. 1 mark

The owners of the Mathsland Concert Hall decide that concerts must not begin before the scheduled starting time. They also make changes to reduce the number of concerts that begin after the scheduled starting time. Following these changes, $M$ is the random variable that represents the number of minutes after the scheduled starting time that concerts begin.

The probability density function for $M$ is

$$f(x) = \begin{cases} \dfrac{8}{(x+2)^3} & x \geq 0 \\ 0 & x < 0 \end{cases}$$

where $x$ is the time, in minutes, after the scheduled starting time.

**e** Calculate the expected value of $M$. 2 marks

**f** **i** Find the probability that a concert now begins more than 15 minutes after the scheduled starting time. 1 mark

**ii** Find the probability that each of the next nine concerts begins no more than 15 minutes after the scheduled starting time and the 10th concert begins more than 15 minutes after the scheduled starting time. Give your answer correct to four decimal places. 2 marks

**iii** Find the probability that a concert begins up to 20 minutes after the scheduled starting time, given that it begins more than 15 minutes after the scheduled starting time. Give your answer correct to three decimal places. 2 marks

**Question 16** (18 marks) ©VCAA 2017N 2BQ3 ●●●

A company supplies schools with whiteboard pens. The total length of time for which a whiteboard pen can be used for writing before it stops working is called its use-time. There are two types of whiteboard pens: Grade A and Grade B. The use-time of Grade A whiteboard pens is normally distributed with a mean of 11 hours and a standard deviation of 15 minutes.

**a** Find the probability that a Grade A whiteboard pen will have a use-time that is greater than 10.5 hours, correct to three decimal places. 1 mark

The use-time of Grade B whiteboard pens is described by the probability density function

$$f(x) = \begin{cases} \dfrac{x}{576}(12 - x)(e^{\frac{x}{6}} - 1) & 0 \le x \le 12 \\ 0 & \text{otherwise} \end{cases}$$

where $x$ is the use-time in hours.

**b** Determine the expected use-time of a Grade B whiteboard pen. Give your answer in hours, correct to two decimal places. 2 marks

**c** Determine the standard deviation of the use-time of a Grade B whiteboard pen. Give your answer in hours, correct to two decimal places. 2 marks

**d** Find the probability that a randomly chosen Grade B whiteboard pen will have a use-time that is greater than 10.5 hours, correct to four decimal places. 2 marks

A worker at the company finds two boxes of whiteboard pens that are not labelled, but knows that one box contains only Grade A whiteboard pens and the other box contains only Grade B whiteboard pens. The worker decides to randomly select a whiteboard pen from one of the boxes. If the selected whiteboard pen has a use-time that is greater than 10.5 hours, then the box that it came from will be labelled Grade A and the other box will be labelled Grade B. Otherwise, the box that it came from will be labelled Grade B and the other box will be labelled Grade A.

**e** Find the probability, correct to three decimal places, that the worker labels the boxes incorrectly. 2 marks

**f** Find the probability, correct to three decimal places, that the whiteboard pen selected was Grade B, given that the boxes have been labelled incorrectly. 2 marks

As a whiteboard pen ages, its tip may dry to the point where the whiteboard pen becomes defective (unusable). The company has stock that is two years old and, at that age, it is known that 5% of Grade A whiteboard pens will be defective.

**g** A school purchases a box of Grade A whiteboard pens that is two years old and a class of 26 students is the first to use them. If every student receives a whiteboard pen from this box, find the probability, correct to four decimal places, that at least one student will receive a defective whiteboard pen. 2 marks

**h** Let $\hat{P}_A$ be the random variable of the distribution of sample proportions of defective Grade A whiteboard pens in boxes of 100. The boxes come from the stock that is two years old. Find $(\hat{P}_A > 0.04 \mid \hat{P}_A < 0.08)$. Give your answer correct to four decimal places. Do not use a normal approximation. 3 marks

**i** A box of 100 Grade A whiteboard pens that is two years old is selected and it is found that six of the whiteboard pens are defective. Determine a 90% confidence interval for the population proportion from this sample, correct to two decimal places. 2 marks

**Question 17** (10 marks) ©VCAA 2021 2BQ4abcdeg ●●●

A teacher coaches their school's table tennis team. The teacher has an adjustable ball machine that they use to help the players practise. The speed, measured in metres per second, of the balls shot by the ball machine is a normally distributed random variable $W$. The teacher sets the ball machine with a mean speed of 10 metres per second and a standard deviation of 0.8 metres per second.

**a** 78% Determine $\Pr(W \geq 11)$, correct to three decimal places. 1 mark

**b** 64% Find the value of $k$, in metres per second, which 80% of ball speeds are below. Give your answer in metres per second, correct to one decimal place. 1 mark

The teacher adjusts the height setting for the ball machine. The machine now shoots balls high above the table tennis table. Unfortunately, with the new height setting, 8% of balls do not land on the table.

Let $\hat{P}$ be the random variable representing the sample proportion of balls that do not land on the table in random samples of 25 balls.

**c** 45% Find the mean and the standard deviation of $\hat{P}$. 2 marks

**d** 49% Use the binomial distribution to find $\Pr(\hat{P} > 0.1)$, correct to three decimal places. 2 marks

The teacher can also adjust the spin setting on the ball machine. The spin, measured in revolutions per second, is a continuous random variable $X$ with the probability density function

$$f(x) = \begin{cases} \dfrac{x}{500} & 0 \leq x < 20 \\ \dfrac{50 - x}{750} & 20 \leq x \leq 50 \\ 0 & \text{elsewhere} \end{cases}$$

**e** 21% Find the maximum possible spin applied by the ball machine, in revolutions per second. 1 mark

**f** 39% Find the standard deviation of the spin, in revolutions per second, correct to one decimal place. 3 marks

**Question 18** (17 marks) ©VCAA 2019 2BQ4 ●●●

The Lorenz birdwing is the largest butterfly in Town A.

The probability density function that described its life span, $X$, in weeks, is given by

$$f(x) = \begin{cases} \frac{4}{625}(5x^3 - x^4) & 0 \leq x \leq 5 \\ 0 & \text{elsewhere} \end{cases}$$

**a** 80% Find the mean life span of the Lorenz birdwing butterfly. 2 marks

**b** 57% In a sample of 80 Lorenz birdwing butterflies, how many butterflies are expected to live longer than two weeks, correct to the nearest integer? 2 marks

**c** 66% What is the probability that a Lorenz birdwing butterfly lives for at least four weeks, given that it lives for at least two weeks, correct to four decimal places? 2 marks

The wingspans of Lorenz birdwing butterflies in Town A are normally distributed with a mean of 14.1 cm and a standard deviation of 2.1 cm.

**d** 81% Find the probability that a randomly selected Lorenz birdwing butterfly in Town A has a wingspan between 16 cm and 18 cm, correct to four decimal places. 1 mark

**e** 61% A Lorenz birdwing butterfly is considered to be **very small** if its wingspan is in the smallest 5% of all the Lorenz birdwing butterflies in Town A. Find the greatest possible wingspan, in centimetres, for a **very small** Lorenz birdwing butterfly in Town A, correct to one decimal place. 1 mark

Each year, a detailed study is conducted on a random sample of 36 Lorenz birdwing butterflies in Town A. A Lorenz birdwing butterfly is considered to be **very large** if its wingspan is greater than 17.5 cm. The probability that the wingspan of any Lorenz birdwing butterfly in Town A is greater than 17.5 cm is 0.0527, correct to four decimal places.

**f** **i** 73% Find the probability that three or more of the butterflies, in a random sample of 36 Lorenz birdwing butterflies from Town A, are **very large**, correct to four decimal places. 1 mark

**ii** 33% The probability that $n$ or more butterflies, in a random sample of 36 Lorenz birdwing butterflies from Town A, are **very large** is less than 1%. Find the smallest value of $n$, where $n$ is an integer. 2 marks

**iii** 49% For random samples of 36 Lorenz birdwing butterflies in Town A, $\hat{P}$ is the random variable that represents the proportion of butterflies that are **very large**. Find the expected value and the standard deviation of $\hat{P}$, correct to four decimal places. 2 marks

**iv** 25% What is the probability that a sample proportion of butterflies that are **very large** lies within one standard deviation of 0.0527, correct to four decimal places? Do not use a normal approximation. 2 marks

**g** 28% The Lorenz birdwing butterfly also lives in Town B.

In a particular sample of Lorenz birdwing butterflies from Town B, an approximate 95% confidence interval for the proportion of butterflies that are very large was calculated to be (0.0234, 0.0866), correct to four decimal places.

Determine the sample size used in the calculation of this confidence interval. 2 marks

# SOLUTIONS

## UNIT 3

## Chapter 1

### Area of Study 1: Functions, relations and graphs

### Short-answer questions

#### Question 1

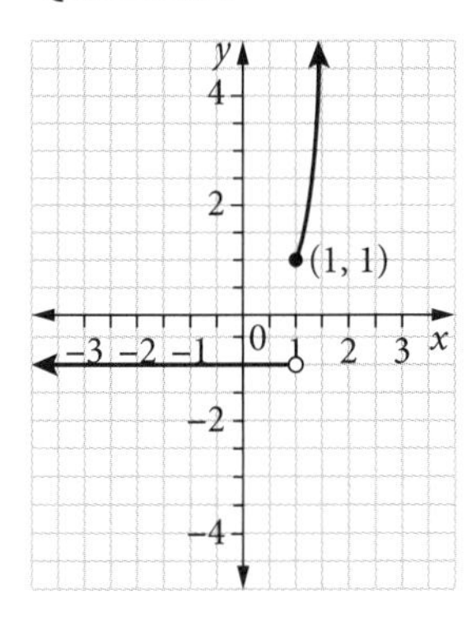


#### Question 2

Set up the equation $y = ax^2 + 4$ and substitute $(1, 0)$.

$0 = a + 4$, so $a = -4$.

Equation is $y = -4x^2 + 4$.

#### Question 3

Set up the equation $y = a\sqrt{2 - x} + 2$ and substitute $(0, -4)$.

$$-4 = a\sqrt{2} + 2$$
$$-6 = a\sqrt{2}$$
$$a = -\frac{6}{\sqrt{2}} = -3\sqrt{2}$$

$$\therefore y = -3\sqrt{2}\sqrt{2 - x} + 2$$

Equation is $y = -3\sqrt{4 - 2x} + 2$.

#### Question 4

**a** $f(x) = 1 - 5x$

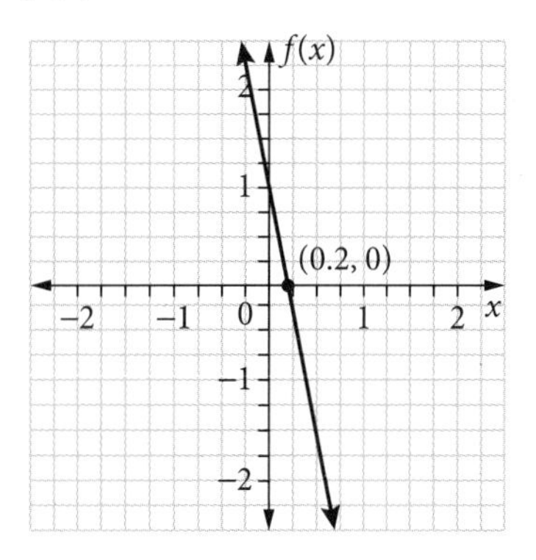


**b** inverse function $f^{-1}$

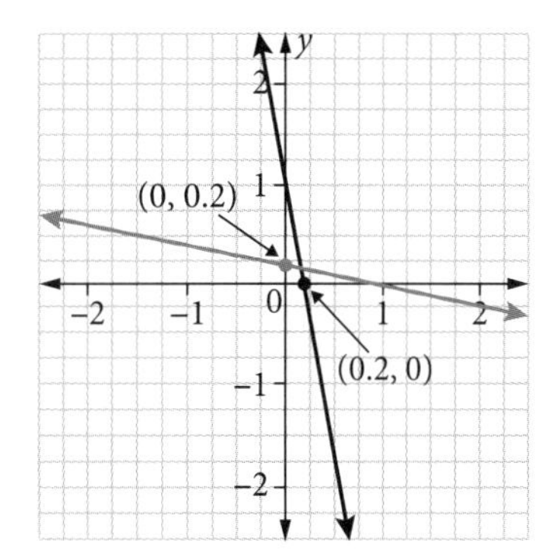


#### Question 5

$$y = 3\sin\left(4x - \frac{\pi}{3}\right)$$

amplitude = 3

$$\text{period} = \frac{2\pi}{4} = \frac{\pi}{2}$$

range = $[-3, 3]$

#### Question 6

$y = \sqrt{x}$

**a** Reflected in the $x$-axis, $y = x$ becomes

$$y = -\sqrt{x}$$

Dilated by a factor of 2 from the $y$-axis, it becomes

$$y = -\sqrt{\frac{x}{2}}$$

Translated 1 unit in the negative direction of the $y$-axis, it becomes

$$y = -\sqrt{\frac{x}{2}} - 1$$

**b** $y = \sqrt{x}$ and $y = -\sqrt{\frac{x}{2}} - 1$

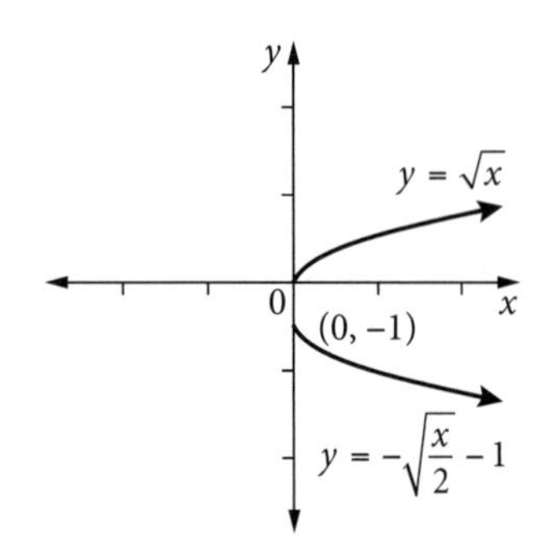


### Question 7

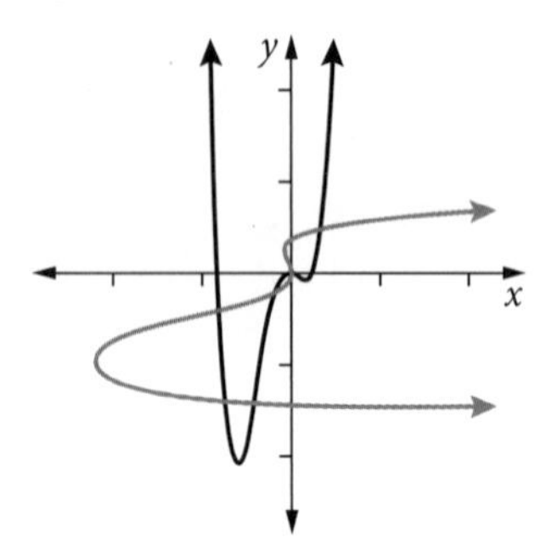


### Question 8

$y = 3e^{x+1} + 1$

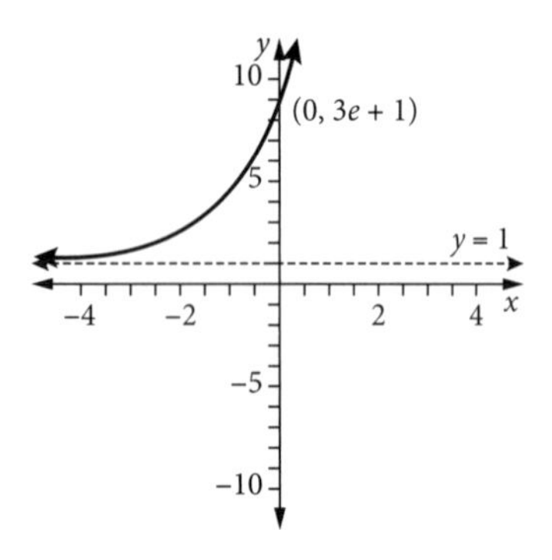


### Question 9

$y = 3\log_e(2 - x)$

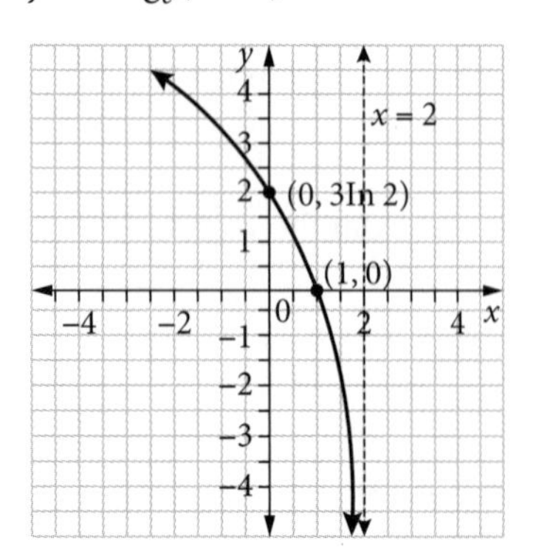


### Question 10

$y = \sin(x)$ and $y = \sin(2x)$ for $0 \le x \le 2\pi$

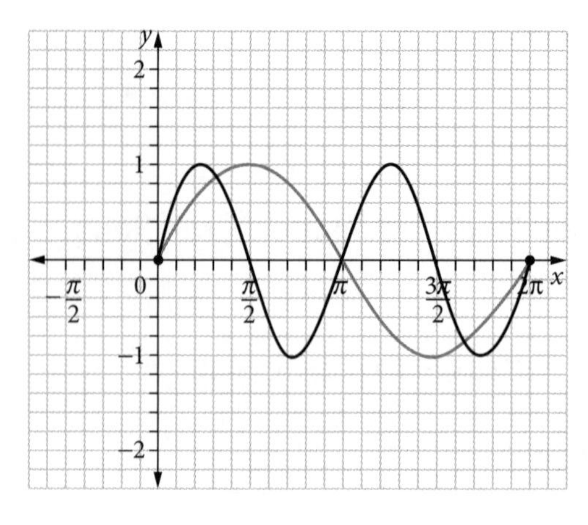


$y = \sin(x) + \sin(2x)$

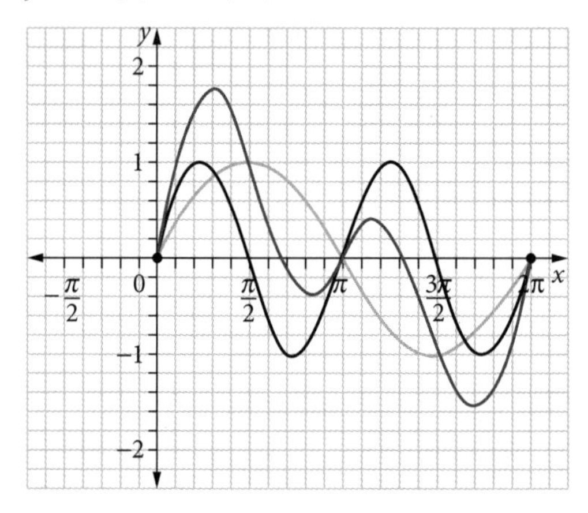


### Question 11

$f(x) = x^2$ and $g(x) = e^{x-2}$

Test for $f(g(x))$: ran (inner) $\subseteq$ dom (outer)

$(0, \infty) \subseteq R$, so $f(g(x))$ exists.

Test for $g(f(x))$: ran (inner) $\subseteq$ dom (outer)

$[0, \infty) \subseteq R$, so $g(f(x))$ exists.

### Question 12

**a** $y = 5\log_2(2 - x) + 1$

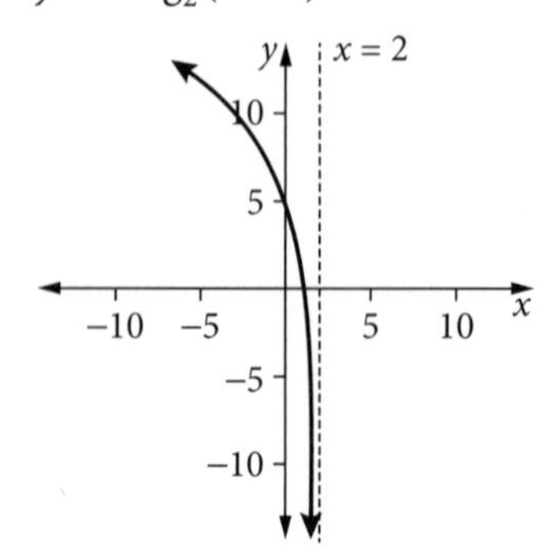


Axial intercepts $(0, 5\log_2(2) + 1)$ and $\left(-2^{-\frac{1}{5}} + 2, 0\right)$.

**b** domain: $(-\infty, 2)$, range: $R$

### Question 13

**a** $y = -2\tan(\pi x), -1 \le x \le 1$

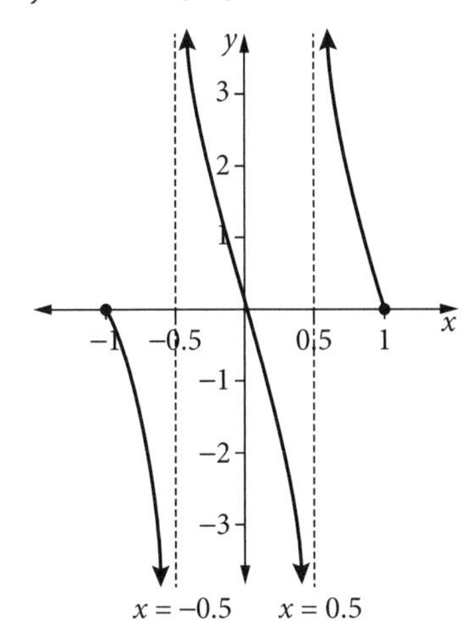


**b** domain: $[-1,1]\backslash\{-0.5, 0.5\}$, range: $R$

### Question 14

**a** $y = 3^{-x}$

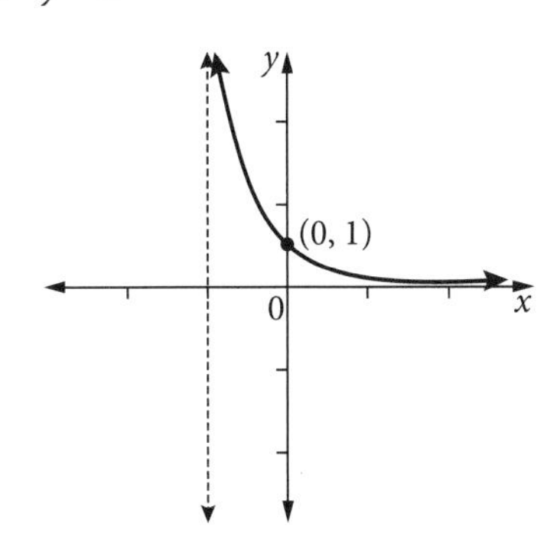


**b** $y = -\log_3(x)$

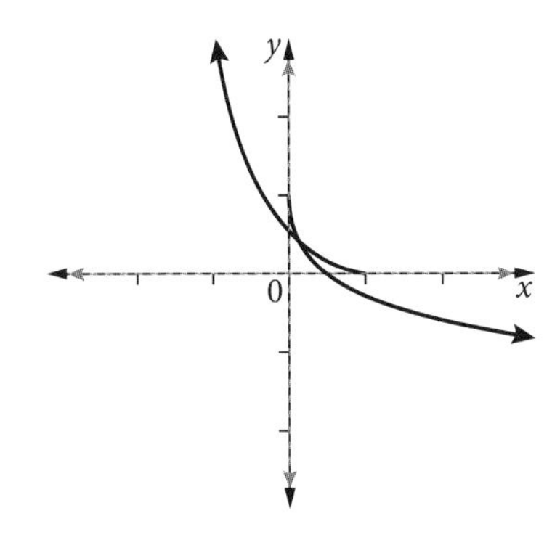


### Question 15

$y = ae^{-x} + b$

Horizontal asymptote at $y = -2$ gives $y = ae^{-x} - 2$.

The graph passes through the origin, so this gives $0 = ae^0 - 2$

So $a = 2$.

The equation of the function is $y = 2e^{-x} - 2$.

### Question 16

**a** $y = a(x - b)(x - c)(x - d)$

The $x$-intercepts at $x = 3$ and $x = 2$ give $y = a(x - 3)(x - 2)(x - d)$

Using the fact that the graph passes through the points $(0, 5)$ and $(1, 1)$ gives the two equations.

$5 = a(-3)(-2)(-d)$

and

$1 = a(-2)(-1)(1 - d)$

Solve to get $a = -\frac{1}{3}$, $d = \frac{5}{2}$

Equation of the function is

$y = -\frac{1}{3}(x - 3)(x - 2)\left(x - \frac{5}{2}\right)$

**b**

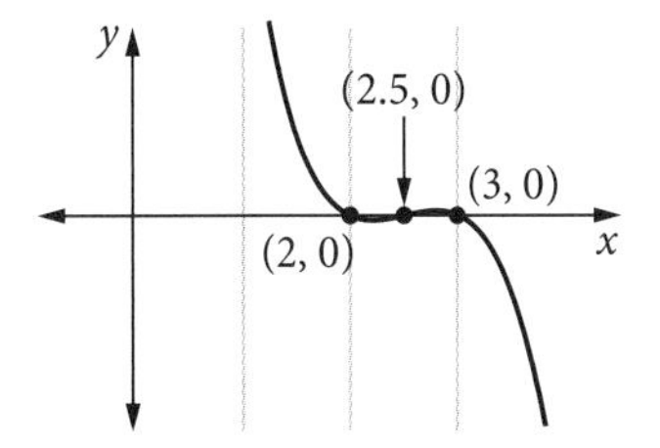


### Question 17

$u(x) = 3x - 1$ and $g(u(x)) = [u(x)]^2$

**a** $[u(x)]^2 = (3x - 1)^2 = h(x)$

**b** $h(2) = 25$

### Question 18

$f(x) = \frac{-2}{(x - 3)^2} + 1$

**a** $y = f(-x)$

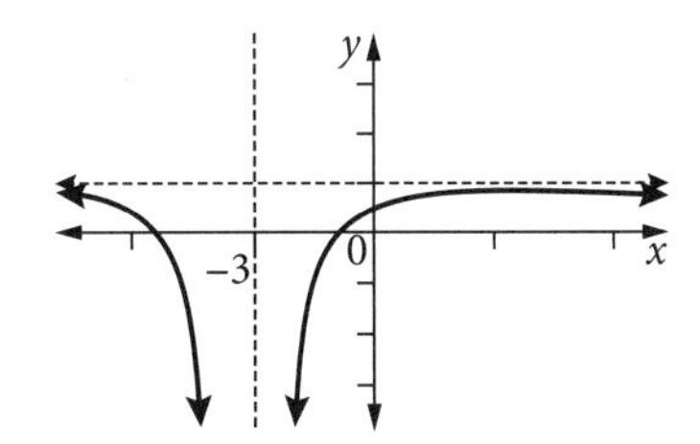


**b** $y = f(1 - x)$

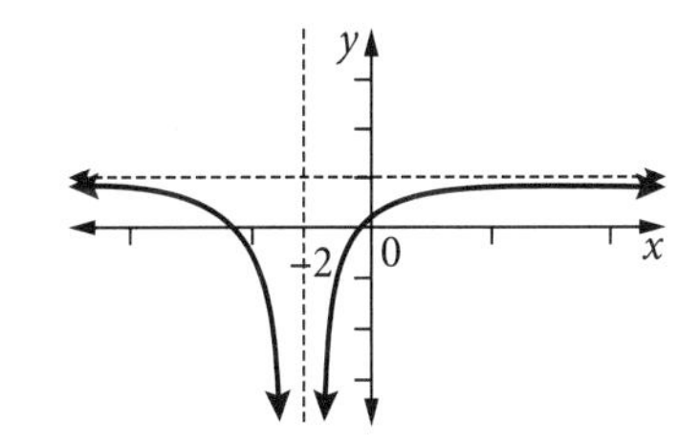

**c** $y = -f(-x)$

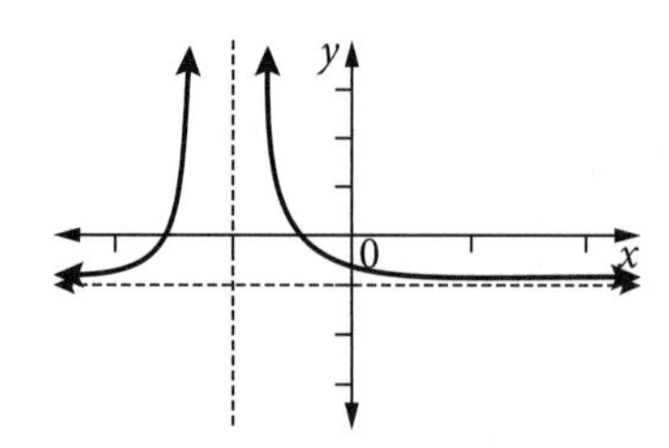


### Question 19

Midpoint of $A$ $(-6, 7)$ and $B$ $(2, -9)$ is $(-2, -1)$.

A line that is perpendicular to the line $y = -4x + 1$ has a gradient of $\frac{1}{4}$.

Equation of the line: $y - y_1 = m(x - x_1)$

$$y + 1 = \frac{1}{4}(x + 2)$$

$$4y = x - 2$$

### Question 20

**a**

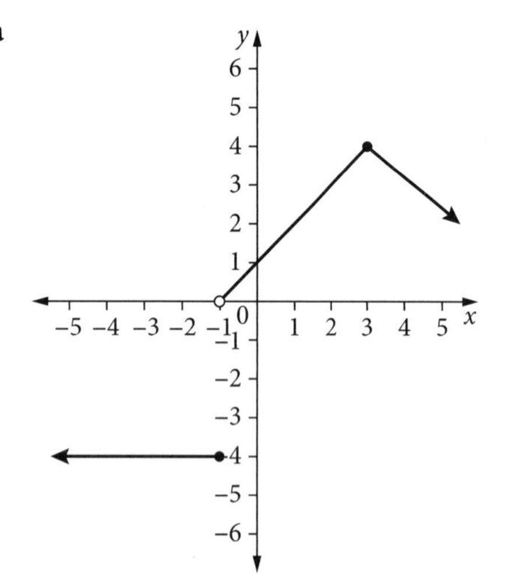


**b** $f(-2) = -4, f(0) = 1, f(6) = 2$

**c** $f(x) = 1$ where $x = 0$ and $x = 7.5$

## Multiple-choice questions

**1** A

gradient $= -\frac{3}{1} = -3$

**2** D

The quadratic graph has a stationary point at $x = 1$.

point $(1, -1)$

**3** E

$y = \pm\sqrt{x + 3}$ is **not** a function. It is one-to-many.

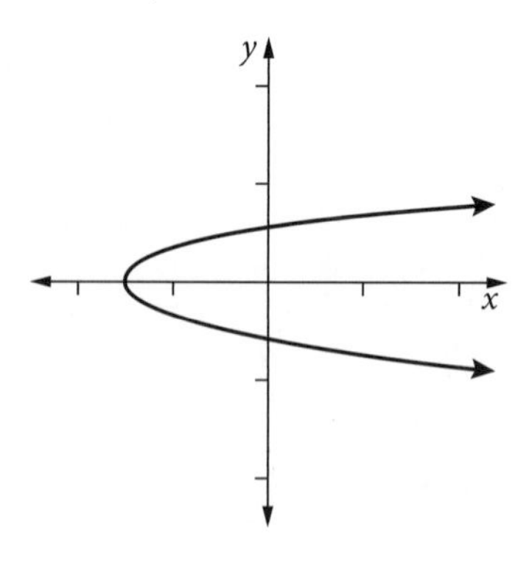


**4** C

The graph has $x$-intercepts at $x = -1$, and a repeated factor at $x = 1$.

$y$-intercept at $y = 4$.

Graph is $y = 4(x + 1)(x - 1)^2$.

**5** C

For the inverse of $\{(-1, 4), (0, 3), (1, 1), (2, -1)\}$, swap the $x$ and $y$ values to get $\{(4, -1), (3, 0), (1, 1), (-1, 2)\}$.

**6** D

$y = 0.5x - 3$, so swap $x$ and $y$ and rearrange to get $f^{-1}(x) = 2x + 6$.

**7** E

$f: (0, 3) \to R$, where $f(x) = 0.5x - 3$, so swap $x$ and $y$ and the domain and range to get $f^{-1}(x) = 2x + 6, x \in (-3, -1.5)$.

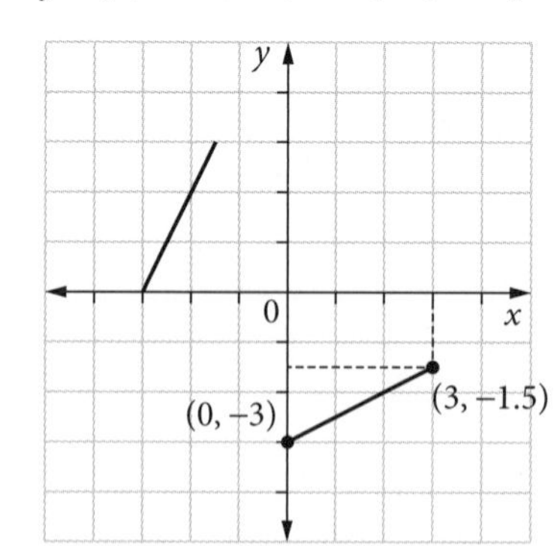


**8** A

Range of lower semicircle $y = -\sqrt{9 - x^2}$ with radius = 3 is $[-3, 0]$.

**9** D

Asymptotes for $y = \frac{1}{(2x - 5)^2} + 2$ are $x = \frac{5}{2}$ and $y = 2$.

**10** D

For $f(x) = \frac{1}{x-1} - 1$, the maximal domain is $R \setminus \{1\}$.

**11** B

$y = -e^{\frac{x}{2}} + 1$ cuts the $x$-axis at $x = 0$.

**12** C

$y = 3e^{-x} + 2$ and $y = 5$ intersect at $(0, 5)$.

**13** E

If $f(x) = x + 2$ and $g(x) = x^2$, then $g(f(x)) = (x+2)^2 = x^2 + 4x + 4$.

**14** C

If $f(x) = 2x + 1$ and $g(x) = 3\ln(x)$, for $f(g(x))$ consider if range $g \subseteq$ dom $f$.

So $R \subseteq R$.

**15** B

$y = 2\sin(\pi x)$, period $= \frac{2\pi}{n} = \frac{2\pi}{\pi} = 2$

**16** A

$y = -3\tan(\pi x)$, period $= \frac{\pi}{n} = \frac{\pi}{\pi} = 1$

**17** E

The parabola has a repeated factor at $x = 1$ and a $y$-intercept of $y = 4$.

It could be modelled by $y = 2a(x-1)^2$.

**18** D

$y = \sqrt{x}$ to $y = 2 + 3\sqrt{x-1}$
The transformations required are

- dilation by a factor of 3 from the $x$-axis
- translation 1 unit to the right
- translation 2 units up.

**19** B

The domain is defined for the intersection of the domains of $g(x)$ and $h(x)$.

Intersection of $x \in [-1, 3]$ and $x \in [0, 4]$ is $x \in [0, 3]$.

**20** A

A piecewise function is given by

$$f(x) = \begin{cases} x + 2 & x \le 0 \\ \frac{x^2}{4} & 0 < x \le 1 \\ -3 & x > 1 \end{cases}$$

$f(x) = -3, x > 1$ only, so $f(3) = -3$.

**21** D

Graph of the inverse function looks like D.

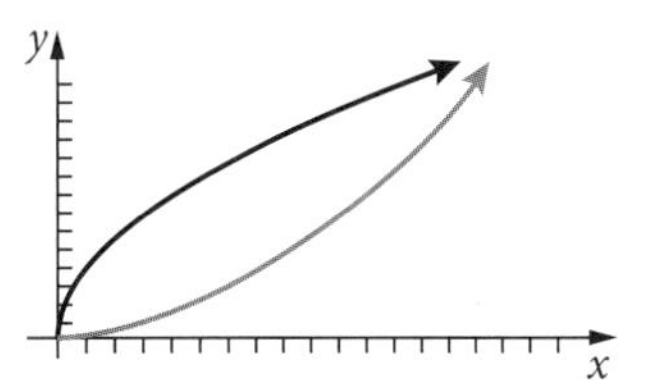


**22** E

If $f(x) = -\sqrt{2(x+2)} - 1$, the range of $f^{-1}$ is the domain of $f = [-2, \infty)$.

**23** A

$y = 2(2^{x-1}) - 8 = 2^x - 8$ has *axial* intercepts at $x = 3$ and $y = -7$.

**24** E

$f(x) = x^2 - 2x$ and $g(x) = \frac{1}{x^2}$

Test: ran (inner) $\subseteq$ dom (outer)

$(0, \infty) \subseteq R$, so option E is correct.

A, C and D are incorrect.

B is incorrect as dom $f(g(x)) = (0, \infty)$ $\neq$ dom $f = R$

**25** D

$2y - 6 = x$ passes through $\left(\frac{1}{2}, b\right)$.

This gives $2b - 6 = \frac{1}{2}$.

So $b = \frac{13}{4}$.

**26** B

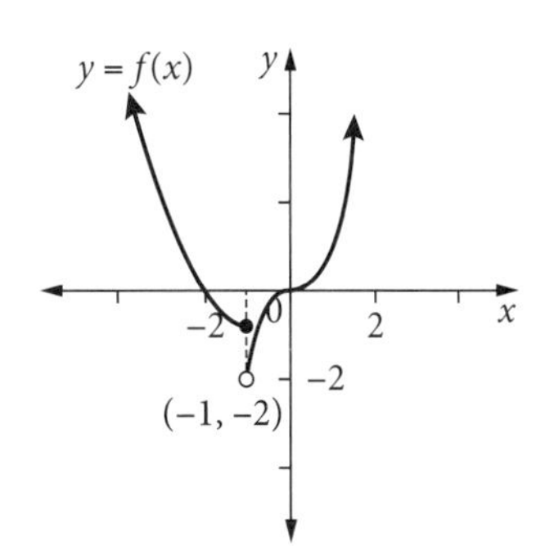


Minimum at $x = -1, y = -2$, range is $(-2, \infty)$.

**27** D

$y = x^{\frac{2}{3}}$, dilate by a factor of 3 from the $y$-axis.

Image is $y = \left(\frac{x}{3}\right)^{\frac{2}{3}}$.

**28** E

$y = -3\sin(2x)$

period $= \frac{2\pi}{n} = \frac{2\pi}{2} = \pi$, amplitude $= 3$

**29** A

$y = \frac{-2}{\sqrt{x}}$ reflected in the $x$-axis is $y = \frac{2}{\sqrt{x}}$.

**30** B

From the graph, the range of $y = \sin(x) - \cos(2x)$ is $[-1.125, 2]$. This can be confirmed by sketching the graph on CAS.

**31** B

$y = 3\tan(3x)$ for $\theta \in [0, \pi]$.

Expect asymptotes at $x = -\frac{\pi}{2}, \frac{\pi}{2}$ for $y = \tan(x)$.

The $(3x)$ means that asymptotes will be multiples of $x = -\frac{\pi}{6}, \frac{\pi}{6}$, with the period $= \frac{\pi}{3}$.

Asymptotes at $x = \frac{\pi}{6}, \frac{\pi}{2}, \frac{5\pi}{6}$.

**32** D

point $(-2, 3)$

Reflection in the $y$-axis gives $(2, 3)$.

Then dilation by a factor of 5 from the $x$-axis gives $(2, 15)$.

**33** A

Graph is a –ve quartic with $x$-intercepts at $x = 0$ (triple factor) and $x = 2$.

equation: $y = -ax^3(x - 2)$

**34** B

$y = a\sin(n(x - b))$

Period $= \frac{2\pi}{n} = \frac{2\pi}{3}$, so $n = 3$ and amp $= 2$.

It is reflected over the $x$-axis and translated by $\pi$ in the +ve $x$-axis.

The equation is $y = -2\sin(3(x - \pi))$.

**35** E

$f(x) = 3\log_e(x - 2)$, so it is not true that $f(x)$ and $f(3x)$ have the same vertical asymptote at $x = 2$.

**36** D

It is a tan graph with period $= \frac{\pi}{n} = \frac{\pi}{3}$, so $n = 3$.

There is a translation of 2 in the positive $y$ direction as it crosses the $x$-axis at $(0, 2)$.

$y = \tan(3x) + 2$

**37** A

$f(x) = 10x - 2$, so

$g(x) = f(f(x)) = 10(10x - 2) - 2 = 100x - 22$

Swap $x$ and $y$ and rearrange to get the inverse

$g^{-1}(x) = \frac{x}{100} + \frac{11}{50}$.

**38** C

The graph looks like a truncus, reflected over the $x$-axis and translated $m$ units vertically.

$y = m - \frac{n}{x^2}$

**39** A

$y = \cos(x)$ is transformed to $y = -\cos\left(4x - \frac{\pi}{2}\right)$ by

- a reflection in the $x$-axis
- dilation by a factor of $\frac{1}{4}$ from the $y$-axis
- a horizontal translation of $\frac{\pi}{8}$ units to the right.

**40** D

The function $f(x) = 3x^2$ is dilated by a factor of 2 from the $x$-axis and reflected in the $y$-axis.

$y = 6x^2 - 1$

**41** E

For $y = \log_e(x)$ to be transformed to $y = \log_e(5 - 2x)$, the transformations required are

- reflection in the $y$-axis
- dilation by a factor of $\frac{1}{2}$ from the $y$-axis
- translation of $\frac{5}{2}$ units to the right.

**42** A

The transformation which maps the curve with equation $y = \sin(x)$ to the curve with equation $y = 2\sin(x-3)+1$ follows the mapping notation $(x, y) \rightarrow (x+3, 2y+1)$.

**43** C

The range of $y = 11 - 4\cos(\pi(3-x))$ is found by using the amplitude of 4 (i.e. $[-4, 4]$), then translating vertically by 11 (i.e. $[-4+11, 4+11]$).

This gives $[7, 15]$.

**44** D

$f(x) = A\sin(nx) + B$

The graph has amplitude = 4.

Period $= 1 = \frac{2\pi}{n}$, so $n = 2\pi$.

There is a translation of 6 in the positive $y$ direction.

$A = 2$, $n = 2\pi$ and $B = 6$.

**45** A

For the transformations in the order

- dilate by 2 units from the $y$-axis
- reflect over the $y$-axis
- translate by 1 unit in the negative direction of the $x$-axis,

the image of the curve with equation $y = x^4$ is $y = \frac{(x+1)^4}{16}$.

**46** B

$y = ax^n e^{-kx} + b$, given $a = 2$ and $n = \frac{1}{2}$.

Solve $y = 2x^{\frac{1}{2}}e^{-kx} + b$ with points $(0, 1)$ and $(2, 2)$ to find $k$ and $b$.

$k = \frac{3}{4}\log_e(2)$, $b = 1$

**47** E

$f^{-1}(x) = \sqrt{2x-1}$, so $f(x) = \frac{1}{2}(x^2+1)$.

Equate to solve or test the solutions.

The intersection will be on the line $y = x$ and is $(1, 1)$.

**48** D

The graph is exponential. It is reflected over the $x$-axis, with a horizontal asymptote at $y = 2$, meaning a vertical translation of 2.

Axial intercept at $(0, -1)$.

Equation of the graph: $y = -3e^x + 2$

**49** A

Set up the equation $y = ax^2 + bx + c$ in CAS and enter the points $(0, 1)$, $(2, 3)$, $(5, 7)$.

The equation is $y = \frac{x^2 + 13x + 15}{15}$.

**50** C

The equation could be of the form $y = x^n$, where $x \in Q$ as the domain is $[0, \infty)$.

Use the approximate point $(2, 3)$ to check the options given.

Only $y = x^{\frac{3}{2}}$ approximately satisfies the point values.

## Extended-answer questions

### Question 1

**a** Let $f(x) = ax^3 + bx^2 + 2x + 10$,

and $f(2) = 0$,

and $f(-1) = -3$.

```
define f(x)=ax^3+bx^2+2x+10
                                done
{f(2)=0
{f(-1)=-3 | a,b
                      {a=5/2, b=-17/2}
```

Solve simultaneously to get $a = \frac{5}{2}$ and $b = -\frac{17}{2}$.

(For a ‘show that’ question, it would be necessary to solve these equations simultaneously by hand.)

**b** $y = P(x) = \frac{5}{2}x^3 - \frac{17}{2}x^2 + 2x + 10$

We know that $f(2) = 0$, so $x - 2$ is a factor.

By division, the quadratic factor is $\frac{5}{2}x^2 - \frac{7}{2}x - 5$.

$$\text{propFrac}\left(\frac{\frac{5}{2}x^3-\frac{17}{2}x^2+2x+10}{x-2}\right)$$
$$\frac{5\cdot x^2}{2}-\frac{7\cdot x}{2}-5$$

$$y = (x-2)\left(\frac{5}{2}x^2 - \frac{7}{2}x - 5\right)$$
$$= \frac{1}{2}(x-2)(5x^2 - 7x - 10)$$

$$\text{rFactor}\left(\frac{5\cdot x^2}{2}-\frac{7\cdot x}{2}-5\right)$$
$$\frac{5\cdot\left(x+\frac{\sqrt{249}}{10}-\frac{7}{10}\right)\cdot\left(x-\frac{\sqrt{249}}{10}-\frac{7}{10}\right)}{2}$$

$$y = \frac{5}{2}(x-2)\left(x - \frac{7 \pm \sqrt{249}}{10}\right)$$

$y = 0$ at $x = 2$, $x = \frac{7 \pm \sqrt{749}}{10}$

Or **solve** can be used.

$$\text{solve}\left(\frac{5}{2}\cdot x^3-\frac{17}{2}x^2+2\cdot x+10=0, x\right)$$
$$\left\{x=2, x=\frac{-\sqrt{249}}{10}+\frac{7}{10}, x=\frac{\sqrt{249}}{10}+\frac{7}{10}\right\}$$

**c**

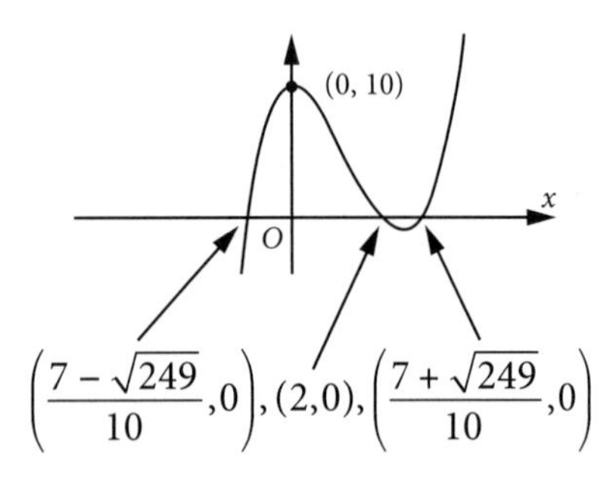


**d**

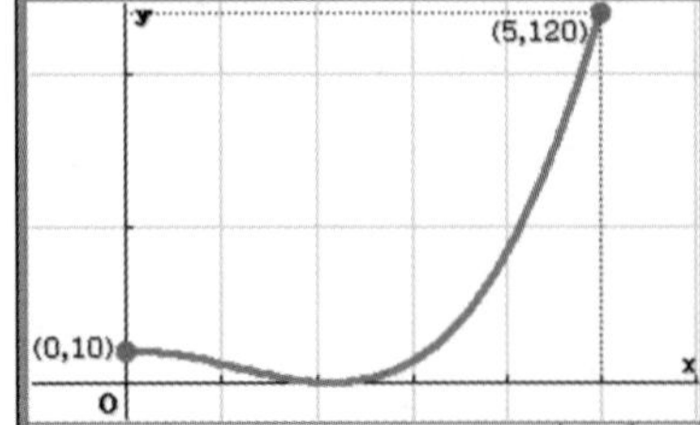


Max. at the end = 120 m

**e** The ski slope ends at the point (0, 10).

Height at the end = 10 m

**f**

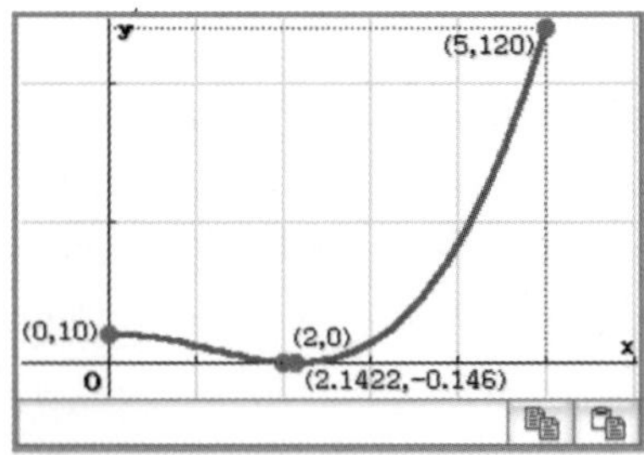


The minimum point (2.14, −0.15) goes underground to a depth of approximately 0.15 m.

The ski slope will not go ahead.

### Question 2

**a**

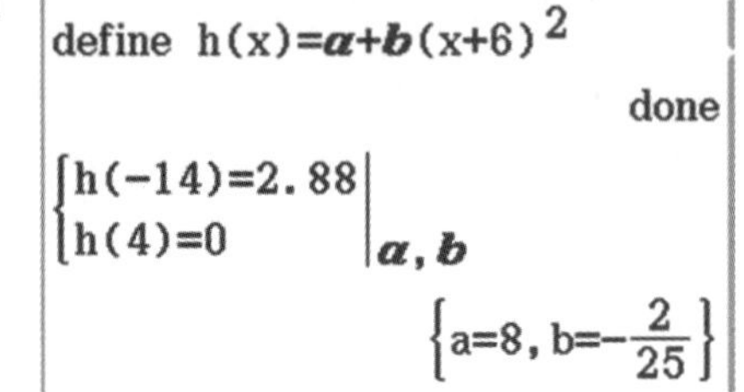


Define the function $h(x) = a + b(x+6)^2$ and use $h(4) = 0$ and $h(-14) = 2.88$.

Solve simultaneously to get $a = 8$ and $b = -\frac{2}{25}$.

**b**

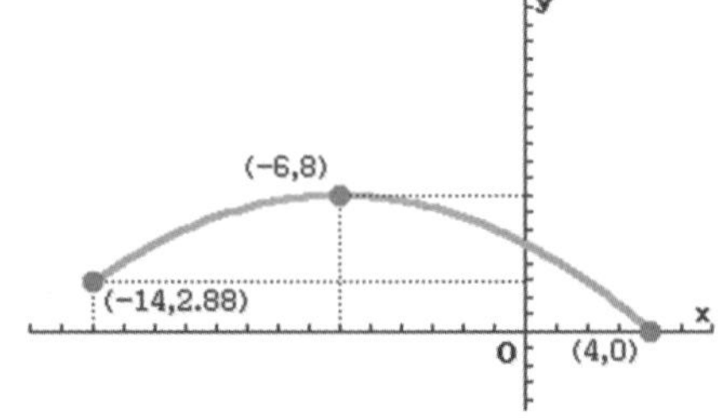


**c** $y$-intercept (0, 5.12)

$5.12 > 5$

Yes, the water will go over the fence.

**d** Maximum point is (−6, 8).

$8 < 10$

Water will not hit the power lines.

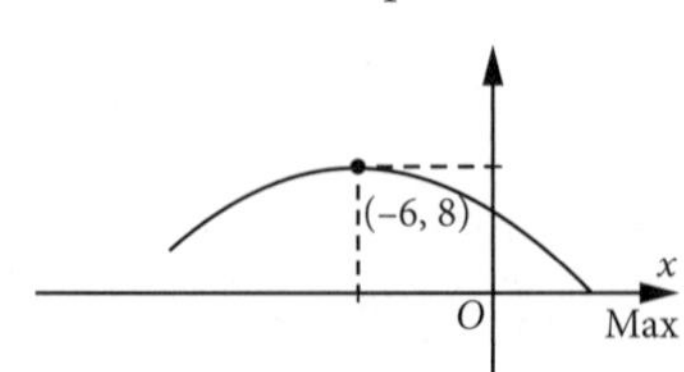

**e** Solve $h(x) = 2$.

This gives $x = -6 + 5\sqrt{3}$.

The only solution in the domain $x \in [-14, 4]$ is $x = -6 + 5\sqrt{3} \approx 2.66$.

Scruffy will reach the water after the maximum point and in the neighbour's garden, approximately 2.66 metres beyond the fence and until it reaches the ground.

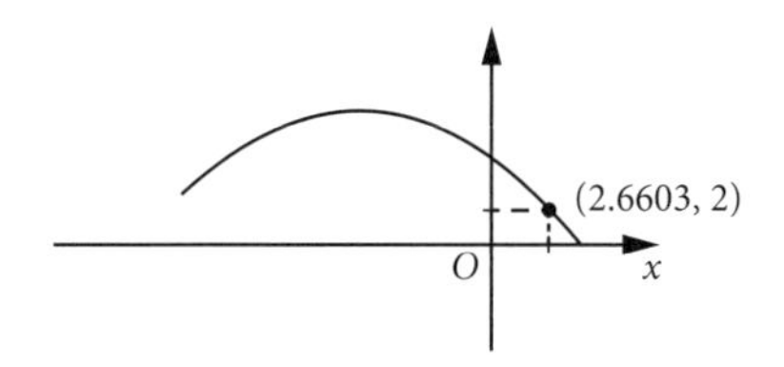


### Question 3

**a** $3x^3 + ax^2 + 5x = 3(x + b)^3 + c$

$= 3(x^3 + 3x^2b + 3xb^2 + b^3) + c$

So $3x^3 + ax^2 + 5x = 3x^3 + 9x^2b + 9xb^2 + 3b^3 + c$.

Equating coefficients:

$a = 9b$ (coefficients of $x^2$)

$5 = 9b^2$ (coefficients of $x^1$)

$0 = 3b^3 + c$ (coefficient of $x^0$)

```
expand(3(x+b)^3+c)
        3·x^3+3·b^3+9·b·x^2+9·b^2·x+c
{a=9b
{5=9b^2
{0=c+3b^3 | a, b, c
{{a=-3·√5, b=-√5/3, c=5·√5/9}
, {a=3·√5, b=√5/3, c=-5·√5/9}}
```

Hence $a = \pm 3\sqrt{5}$, $b = \pm\dfrac{\sqrt{5}}{3}$, $c = \mp\dfrac{5\sqrt{5}}{9}$

$$\therefore\ 3x^3 + ax^2 + 5x = 3\left(x - \frac{\sqrt{5}}{3}\right)^3 + \frac{5\sqrt{5}}{9}$$

or

$$3x^3 + ax^2 + 5x = 3\left(x + \frac{\sqrt{5}}{3}\right)^3 - \frac{5\sqrt{5}}{9}$$

**b** $b < 0$ and $c > 0$

$$f(x) = 3\left(x - \frac{\sqrt{5}}{3}\right)^3 + \frac{5\sqrt{5}}{9}$$

$f(x) = 0$ gives $\left(x - \dfrac{\sqrt{5}}{3}\right)^3 = \dfrac{5\sqrt{5}}{27}$.

$$x = -\frac{\sqrt{5}}{3} + \frac{\sqrt{5}}{3}$$

$x = 0$

**c** $b > 0$ and $c < 0$

$$f(x) = 3\left(x + \frac{\sqrt{5}}{3}\right)^3 - \frac{5\sqrt{5}}{9}$$

$f(x) = 0$ gives $\left(x + \dfrac{\sqrt{5}}{3}\right)^3 = \dfrac{5\sqrt{5}}{27}$.

$$x = \frac{\sqrt{5}}{3} - \frac{\sqrt{5}}{3}$$

$x = 0$

**d** Point of intersection at $(0, 0)$.

**e**

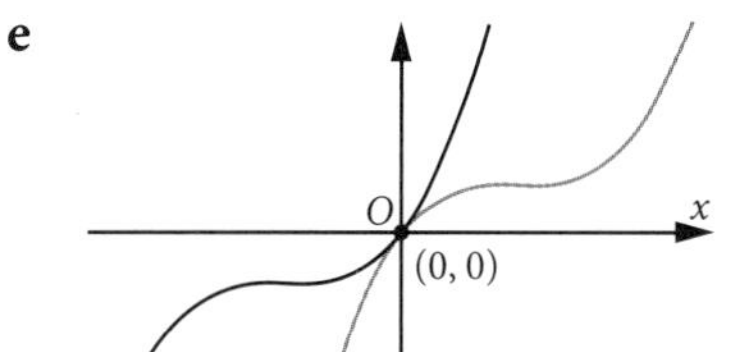


**f**

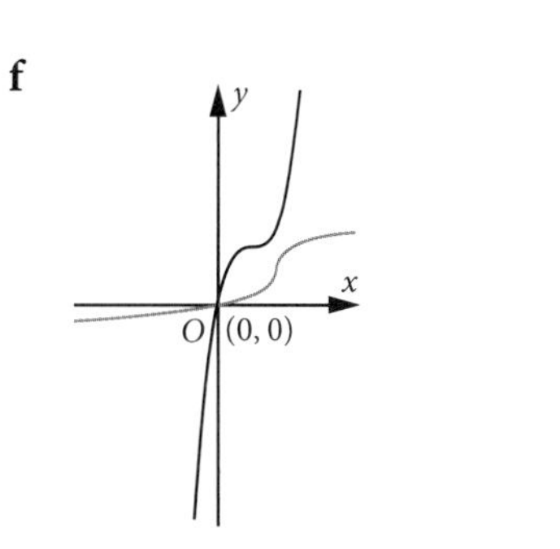

### Question 4

**a** $T_0 = 90°\text{C}$

**b** $T = T_0 \times 2^{-kt}$

It takes 30 minutes for the temperature to halve, so $t = \frac{1}{2}$ and $T = \frac{1}{2}T_0$.

$\Rightarrow \frac{1}{2}T_0 = T_0 \times 2^{-0.5k}$

$\Rightarrow \frac{1}{2} = 2^{-0.5k}$

$2^{-1} = 2^{-0.5k}$

This gives $\frac{1}{2}k = 1$.

$k = 2$

After 60 minutes, i.e. at $t = 1$:

$T = T_0 \times 2^{-2 \times 1}$

$T = T_0 \times \frac{1}{4}$

The fraction of the original temperature after 60 minutes $= \frac{1}{4}$.

**c** Using the results from part **b**,

$T = 90 \times \frac{1}{4} = 22.5°\text{C}$

**d** $T = T_0 \times 2^{-kt} + 20$

$t = 0$ gives $T = T_0 + 20$.

Original temperature of the cup of tea is 90°C.

$90 = T_0 + 20$, so $T_0 = 70$

It takes 20 minutes for the temperature to halve, so $t = \frac{1}{3}$.

$\Rightarrow \frac{1}{2} \times 90 = 70 \times 2^{-k\frac{1}{3}} + 20$

$45 = 70 \times 2^{-k\frac{1}{3}} + 20$

This gives $k \approx 4.46$.

$\text{solve}\left(45=70\cdot 2^{-\frac{k}{3}}+20, k\right)$

$\{k=4.456280482\}$

**e**

| $T = T_0 \times 2^{-kt}$ | |
|---|---|
| Step 1: translated 15 units in the negative direction of the $t$-axis | $T_1 = T_0 \times 2^{-k(t+15)}$ |
| Step 2: translated 7 units in the negative direction of the $T$-axis | $T_2 = T_0 \times 2^{-k(t+15)} - 7$ |
| Step 3: dilated by a factor of 5 from the $t$-axis | $T_3 = 5T_0 \times 2^{-k(t+15)} - 35$ |
| Step 4: dilated by a factor of $\frac{1}{2}$ from the $T$-axis | $T_4 = 5T_0 \times 2^{-k(2t+15)} - 35$ |
| Step 5: reflected in the $T$-axis | $T_5 = 5T_0 \times 2^{k(2t-15)} - 35$ |
| New rule for $T(t)$ | $T(t) = 5T_0 \times 2^{k(2t-15)} - 35$ |

### Question 5

**a** $f$ is a one-to-one function where dom $f = R$ and ran $f = (c, \infty)$.

**b** $a = \frac{2e^4}{e^2 - 1}$, $c = \frac{-2}{e^2 - 1}$

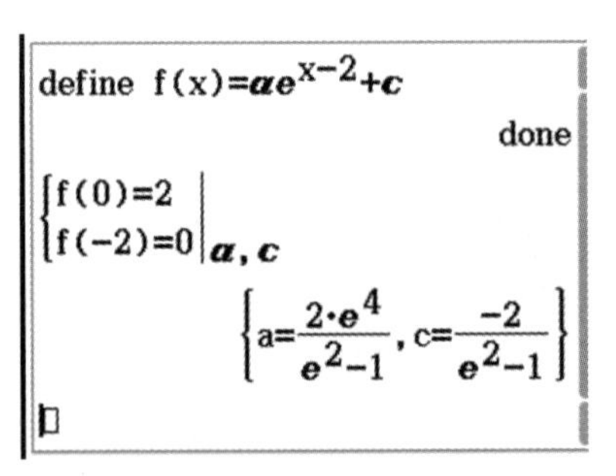


**c** $R \to R, f(x) = \frac{2e^4}{e^2 - 1}e^{x-2} - \frac{2}{e^2 - 1}$

**d**

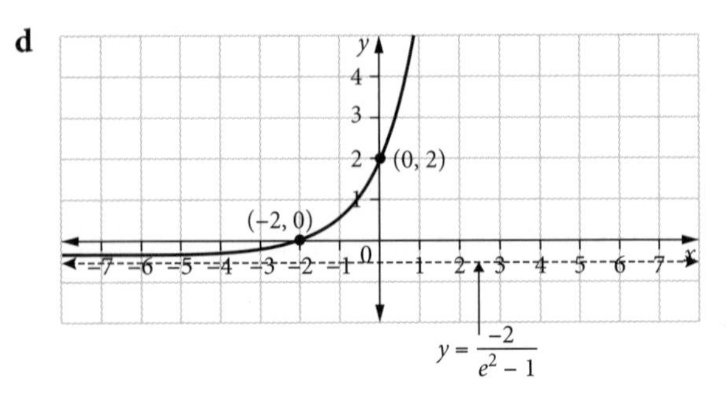

**e** $f^{-1}(x) = \ln\left(\frac{xe^2}{2} - \frac{x}{2} + 1\right) - 2$

domain $f^{-1}(x)$ = range $f = (c, \infty)$

$= \left(\frac{-2}{e^2 - 1}, \infty\right).$

**f** Solving $f^{-1}(x) = f(x)$ gives no points of intersection.

**g**

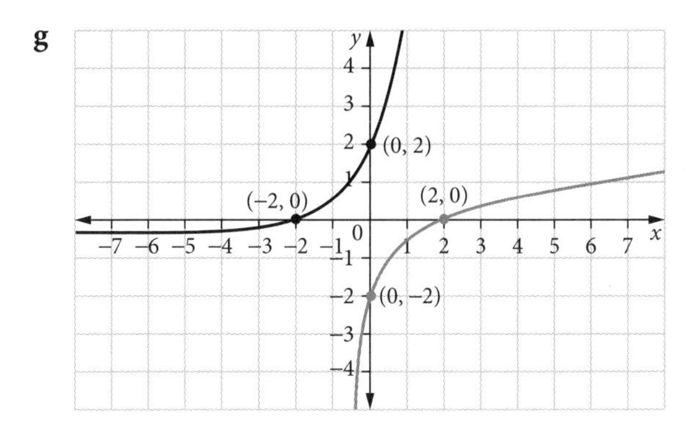


# Chapter 2

## Area of Study 2: Algebra, number and structure

## Short-answer questions

**Question 1**

$\sqrt{9x - 4} = 2$

$9x - 4 = 4$

$9x = 8$

$x = \frac{8}{9}$

**Question 2**

**a** 0

**b** Set up an equation $y = ax^2 + 4$ and substitute $(1,0)$.

$0 = a + 4$, so $a = -4$.

The equation is $y = -4x^2 + 4$.

Swap $x$ and $y$ and rearrange to find the inverse.

$x = -4y^2 + 4$

$4y^2 = 4 - x$

$y = \pm\sqrt{\frac{4 - x}{4}} = \pm\frac{\sqrt{4 - x}}{2}$

$\therefore y^{-1} = \frac{\sqrt{4 - x}}{2}$

**Question 3**

$P(x) = 55 - 18x + 2x^2 - 3x^3 + x^4$

$P(-2) = 55 + 36 + 8 + 24 + 16 = 139$

**Question 4**

$\log_e(y) = \log_e(x) + \log_e(p)$

$\Leftrightarrow \log_e(y) = \log_e(xp)$

Equating terms,

$y = xp$

**Question 5**

$\frac{\log_3(27)}{\log_3(9)} = \frac{\log_3(3^3)}{\log_3(3^2)} = \frac{3\log_3(3)}{2\log_3(3)} = \frac{3}{2}$

**Question 6**

$\sin(x) = \frac{1}{\sqrt{2}}$ for $0° \le x° \le 180°$

$x = 45°, 180° - 45°$

$x = 45°, 135°$

**Question 7**

$y = a\cos(kt) + c$

When $t = 0$, the height is 20 metres.

When $t = 10$, the height is 60 metres.

When $t = 20$, the height is 20 metres for the 1st time since $t = 0$.

**a** period $= \frac{2\pi}{k} = 20$

So $k = \frac{\pi}{10}$.

**b** $a = -20$

**c** $c = 40$

**Question 8**

$\log_e(x^2) = 2 \Leftrightarrow e^2 = x^2$

$\therefore x = e$

**Question 9**

$f(x) = 3 + \log_e(x + 2)$

Swap $x$ and $y$ and rearrange to find the inverse.

$x = 3 + \log_e(y + 2) \Leftrightarrow y + 2 = e^{x-3}$

$\therefore f^{-1}(x) = e^{x-3} - 2$

### Question 10

If $g: [D, \infty) \to R, g(x) = (x-2)^2 + 3$,

**a** $D \geq 2$ for $g^{-1}$ to exist.

**b** Swap $x$ and $y$ and rearrange to find the inverse.

$x = (y-2)^2 + 3 \Leftrightarrow (y-2)^2 = x - 3$

$\therefore y - 2 = \pm\sqrt{x-3}$

Select the positive branch to suit restricted domain.

$g^{-1}(x) = \sqrt{x-3} + 2$

domain $g^{-1}(x) =$ range $g(x) = [3, \infty)$

$g^{-1}: [3, \infty) \to R, g^{-1}(x) = \sqrt{x-3} + 2$

### Question 11

$f(x) = x^2$ and $g(x) = \log_e(x-2)$

Test for $f(g(x))$: ran (inner) $\subseteq$ dom (outer)

$R \subseteq R$

So $f(g(x))$ exists.

Test for $g(f(x))$: ran (inner) $\subseteq$ dom (outer)

$[0, \infty) \subseteq (2, \infty)$

So $g(f(x))$ does not exist.

### Question 12

Let $P(x) = x^3 - x^2 - x + 1$.

$P(1) = 1^3 - 1^2 - 1 + 1 = 0$

So $x - 1$ is a factor.

Dividing gives the quadratic factor as $(x^2 - 1) = (x-1)(x+1)$.

Solution to equation $x^3 - x^2 - x + 1 = 0$: $x = 1, x = -1$

### Question 13

$9^x - 2(3^{x+1}) + 9 = 0$

$\Leftrightarrow 3^{2x} - 2(3^x \times 3^1) + 9 = 0$

$\Leftrightarrow 3^{2x} - 6(3^x) + 9 = 0$

Letting $A = 3^x$ gives $A^2 - 6A + 9 = 0$.

$(A-3)^2 = 0$ so $A = 3$.

$3^x = 3$ gives $x = 1$.

### Question 14

If $f(x) = x^3 + 1$ and $g(x) = \sqrt{x-1}$, define $g(f(x))$ and $f(g(x))$ if possible.

For $g(f(x))$, test ran $(f) \subseteq$ dom $(g)$.

$R \not\subseteq [1, \infty)$, so $g(f(x))$ does not exist.

For $f(g(x))$, test ran $(g) \subseteq$ dom $(f)$.

$[0, \infty) \subseteq R$ so $f(g(x))$ exists.

rule: $f(g(x)) = (x-1)^{\frac{3}{2}} + 1$

dom $[f(g(x))] =$ dom $[g(x)] = [1, \infty)$

### Question 15

$2\sin^2(x) - \sin(x) - 1 = 0$

Let $A = \sin(x)$.

This gives

$2A^2 - A - 1 = 0$

$(2A+1)(A-1) = 0$

$\therefore A = -\frac{1}{2}, A = 1$

So $\sin(x) = -\frac{1}{2}$, $\sin(x) = 1$ for $0 \leq x \leq 2\pi$.

So $x = \frac{7\pi}{6}, \frac{11\pi}{6}$, and $x = \frac{\pi}{2}$.

### Question 16

$T = a\sin(nt) + b$

Maximum of 100°F at $t = 1.5$.

Minimum of 40°F at $t = 4.5$.

Amplitude $= a = 30$

Period $= \frac{2\pi}{n} = 6$ for 2 times 3 months, $n = \frac{\pi}{3}$

Vertical translation will be 70.

Thus $T = 30\sin\left(\frac{\pi}{3}t\right) + 70$.

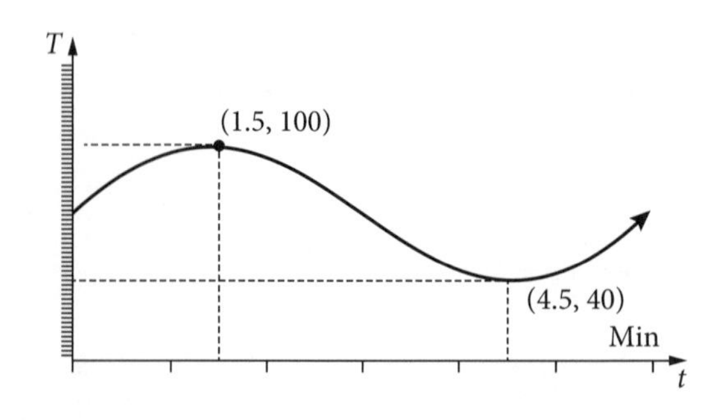

### Question 17

**a** $\sin\left(2\left(x - \frac{\pi}{2}\right)\right) = \frac{1}{\sqrt{2}}$

A CAS screen will help you to see the patterns.

```
solve(sin(2·(x-π/2))=1/√2, x)
        {x=π·constn(1)+5·π/8, x=π·constn(2)+7·π/8}
```

Or, by hand,

$\sin\left(2\left(x - \frac{\pi}{2}\right)\right) = \frac{1}{\sqrt{2}}$

gives

$\left(2\left(x - \frac{\pi}{2}\right)\right) = \frac{\pi}{4}, \pi - \frac{\pi}{4}, 2\pi + \frac{\pi}{4}, 3\pi - \frac{\pi}{4}\ldots$

$\left(2\left(x - \frac{\pi}{2}\right)\right) = \frac{\pi}{4}, \frac{3\pi}{4}, \frac{9\pi}{4}, \frac{11\pi}{4}\ldots$

$\therefore \left(x - \frac{\pi}{2}\right) = \frac{\pi}{8}, \frac{3\pi}{8}, \frac{9\pi}{8}, \frac{11\pi}{8}\ldots$

$\therefore x = \frac{5\pi}{8}, \frac{7\pi}{8}, \frac{13\pi}{8}, \frac{15\pi}{8}\ldots$

So the pattern is $\frac{5\pi}{8} + \pi n$ or $\frac{7\pi}{8} + \pi n$, where $n \in Z$.

$x = \frac{5\pi}{8} + \pi n, x = \frac{7\pi}{8} + \pi n, n \in Z.$

**b** Exact solutions for $\left(-\frac{\pi}{2}, \frac{\pi}{2}\right)$.

$x = -\frac{3\pi}{8}, x = -\frac{\pi}{8}$

| | $n = -1$ | $n = 0$ | $n = 1$ |
|---|---|---|---|
| $x = \frac{5\pi}{8} + \pi n$ | $x = -\frac{3\pi}{8}$ | $x = \frac{5\pi}{8}$ out of domain | out of domain |
| $x = \frac{7\pi}{8} + \pi n$ | $x = -\frac{\pi}{8}$ | $x = \frac{7\pi}{8}$ out of domain | out of domain |

### Question 18

The system of equations

$(k - 1)x - 2y = 6$
$-x + ky = 3$

$y = \frac{(k-1)x - 6}{2}$

and

$y = \frac{x + 3}{k}$

Equate gradients:

$\frac{k-1}{2} = \frac{1}{k}$ gives $k(k - 1) = 2$

$k^2 - k - 2 = 0$

Solving $(k - 2)(k + 1) = 0$ gives $k = 2, k = -1$.

Test the values of $k$ in the system of equations.

When $k = 2$,
$x - 2y = 6$
$-x + 2y = 3$

These are parallel lines, so there are no solutions for $k = 2$.

When $k = -1$,
$-2x - 2y = 6$
$-x - y = 3$

These are the same line, so there are infinite solutions for $k = -1$.

**a** $R\backslash\{-1, 2\}$

**b** $k = 2$

**c** $k = -1$

### Question 19

Divide $Q(x) = x^3 - 9x^2 + 7x - 3$ by $x - 3$.

Using synthetic division,

| 3 | 1 | −9 | 7 | −3 |
|---|---|---|---|---|
| | | 3 | −18 | −33 |
| | 1 | −6 | −11 | −36 |

quotient $= x^2 - 6x - 11$, remainder $= -36$

### Question 20

$3x^2 - 4x + 1 = a(x + b)^2 + c$
$3x^2 - 4x + 1 = ax^2 + 2abx + ab^2 + c$

Equate coefficients:

$a = 3$ … [1]

$-4 = 2ab$ … [2]

$1 = ab^2 + c$ … [3]

[1] and [2] give $-4 = 6b$, so $b = -\frac{2}{3}$.

Substituting $a$ and $b$ into [3]:

$1 = 3\left(-\frac{2}{3}\right)^2 + c$, so $c = -\frac{1}{3}$.

$a = 3, b = -\frac{2}{3}, c = -\frac{1}{3}$

**Question 21**

The value(s) of $k$ for which the following simultaneous linear equations have a unique solution.

$kx - 9y = 5$
$x - ky = k$

Rearrange to find and then equate gradients gives

$\frac{k}{9} = \frac{1}{k}$

gives $k = 3, k = -3$.

The solution is $k \in R \setminus \{\pm 3\}$.

**Question 22**

Find the value(s) of $k$ for which the following simultaneous linear equations have infinite solutions.

$kx - 9y = 5$
$x - ky = k$

Test each value of $k$ in equations.

Gives same gradient, different $y$-intercepts. Hence parallel lines.

There are no values of $k$ for infinite solutions.

## Multiple-choice questions

**1** D

Graph has gradient = −3 and $y$-intercept of $y = 3$.

$y = -3x + 3$

**2** C

Graph has stationary point $(1, -1)$ and $y$-intercept of $y = 3$.

$y = 4(x - 1)^2 - 1$

**3** E

A polynomial has positive integer powers.

$\therefore f(x) = -\frac{1}{x} + x - 2$ is **not** a polynomial.

**4** A

$P(x) = -2x^3 + 7x - 5$

$P(1) = -2 + 7 - 5 = 0$

$\therefore P(1) = 0$ is the **correct** statement.

**5** B

$Q(x) = -x^4 - 2x^3 + 3x^2 + 4x - 4$ is a −ve quartic with repeated factors at $x = -2$ and $x = 1$.

**6** D

A factor of $Q(x) = -x^4 - 2x^3 + 3x^2 + 4x - 4$ is $(1 - x)$.

Dividing $Q(x)$ by $(1 - x)$, we get $x^3 + 3x^2 - 4$

$$\therefore Q(x) = (1 - x)(x^3 + 3x^2 - 4) = (1 - x)(x - 1)(x + 2)^2$$

The other factors are $(x - 1)$, $(x + 2)$, $(x + 2)$.

**7** A

$P(x) = -2(x - 3)(x + 2)(x - 5)$

$P(0) = -2(-3)(2)(-5) = -60$

$y$-intercept is $(0, -60)$.

**8** E

$$4x^2 - 3x + 1 = a(x - b)^2 + c = ax^2 + ab^2 - 2abx + c$$

Equate terms to get

$a = 4, b = \frac{3}{8}, c = \frac{7}{16}$

**9** B

If $(ax + b)$ is a factor of $P(x)$, then from the remainder theorem, $P\left(-\frac{b}{a}\right) = 0$.

**10** E

$P(x) = x^3 + x^2 - 5x + 7$ divided by $(x + 1)$.

remainder $= P(-1) = -1 + 1 + 5 + 7 = 12$

**11** D

$f(x) + f(y) = f(xy)$

Try option D.

$f(x) = \log_e(x)$

LHS $= f(x) + f(y) = \log_e(x) + \log_e(y)$

RHS $= f(xy) = \log_e(xy) = \log_e(x) + \log_e(y)$

$f(x) = \log_e(x)$ works.

**12** C

$f(x) \times f(y) = f(x + y)$

Try option C:

$f(x) = e^x$

LHS $= f(x) \times f(y) = e^x \times e^y$

RHS $= f(x + y) = e^{x+y} = e^x \times e^y$

$f(x) = e^x$ works.

**13** B

$f: (-\infty, A] \rightarrow R, f(x) = x^2 + 4x$

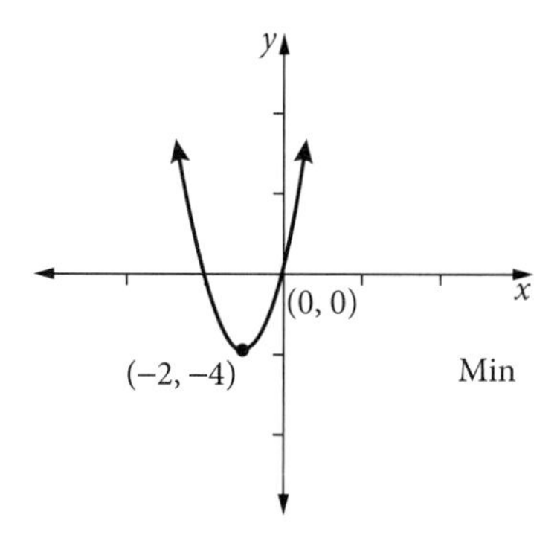


Needs to be one-to-one and not many-to-one.

For the inverse $f^{-1}$ to exist, $A \leq -2$.

**14** D

$f: (-\infty, 1) \rightarrow \mathrm{R}, f(x) = x^2 - 2x$

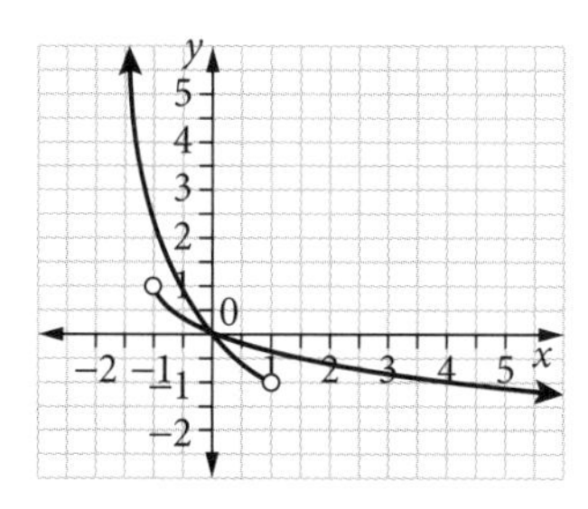


$f^{-1}$ is the lower half of the square root graph.

Swap $x$ and $y$ to get
$f^{-1}(x) = 1 - \sqrt{x+1}, x \in (-1, \infty)$.

**15** A

Equate $f$ and $f^{-1}$.

Swap $x$ and $y$ to get $f^{-1}(x) = -\sqrt{x+4}$.

Solve $x^2 - 4 = -\sqrt{x+4}$.

In domain $(-\infty, 0]$, point of intersection
$= (-1.56, -1.56)$.

**16** A

Equate $y = 2a^2(x-1)^2 = 2$ to find
the intersections.

Solving for $x$, we get $\left(\frac{a-1}{a}, 2\right), \left(\frac{a+1}{a}, 2\right)$.

**17** B

$25° = \frac{25 \times \pi}{180}$ radians $= \frac{5\pi}{36}$ radians

**18** B

$\frac{x}{12} = \frac{x}{12} \times \frac{180}{\pi} = 15°$

**19** B

$\sin\left(\frac{7\pi}{2}\right) = \cos\left(-\frac{\pi}{2}\right) = -1$

**20** C

$\cos\left(\frac{7\pi}{2}\right) = \cos\left(-\frac{\pi}{2}\right) = 0$

**21** C

Equate $f(x) = -2\sqrt{x+2} - 1$ and $g(x) = x + 1$ to find intersection point.

Solve $-2\sqrt{x+2} - 1 = x + 1$ to get $(-2, -1)$.

**22** B

$y = 2(2^{x-1}) - 8$

Swap $x$ and $y$ to find $y^{-1}$.

$x + 8 = 2(2^{y-1})$

$\frac{x+8}{2} = 2^{y-1} \Leftrightarrow \log_2\left(\frac{x+8}{2}\right) = y - 1$

$y^{-1} = \log_2\left(\frac{x+8}{2}\right) + 1 = \log_2\left(\frac{x+8}{2}\right) + \log_2(2)$

So $y^{-1} = \log_2(x+8)$.

**23** B

$f(x) = x^2 - 2x$ and $g(x) = \frac{1}{x^2}$

For $g(f(x))$, test ran (inner) = dom (outer)

$[-1, \infty) \not\subset R \setminus \{0\}$

Restrict $f(x)$ to the range $[-1, 0) \cup (0, \infty)$.

So domain $g(f(x)) = \text{dom } f(x)$
$= (-\infty, 0) \cup (0, 2) \cup (2, \infty)$ or $R \setminus \{0, 2\}$.

So the rule of $g(f(x))$ is

$g(f(x)) = \frac{1}{(x^2 - 2x)^2}$ for $x \in R \setminus \{0, 2\}$.

**24** B

Rearrange $2y - 6x = 3$ to get $y = 3x + \frac{3}{2}$

Solve for $x$: $0 = 3x + \frac{3}{2}$

$x = -\frac{1}{2}$

**25** E

$\sin\left(\frac{7\pi}{6}\right) = \sin\left(\pi + \frac{\pi}{6}\right) = -\sin\left(\frac{\pi}{6}\right)$

**26** A

$\log_x(25) = 2 \Leftrightarrow x^2 = 25$

**27** A

$9^{x-2} \times 3^x = 9 \Leftrightarrow 3^{2(x-2)} \times 3^x = 3^2$

$3^{3x-4} = 3^2$

Equating powers and solving gives $x = 2$.

**28** A

For $f(x) = \frac{-2}{\sqrt{x}}$, you need $x > 0$.

**29** E

$y = \sin(x) - \cos(2x)$

$y = 0$ means $x$-intercepts for the domain $(\pi, 2\pi)$.

$x = \frac{3\pi}{2}$

**30** B

$2\log_3(x) - \log_3(x+1) + 2\log_3(x+4)$
$= \log_3(x)^2 - \log_3(x+1) + \log_3(x+4)^2$
$= \log_3\left(\frac{x^2(x+4)^2}{x+1}\right)$

**31** C

$ax^2 + b = cd \Leftrightarrow d = \frac{ax^2+b}{c}$

**32** A

$x + y = 5$
$2x - y = 1$

Adding the two equations gives $3x = 6 \Rightarrow x = 2$. Substituting into the first equation gives $y = 3$.

**33** D

$f(x) = -2\sin(3(x - \pi))$.

Line $y = 2$ in the domain $(-\pi, \pi)$ touches the graph 3 times.

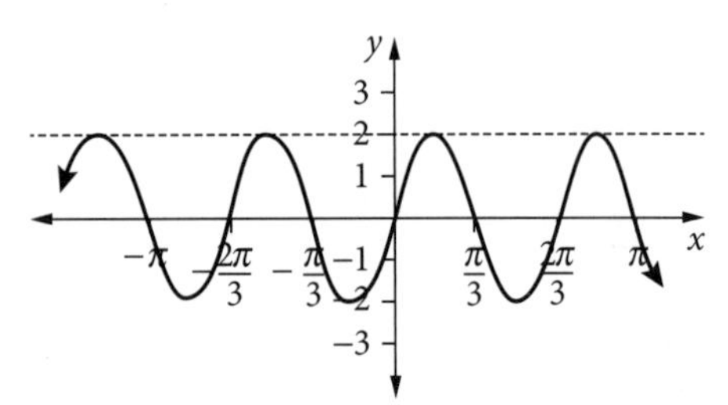


**34** C

For $y = -e^{x+2} - 3$, swap $x$ and $y$ and rearrange to find the inverse.

$x = -e^{y+2} - 3 \Leftrightarrow e^{y+2} = -3 - x$
$\Leftrightarrow \log_e(-3 - x) = y + 2$

So $f^{-1}(x) = \log_e(-3 - x) - 2$
and dom of $f^{-1}$ = ran of $f$,
dom of $f^{-1} = (-\infty, -3)$

Hence $f^{-1}: (-\infty, -3) \rightarrow R$,
$f^{-1}(x) = \log_e(-3 - x) - 2$.

**35** B

$f(x) = e^x$

Try option B:

$f(x)f(y) = f(x + y)$

LHS $= e^x \times e^y = e^{x+y}$

RHS $= e^{x+y}$

$\Rightarrow f(x)f(y) = f(x + y)$ is true.

**36** D

$\log_a(y) = \frac{1}{2}$
$2\log_a(y^3) = 3 \times 2\log_a(y) = 6 \times \frac{1}{2} = 3$

**37** A

$[\log_5(x)]^2 = \log_5(x^2)$
$\Leftrightarrow \log_5(x) \times \log_5(x) = 2\log_5(x)$
$\Leftrightarrow \log_5(x) \times \log_5(x) - 2\log_5(x) = 0$
$\Leftrightarrow \log_5(x)\,[\log_5(x) - 2] = 0$

This means $\log_5(x) = 0$ or $\log_5(x) = 2$.

$\Leftrightarrow 5^0 = x$ or $5^2 = x$

$x$ is equal to 1 or 25.

**38** A

$N = Ae^{-kt}$

$N = 4.12$ when $t = 2$ gives $4.12 = Ae^{-2k}$.

$N = 2.62$ when $t = 5$ gives $2.62 = Ae^{-5k}$

Solve simultaneously to get $A = 5.6$, $k = 0.2$.

**39** D

$\sin^2(2\pi x) = \frac{3}{4}$ for $x \in \left[0, \frac{1}{2}\right]$

$\Rightarrow \sin(2\pi x) = \frac{\sqrt{3}}{2}$

$\Rightarrow \quad 2\pi x = \frac{\pi}{3}, \frac{2\pi}{3}$

$x = \frac{1}{6}, \frac{1}{3}$

**40** B

$x^{-3} + x^{-1} = \frac{1}{x^3} + \frac{1}{x} = \frac{1 + x^2}{x^3}$

**41** C

A system of linear equations is

$3x + 4y + 1 = 0$

$6x + 8y + 2 = 0$

Graphs are the same line (same gradient and same $y$-intercept) so there are an infinite number of points of intersection.

**42** D

Rearrange the equations

$kx + 4y = 1$ and $(k - 1)x + ky = 2k$

to get

$y = \frac{1 - kx}{4}$ and $y = \frac{2k - (k - 1)x}{k}$.

Equate the gradients to find where there is **no** unique solution.

$\Rightarrow -\frac{k}{4} = \frac{-(k-1)}{k}$
$\Rightarrow k = 2$

So we have a unique solution for $k \in R \setminus \{2\}$.

Test $k = 2$:

$2x + 4y = 1$
$x + 2y = 4$

These are parallel lines (same gradient, different y-intercept), so there are no solutions for $k = 2$.

**43** D

$f(x) = 3\log_e(x - b)$
$f(2) = 6$ gives $6 = 3\log_e(2 - b)$.
$\Leftrightarrow e^2 = 2 - b$
So $b = 2 - e^2$.

**44** A

$5^{x-2} = 10 \Leftrightarrow \log_5(10) = x - 2$

So $x = \log_5(10) + 2$.

Applying a change of base gives

$x = \frac{\log_{10}(10)}{\log_{10}(5)} + 2 = \frac{1}{\log_{10}(5)} + 2$

Then $x = \frac{1}{\log_{10}(5)} + 2$.

**45** A

Rearrange the equations
$(k - 3)x - 5y = -2$ and $2x - (2k + 2)y = -4$
to get $y = \frac{(k-3)x + 2}{5}$ and $y = \frac{2x + 4}{2k + 2}$.

Equate gradients to find where there is no unique solution.

$\Rightarrow \frac{k-3}{5} = \frac{2}{2k+2} = \frac{1}{k+1}$
$\Rightarrow k = -2, 4$

So we have a unique solution for $k \in R \setminus \{-2, 4\}$.

Test the values in the system of equations.

For $k = -2$,
$-5x - 5y = -2$
$2x + 2y = -4$
we get parallel lines.

For $k = 4$,
$x - 5y = -2$
$2x - 10y = -4$
we get the same line.

So there are infinitely many solutions for $k = 4$.

**46** B

$\sin(x) = 1$
$\Rightarrow x = \frac{\pi}{2}, \frac{5\pi}{2}, \frac{9\pi}{2} \ldots$
$\Rightarrow x = \frac{\pi}{2} + 2n\pi, n \in \mathbf{Z}$

**47** A

This is an exponential graph, reflected over the $x$-axis, with a horizontal asymptote at $y = 2$, meaning a vertical translation of 2.

Axial intercept at $(0, -1)$.

Equation of the graph: $y = -3e^x + 2$

Swap $x$ and $y$ and rearrange to find the inverse.

$x = -3e^y + 2$
$\Leftrightarrow y = \log_e\left(\frac{2 - x}{3}\right)$

**48** E

$T = T_0 \times 2^{-kt}$

20 minutes for the temperature to halve gives

$0.5T_0 = T_0 \times 2^{-20k}$
$0.5 = 2^{-20k} \Leftrightarrow \log_2(0.5) = -20k$
$\Leftrightarrow -\frac{1}{20}\log_e(0.5) = k$
$\therefore k = \frac{1}{20}\log_2(2) = \frac{1}{20}$

To find the temperature after 60 minutes, substitute $k$ and 60 into the equation to get

$T = T_0 \times 2^{-3} = \frac{1}{8}T_0$

This gives the fraction as $\frac{1}{8}$.

# Extended-answer questions

### Question 1

**a** Two lines are defined by $y = 3x - k$ (line 1) and $y = 3kx + k$ (line 2).

The gradient of line 1 is 3.

The gradient of line 2 is $3k$.

Solving $3 = 3k$ gives $k = 1$.

**b** Perpendicular lines for $m_1 \times m_2 = -1$

$3 \times 3k = -1$

$k = -\frac{1}{9}$

**c** $y = 3x - k$ and $y = 3kx + k$ with $k = -\frac{1}{9}$

becomes $y = 3x + \frac{1}{9}$ and $y = -\frac{1}{3}x - \frac{1}{9}$.

Solve for the intersection point.

The intersection point is $\left(-\frac{1}{15}, -\frac{4}{45}\right)$.

**d**

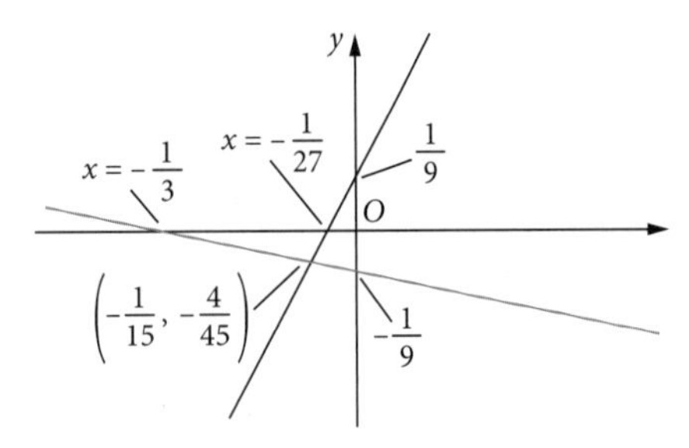


**e** $x$-intercept of line with negative gradient, $x = -\frac{1}{3}$

$x$-intercept of line with positive gradient, $x = -\frac{1}{27}$

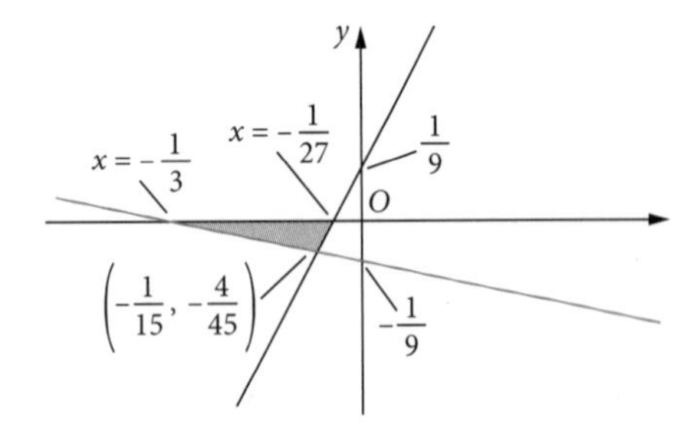


$$\text{Area of triangle} = \frac{1}{2}\text{base} \times \text{height}$$

$$= \frac{1}{2} \times \left(\frac{1}{3} - \frac{1}{27}\right) \times \frac{4}{45}$$

$$= \frac{16}{1215} \text{ square units}$$

### Question 2

**a** $f(x) = ax^4 - 2bx^2 + c$

$f(2) = 0$

$f\left(-\frac{1}{2}\right) = 0$

$f(-1) = 6$

Solve simultaneously.

Thus $a = -\frac{8}{3}$, $b = -\frac{17}{3}$, $c = -\frac{8}{3}$

**b** $f(x) = ax^4 - 2bx^2 + c$

$$f(x) = -\frac{8}{3}x^4 + \frac{34}{3}x^2 - \frac{8}{3}$$

$$= -\frac{2}{3}(4x^4 - 17x^2 + 4)$$

$$= -\frac{2}{3}(x^2 - 4)(4x^2 - 1)$$

$$= -\frac{2}{3}(x + 2)(x - 2)(2x + 1)(2x - 1)$$

**c**

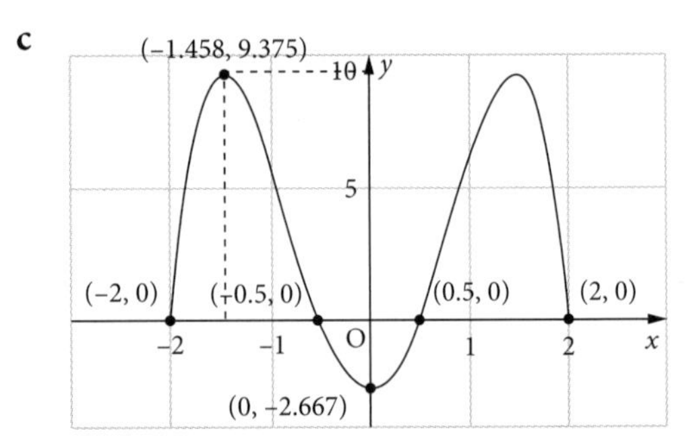


**d** $f(x - 2) + \frac{8}{3} = -\frac{2}{3}[4(x - 2)^4 - 17(x - 2)^2 + 4] + \frac{8}{3}$

$$g(x) = -\frac{2}{3}\left[4\left(\frac{1}{2}x - 2\right)^4 - 17\left(\frac{1}{2}x - 2\right)^2 + 4\right] + \frac{8}{3}$$

**e** $g(x) = -\frac{1}{6}(x - 4)^2(x^2 - 8x - 1)$

$g(x) = -\frac{1}{6}(x - 4)^2(x - 4 + \sqrt{17})(x - 4 - \sqrt{17})$

Rational root $x = 4$, irrational roots $x = 4$ Pr(H

**f**

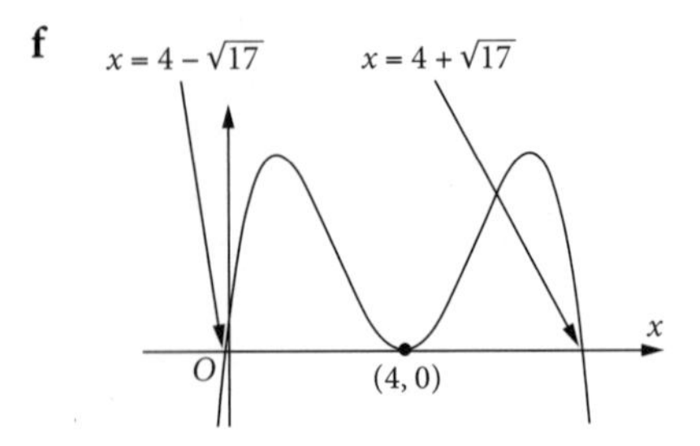


### Question 3

**a** $< 19$ and $19 \leq$ does not repeat the value at $x = 19$ so the hybrid graph is a many-to-one function.

**b** $f(10) = 3$

**c** $f(20) = 8$

**d** $A = 40.56$

**e**

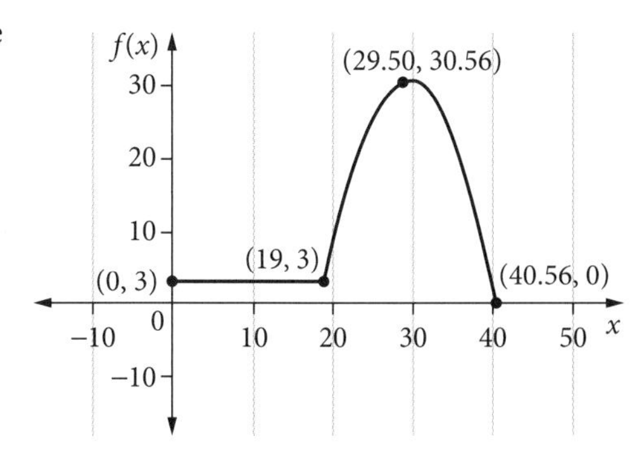


## Question 4

**a** **i** $r(0) = 2500$

**ii** $r_{min} = 2500 - 1700 = 800$,
$r_{max} = 2500 + 1700 = 4200$

**iii** period $= \dfrac{2\pi}{\frac{\pi}{80}} = 160$ weeks

**b** $a = \dfrac{2500 - 700}{2} = 900$

$f(t) = a\sin(b(t - 60)) + 1600$

$\dfrac{2\pi}{b} = 160,\ b = \dfrac{\pi}{80}$

**c** $r(t) + f(t) = -900\sin\left(\dfrac{\pi t}{80} + \dfrac{\pi}{4}\right) + 1700\sin\left(\dfrac{\pi t}{80}\right) + 4100$

The maximum combined population is 5339 correct to the nearest whole number.

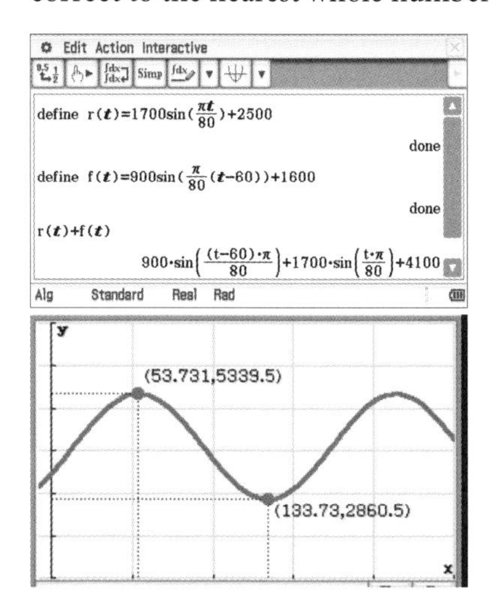


**d** period $= \dfrac{2\pi}{\frac{\pi}{80}} = 160$ weeks

## Question 5

**a** The range is $R$.

**b** **i** $f(x) = \log_e\left(x + \dfrac{1}{2}\right) - \log_e\left(\dfrac{1}{2} - x\right)$

$f'(0) = 4$

```
define f(x)=ln(x+1/2)-ln(1/2-x)
                                done
diff(f(x),x,1,0)
                                   4
```

**ii** $f$ is strictly increasing over the domain $\left(-\dfrac{1}{2}, \dfrac{1}{2}\right)$.

**c** $f(x) + f(-x)$

$= \log_e\left(x + \dfrac{1}{2}\right) - \log_e\left(\dfrac{1}{2} - x\right) + \log_e\left(-x + \dfrac{1}{2}\right) - \log_e\left(\dfrac{1}{2} + x\right)$

$= \log_e\left(x + \dfrac{1}{2}\right) - \log_e\left(\dfrac{1}{2} + x\right) - \log_e\left(\dfrac{1}{2} - x\right) + \log_e\left(-x + \dfrac{1}{2}\right)$

$= 0$

**d** Let $y = f(x)$, inverse swap $x$ and $y$.

Solve $x = \log_e\left(y + \dfrac{1}{2}\right) - \log_e\left(\dfrac{1}{2} - y\right)$ for $y$.

$f^{-1}(x) = \dfrac{1}{2}\tanh\left(\dfrac{x}{2}\right) = \dfrac{1}{2} - \dfrac{1}{e^x + 1} = \dfrac{e^x - 1}{2(e^x + 1)}$

The domain is $R$. It is the same as the range of $f$.

```
define f(x)=ln(x+1/2)-ln(1/2-x)
                                done
solve(f(y)=x, y)
                    {y=(e^x-1)/(2·(e^x+1))}
```

## Question 6

**a** $4\pi$

**b** $-1.722$

**c** $2\pi$

**d** 2

# Chapter 3

## Area of Study 3: Calculus

## Short-answer questions

### Question 1

$y = 2x\sin(3x)$

$\frac{dy}{dx} = 2\sin(3x) + 6x\cos(3x)$

### Question 2

**a** Average rate of change is zero.

**b** Average rate of change is $\frac{0-4}{1-0} = -4$.

**c** The equation of the parabola is $y = -4x^2 + 4$.

$\frac{dy}{dx} = -8x$

The rate of change at $x = 1$ is $-8$.

### Question 3

$y = x\sin(x)$

**a** $\frac{dy}{dx} = \sin(x) + x\cos(x)$

**b** $\int \sin(x)\,dx + \int x\cos(x)\,dx = x\sin(x)$

$\int x\cos(x)\,dx = x\sin(x) - \int \sin(x)\,dx$

$\int 2x\cos(x)\,dx = 2x\sin(x) - 2\int \sin(x)\,dx$

$\therefore \int 2x\cos(x)\,dx = 2x\sin(x) + 2\cos(x)$

### Question 4

If $y = x\log_e(x)$

**a** $\frac{dy}{dx} = \log_e(x) + 1$

**b** $\int (\log_e(x) + 1)dx = x\log_e(x) + c$

$\therefore \int_1^2 (\log_e(x) + 1)dx = [x\log_e(x)]_1^2$

$\int_1^2 (\log_e(x))dx + \int_1^2 1\,dx = [x\log_e(x)]_1^2$

$\int_1^2 (\log_e(x))dx = [x\log_e(x)]_1^2 - \int_1^2 1\,dx$

$\int_1^2 2\log_e(x)dx = 2[x\log_e(x)]_1^2 - 2\int_1^2 1\,dx$

$\therefore \int_1^2 2\log_e(x)dx = 4\log_e(2) - 4$

### Question 5

$y = \frac{\log_e(2x)}{x^2}$

$\frac{dy}{dx} = \frac{x^2 \times \frac{1}{x} - 2x\log_e(2x)}{x^4}$

$= \frac{1 - 2\log_e(2x)}{x^3}$

### Question 6

$y = -x^2 + 2x$

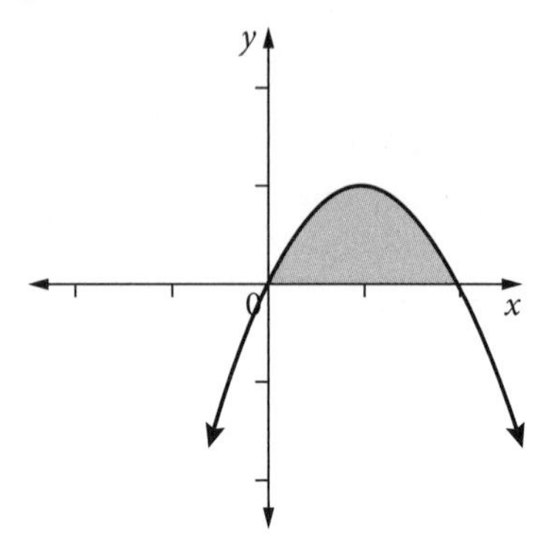


$\text{area} = \int_0^2 (-x^2 + 2x)\,dx$

$= \left[\frac{-x^3}{3} + x^2\right]_0^2$

$= \frac{-2^3}{3} + 4$

$= \frac{4}{3}$ sq. units

### Question 7

$y = \sin(kt)$

$\frac{dy}{dx} = k\cos(kt) = -\pi$ at $t = 1$.

$k\cos(k) = -\pi$

$\cos(k) = -\frac{\pi}{k}$

$k = \pi$ to give $\cos(\pi) = -1$.

### Question 8

**a** $\text{area} = -\int_0^2 (x^2 - 2x)\,dx + \int_2^3 (x^2 - 2x)\,dx$

$= -\left[\frac{x^3}{3} - x^2\right]_0^2 + \left[\frac{x^3}{3} - x^2\right]_2^3$

$= -\left(\frac{8}{3} - 4\right) + \left(\frac{27}{3} - 9\right) - \left(\frac{8}{3} - 4\right)$

$= \frac{8}{3}$ sq. units

**b** $\int_0^3 (x^2 - 2x)\,dx = 0$

### Question 9

$v(t) = t^2 - 6t + 8$
$= (t - 4)(t - 2)$

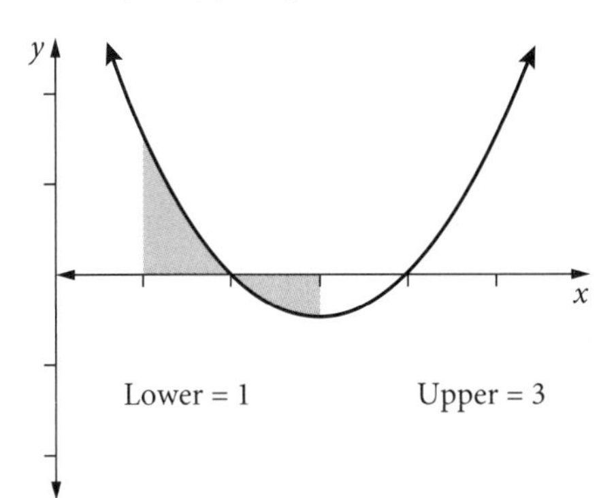


$$\text{distance} = \int_1^2 (t^2 - 6t + 8)\,dt - \int_2^3 (t^2 - 6t + 8)\,dt$$
$$= \frac{4}{3} + \frac{2}{3} = 2$$

Distance travelled between 1 and 3 seconds is 2 metres.

### Question 10

If $g: [k, 4] \to R, g(x) = -(x - 2)^2 + 3$

**a** $g^{-1}$ exists for $k \geq 2$.

**b** $g^{-1}(x) = 2 + \sqrt{3 - x}$

Derivative of $g^{-1}(x) = -\frac{1}{2\sqrt{3 - x}}$

### Question 11

$$y = \sqrt{2x^2 + 4} = (2x^2 + 4)^{\frac{1}{2}}$$
$$\frac{dy}{dx} = \frac{1}{2}(2x^2 + 4)^{-\frac{1}{2}} \times 4x = \frac{2x}{\sqrt{2x^2 + 4}}$$

At $x = 2$,

$$\frac{dy}{dx} = \frac{4}{\sqrt{8 + 4}} = \frac{4}{\sqrt{12}} = \frac{4}{2\sqrt{3}} = \frac{2}{\sqrt{3}} = \frac{2\sqrt{3}}{3}$$

### Question 12

$y = x^3 - x^2 - x - 2$

$\frac{dy}{dx} = 3x^2 - 2x - 1 = 0$ for stationary points

$(3x + 1)(x - 1) = 0$

$x = 1, x = -\frac{1}{3}$

$(1, -3), \left(-\frac{1}{3}, -\frac{49}{27}\right)$

### Question 13

**a** Using the given point, $f(x) = x^3 + ax^2 + b$ gives the equation

$$-3 = 8 + 4a + b$$
$$\Rightarrow 4a + b = -11 \quad \ldots [1]$$

$f'(x) = 3x^2 + 2ax$ gives the equation

$$0 = 12 + 4a$$
$$\Rightarrow a = -3$$

Substituting the value for $a$ into [1], we get $b = 1$.

**b** $f'(x) = 3x^2 - 6x = 0 \Rightarrow x = 0, 2$

There is another stationary point at $(0, 1)$.

### Question 14

$$f(x) = \sin\left(\frac{x}{5}\right)$$
$$f'(x) = \frac{1}{5}\cos\left(\frac{x}{5}\right) \Rightarrow f'\left(\frac{5\pi}{4}\right) = \frac{1}{5}\cos\left(\frac{\pi}{4}\right)$$
$$f'\left(\frac{5\pi}{4}\right) = \frac{1}{5\sqrt{2}}$$

### Question 15

$f(x) = x^3 + 1$ and $g(x) = \sqrt{x - 1}$

**a** For $g(f(x))$, test ran $(f) \subseteq$ dom $(g)$.

$R \not\subseteq [1, \infty)$, so $g(f(x))$ does not exist.

For $f(g(x))$, test ran $(g) \subseteq$ dom $(f)$.

$[0, \infty) \subseteq R$

rule: $f(g(x)) = (\sqrt{x - 1})^3 + 1$

dom $[f(g(x))] =$ dom $g(x) = [1, \infty)$

**b** Let $h(x) = f(g(x))$

$$= (\sqrt{x - 1})^3 + 1$$
$$= (x - 1)^{\frac{3}{2}} + 1$$
$$h'(x) = \frac{3}{2}(x - 1)^{\frac{1}{2}}$$

### Question 16

**a** $y = (5x + 2)^4$

$$\frac{dy}{dx} = 20(5x + 2)^3$$

At $x = 0$, $\frac{dy}{dx} = 20 \times 8 = 160$.

**b** Using $y - y_1 = m(x - x_1)$,

$$y - 16 = 160(x - 0)$$
$$y = 160x + 16$$

**Question 17**

$y = f(x)$

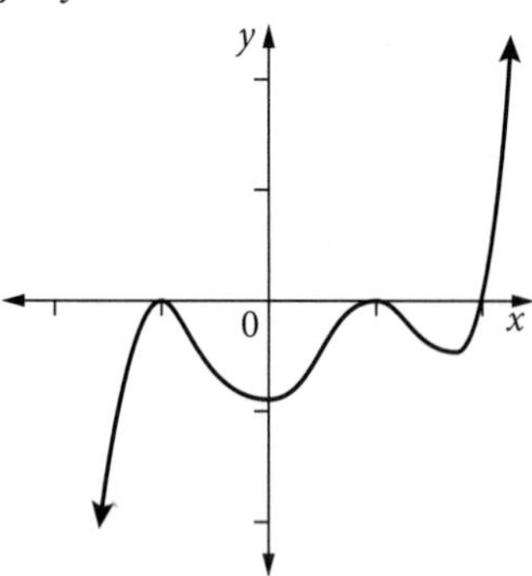


$y = f'(x)$

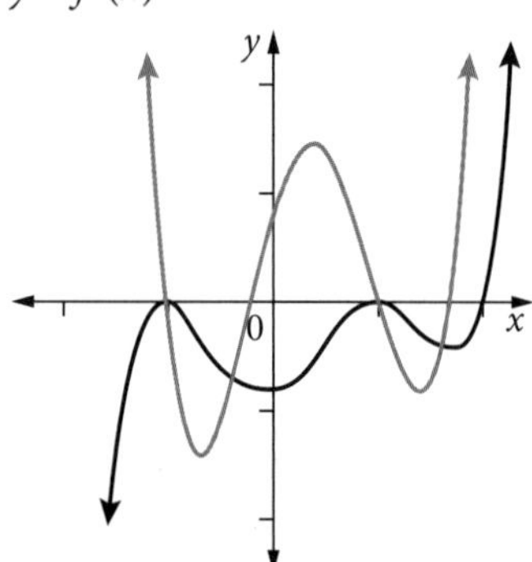


**Question 18**

**a** $f(x) = 3x^2 + 2$

Gradient of the chord $PQ$

$$= \frac{f(-2+h) - f(-2)}{h}$$
$$= \frac{3(-2+h)^2 + 2 - 14}{h}$$
$$= \frac{3(h^2 - 4h + 4) - 12}{h}$$
$$= \frac{3h^2 - 12h + 12 - 12}{h}$$
$$= \frac{h(3h - 12)}{h}$$
$$= 3h - 12$$

**b** Gradient of the tangent at $P = -12$.

**Question 19**

$y = f'(x)$

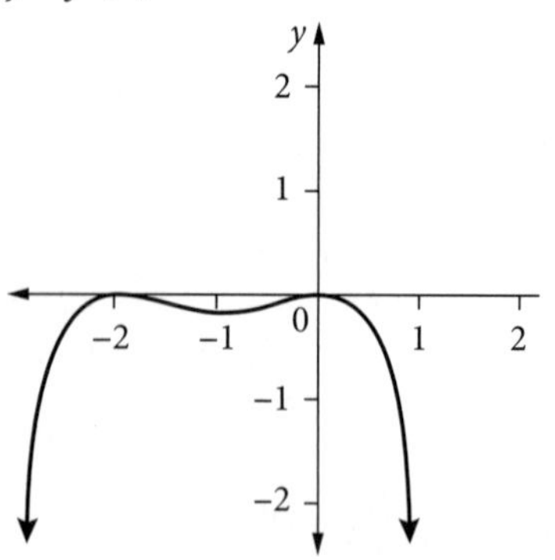


This is a negative polynomial of degree 3.

We expect the anti–derivative graph, $y = f(x)$, to be a negative polynomial of degree 4.

It might look like

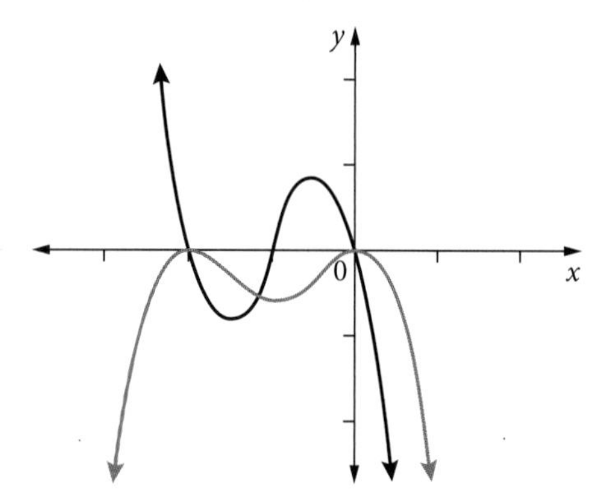


**Question 20**

**a** $T = -0.002h^3 + 30$

$$\text{average rate of change} = \frac{f(10) - f(0)}{10}$$
$$= \frac{-2 + 30 - 30}{10}$$
$$= -0.2$$

**b** $\frac{dT}{dh} = -0.006h^2$

At $h = 10$,

$$\frac{dT}{dh} = -0.006h^2 = -0.6$$

## Multiple-choice questions

**1** E

If $f(x) = 2x^2 + 4x$, then $f'(x) = 4x + 4$.

**2** C

If $f(x) = 2x^2 + 4x$, then an anti–derivative is

$$y = \frac{2x^3}{3} + 2x^2$$

**3** A

If $f(x) = 3\sqrt{x} = 3x^{\frac{1}{2}}$, then $f'(x) = \frac{3}{2}x^{-\frac{1}{2}} = \frac{3}{2\sqrt{x}}$.

So $f'(3)$ equals $\frac{3}{2\sqrt{3}} = \frac{\sqrt{3}}{2}$.

**4** A

If $f(x) = 2\sqrt{x} = 2x^{\frac{1}{2}}$, then $f'(x) = x^{-\frac{1}{2}} = \frac{1}{\sqrt{x}}$.

**5** D

$g(x) = -\frac{1}{4}(x^4 - 6x^2 + 8x)$

So $g'(x) = -\frac{1}{4}(4x^3 - 12x + 8) = -(x^3 - 3x + 2)$.

This gives stationary points at $x = -2$ and $x = 1$.

This is a negative quartic. The derivative graph will be a negative cubic with stationary points at $x = -2$ and $x = 1$ and a repeated factor at $x = 1$.

So the graph that best represents $g'(x)$ is D.

**6** E

If $f(x) = -x^4 - 2x^3 + 3x^2 + 4x - 4$, then

$f'(x) = -4x^3 - 6x^2 + 6x + 4$.

This can be factorised to get

$f'(x) = -2(x + 2)(x - 1)(2x + 1)$.

**7** A

For $g(x) = -2(x - 3)(x + 2)(x - 5)$
$= -2x^3 + 12x^2 + 2x - 60$

a possible anti–derivative is

$y = -\frac{x^4}{2} + 4x^3 + x^2 - 60x + 10$

where 10 is a possible constant.

**8** C

$f(x) = 2x^3 - 3x^2 + 4$

$f'(x) = 6x^2 - 6x$

Consider the gradient graph

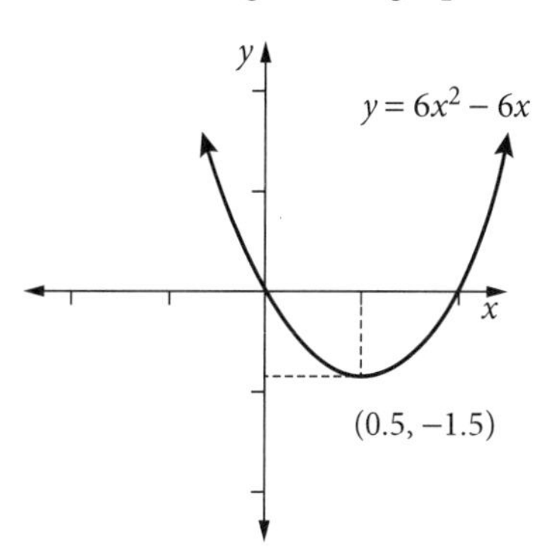


and note that the minimum is at $x = \frac{1}{2}$.

**9** D

$\frac{d}{dx}(-\sin(4x)) = -4\cos(4x)$

**10** C

If $f(x) = e^{-3x}$, then $f'(x) = -3e^{-3x}$.

**11** C

If $f(x) = e^{-5x}$, then $f'(x) = -5e^{-5x}$.

So $f'(5) = -5e^{-25}$.

**12** D

$v = t^3 - t^2 + 4$ so $v'(t) = 3t^2 - 2t$.

At $t = 3$, $v'(t) = 21$.

**13** B

The graph shown has one local minimum and one stationary point of inflection.

The graph has two stationary points.

**14** D

$f: (-4, 2] \rightarrow R, f(x) = x^2 + 4x$. Sketch the graph of $f$ for the restricted domain.

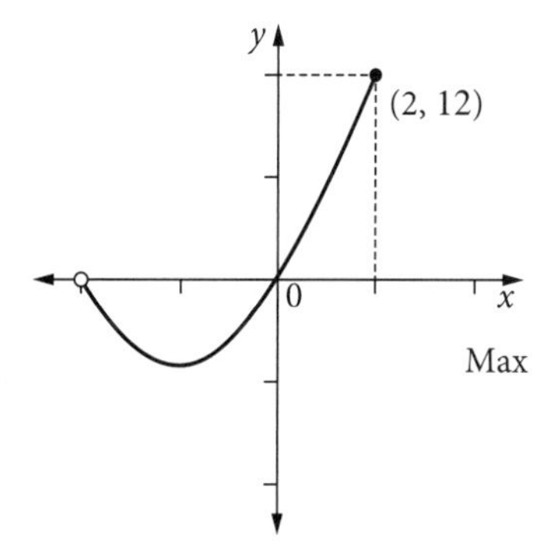


Maximum value is 12.

**15** A

$f: (-4, 2] \to R, f(x) = x^2 + 4x$. Sketch the graph of $f$ for the restricted domain.

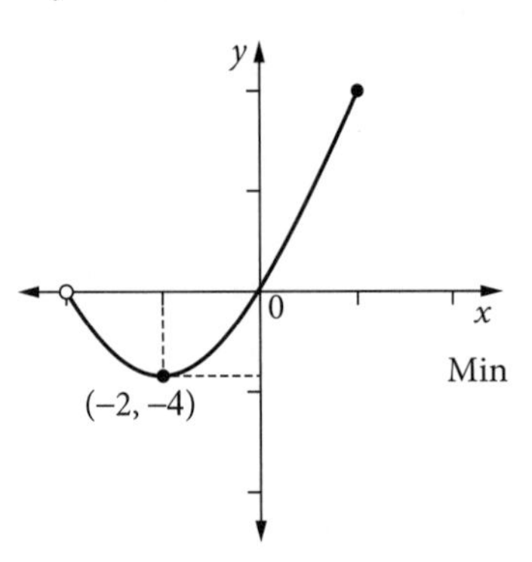


Minimum value is −4.

**16** B

$h(t) = -t^2 + 2t + 1$. Sketch the graph of $h$ or differentiate and equate to zero.

$h'(t) = -2t + 2 = 0$ gives $t = 1$.

$h(1) = -(1)^2 + 2(1) + 1 = 2$

**17** D

$y = 4(x - 1)^2$

The curve and the line $y = 1$ meet at

$x = \frac{1}{2}, x = \frac{3}{2}$.

area $= \int_{\frac{1}{2}}^{\frac{3}{2}}$(upper − lower)

$dx = \int_{\frac{1}{2}}^{\frac{3}{2}} 1 - 4(x - 1)^2 dx$

**18** C

$P(t) = 1000e^{1-t}$

$P'(t) = -1000e^{1-t}$

$P'(10) = -1000e^{-9}$

The rate at which the population is decreasing after 10 years is approximated by 0.123. Note that the word decreasing indicates a −ve exponential power.

**19** A

$y = 2\sin^2(3x)$

Using the chain rule, $\frac{dy}{dx} = 12\sin(3x)\cos(3x)$.

**20** A

$y = 4x^3 - 3x^2 - 2$ so $\frac{dy}{dx} = 12x^2 - 6x$.

$12x^2 - 6x = 0$ gives $x = 0$ and $x = \frac{1}{2}$.

Coordinates are $(0, -2), \left(\frac{1}{2}, -\frac{9}{4}\right)$.

**21** B

$f(x) = ax^2 - bx$ and $f'(x) = 2ax - b$

Substituting $(1, 1)$ into $f(x) = ax^2 - bx$ gives the equation $1 = a - b$.

Substituting a gradient $= -1$ at $x = 1$ into $f'(x) = 2ax - b$ gives the equation $-1 = 2a - b$.

Simultaneously solving the equations $1 = a - b$ and $-1 = 2a - b$, we get $a = -2$ and $b = -3$.

**22** C

If $f(x) = -2\sqrt{x + 2} - 1$, then $f'(x) = -\frac{1}{\sqrt{x + 2}}$.

$f'(2) = -\frac{1}{\sqrt{2 + 2}} = -0.5$

**23** B

$y = \log_e(3x + 3)$, so $\frac{dy}{dx} = \frac{3}{3x + 3} = \frac{1}{x + 1}$.

$\frac{dy}{dx} = 1 = \frac{1}{x + 1}$

$\Rightarrow x = 0$

**24** D

$f(x) = x^2 - 2x$ and $g(x) = \frac{1}{x^2}$

For $g(f(x))$, test ran (inner) = dom (outer)

$[-1, \infty) \not\subset R \setminus \{0\}$

Restrict $f(x)$ so that it has the range $[-1, 0) \cup (0, \infty)$.

So domain $g(f(x)) = \text{dom} f(x) = (-\infty, 0) \cup (0, 2) \cup (2, \infty)$.

$h(x) = g(f(x)) = \frac{1}{(x^2 - 2x)^2}$

$h'(x) = \frac{-4x + 4}{(x^2 - 2x)^3}$ for $x \in (-\infty, 0) \cup (0, 2) \cup (2, \infty)$.

**25** D

$2y - 6x = 3$ can be rearranged as $y = 3x + \frac{3}{2}$.

gradient = 3

**26** C

$y = x^2 - 2x$

average rate of change $= \dfrac{f(3) - f(1)}{3 - 1} = 2$

**27** D

$y = 2x^3$ so $\dfrac{dy}{dx} = 6x^2$

At $x = 2$, gradient = 24.

Use $y - y_1 = m(x - x_1)$, so $y - 16 = 24(x - 2)$.

$y = 24x - 32$

**28** C

Gradient of the secant from $x = 2$ to $x = 2 + h$ for $y = 3x^2 + 2x - 1$ is $\dfrac{f(2+h) - f(2)}{h}$

**29** A

$y = \tan(x)$, so $\dfrac{dy}{dx} = \sec^2(x)$

At $x = a$, $\dfrac{dy}{dx} = \sec^2(a)$.

**30** C

Absolute maximums are at $x = -\dfrac{3\pi}{2}, \dfrac{\pi}{2}$.

**31** D

$g(x) = \sqrt{x}(x^p - 1)$

Using the product rule,

$$g'(x) = (x^p - 1) \times \frac{1}{2\sqrt{x}} + \sqrt{x} \times px^{p-1}$$

$$= \frac{x^{p-1}}{2\sqrt{x}} - \frac{1}{2\sqrt{x}} \times px^{p-1}\sqrt{x}$$

$$= \frac{x^{p-\frac{1}{2}}}{2} - \frac{1}{2\sqrt{x}} + px^{p-\frac{1}{2}}$$

$$= px^{p-\frac{1}{2}} + \frac{x^{p-\frac{1}{2}}}{2} - \frac{1}{2\sqrt{x}}$$

$$= \frac{1}{2}\left(2px^{p-\frac{1}{2}}\right) + \frac{x^{p-\frac{1}{2}}}{2} - \frac{1}{2\sqrt{x}}$$

$$= \frac{1}{2}(2p + 1)x^{p-\frac{1}{2}} - \frac{1}{2\sqrt{x}}$$

**32** D

$f(x) = \sin^n(x)$ so

$f'(x) = n(\sin(x))^{n-1}\cos(x)$
$= n\cos(x)\sin^{n-1}(x)$

**33** D

Sketch the graph of $f(x) = x^{\frac{1}{8}}$

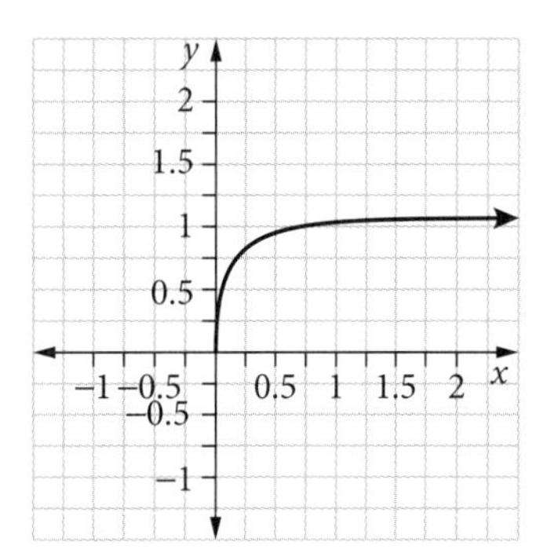


It is **not** true that the function is strictly decreasing for all $x \in R^+$.

It is strictly increasing.

**34** D

$y = x\log_{10}(2x^2 - 1)$

Because this is base 10, we need to use CAS for this derivative.

```
diff(x·log(10,2·x^2−1),x,1,2)
                                  1.837771142
```

The gradient at $x = 2$ is closest to 1.84.

**35** E

$f(x) = -2\sin(3(x - \pi))$. There are 6 stationary points in the domain $(-\pi, \pi)$.

**36** B

$y = x_2 + 1$

The tangent at $(1, 2)$ is $y = 2x$.

The gradient of the curve at $x = 1$ is 2.

**37** A

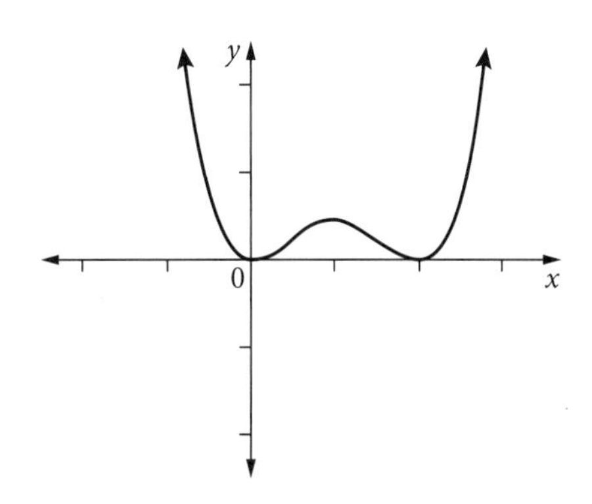


The graph given looks like a positive polynomial of degree 4.

To find the gradient function graph, consider stationary points and $x$-intercepts.

The gradient graph will be a positive polynomial of degree 3.

The graph of $f'(x)$ looks like

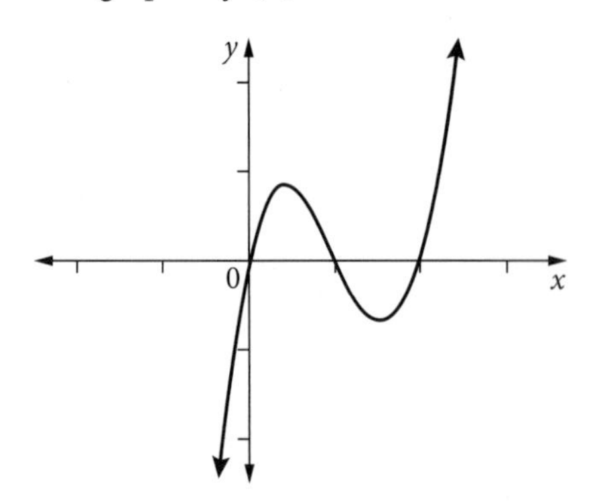


**38** B

average rate of change $= \dfrac{f(x+h) - f(x)}{h}$

**39** D

For $N = Ae^{-kt}$, an anti-derivative is $y = -\dfrac{A}{k}e^{-kt}$.

**40** A

$h(x) = \tan(2\pi x) + 1$, so $h'(x) = 2\pi \sec^2(2\pi x)$.

At $x = 0$, gradient $= 2\pi$.

Use $y - y_1 = m(x - x_1)$, so $y - 1 = 2\pi(x - 0)$

$y = 2\pi x + 1$

**41** C

$f(x) = x^{-2} + x^{-1}$ so $f'(x) = -2x^{-3} - x^{-2}$

$= \dfrac{-x-2}{x^3}$

There are asymptotes at $y = 0$ and $x = 0$.

**42** B

$g(x) = 2x + x^2$ so the average rate of change

$= \dfrac{f(1) - f(0)}{1 - 0} = 3$

**43** E

$y = e^{3x}\sin(x)$

Using the product rule,

$$\frac{dy}{dx} = 3e^{3x}\sin(x) + e^{3x}\cos(x)$$
$$= e^{3x}(\cos(x) + 3\sin(x))$$

**44** B

$f(x) = 3\log_e(x - b)$

$f'(x) = \dfrac{3}{x - b}$

$\therefore f'(4) = \dfrac{3}{4 - b}$

**45** C

$$y = \frac{x}{\log_e(x)}$$

Using the quotient rule,

$$\frac{dy}{dx} = \frac{\log_e(x) \times 1 - x \times \frac{1}{x}}{(\log_e(x))^2}$$
$$= \frac{\log_e(x) - 1}{(\log_e(x))^2}$$

**46** C

To use the chain rule $\dfrac{dy}{dx} = \dfrac{dy}{du} \times \dfrac{du}{dx}$ for the function $y = \cos(3x^2 + 1)$, set $u$ as $3x^2 + 1$ to give $y = \cos(u)$.

**47** D

To use the chain rule $\dfrac{dy}{dx} = \dfrac{dy}{du} \times \dfrac{du}{dx}$ for the function $y = \cos(3x^2 + 1)$, set $u = 3x^2 + 1$ to give $y = \cos(u)$.

$u = 3x^2 + 1$

$\dfrac{du}{dx} = 6x$

**48** B

$$\int_0^1 (-3f(x) + 2x)\,dx = \int_0^1 -3f(x)\,dx + \int_0^1 2x\,dx$$
$$= -3\int_0^1 f(x)\,dx + [x^2]_0^1$$

$= -3 \times 2 + 1$

$= -5$

**49** A

$f(x) = x\log_e(x + 1)$

Graph the function on a CAS to see the shape of the graph.

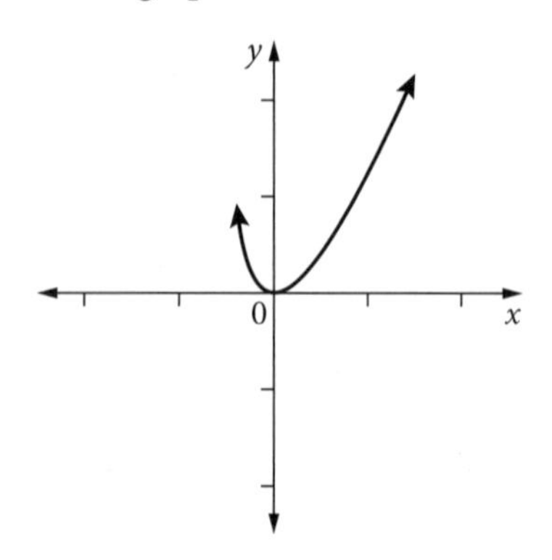


**50** D

By definition, $\lim\limits_{\delta x_i \to 0} \sum\limits_{i=1}^{n} f(x_i)\,\delta x_i = \int_a^b f(x)\,dx$.

# Extended-answer questions

### Question 1

**a** $y = 3x\cos(x)$

$$\frac{dy}{dx} = 3\cos(x) - 3x\sin(x)$$

**b** From part **a**,

$$\int 3\cos(x) - 3x\sin(x)\,dx = 3x\cos(x)$$

This gives

$$3\int x\sin(x)\,dx = 3\int\cos(x)\,dx - 3x\cos(x)$$
$$\int x\sin(x)\,dx = \int\cos(x)\,dx - x\cos(x)$$
$$= \sin(x) - x\cos(x)$$

**c** $$\int x\sin(x)\,dx = \sin(x) - x\cos(x)$$
$$\int_0^{\pi} x\sin(x)\,dx = [\sin(x) - x\cos(x)]_0^{\pi}$$
$$= (\sin(\pi) - \pi\cos(\pi)) - (0 - 0)$$
$$= \pi$$

**d**

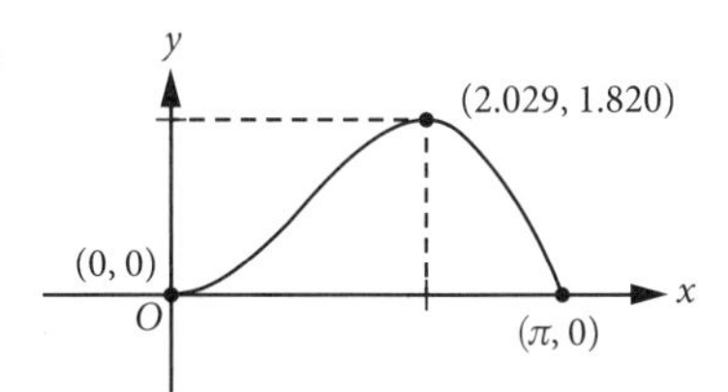


**e** Graph the curve $f(x) = x\sin(x)$ for the domain $x \in [0, \pi]$.

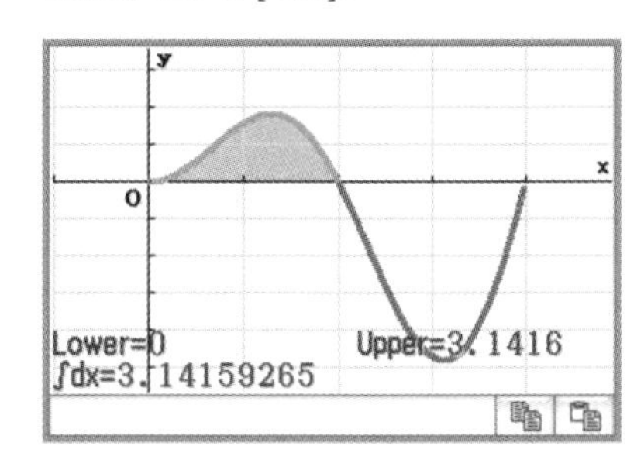


The area required is all above the $x$-axis, so area = $\pi$.

**f** $f(x) = x\sin(x)$

$f'(x) = \sin(x) + x\cos(x) = 0$

gives $x = 0$, $x = 2.029$

Solving with CAS, within the domain, turning points are at $x = 0$, $x = 2.029$

So the local maximum is at $x = 2.029$, matching the graph in part **d**.

### Question 2

**a** $f(t) = 10\sin(\pi t) + 100$

$t = 0$ to $t = 4$

$$\text{average value} = \frac{1}{4-0}\int_0^4 (10\sin(\pi t) + 100)\,dt$$
$$= 100\text{ dB}$$

**b** $t = 0$ to $t = 4.75$

average value

$$= \frac{1}{4.75-0}\int_0^{4.75} (10\sin(\pi t) + 100)dt$$
$$= 101.14\text{ dB}$$

**c** amplitude = 10

$$\text{period} = \frac{2\pi}{\pi} = 2$$

**d**

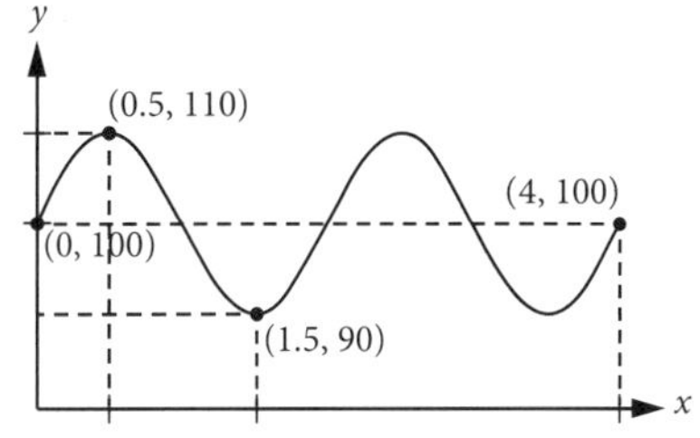


**e** dangerous ≥ 95

Solve $f(t) \geq 95$.

Dangerous for $0 \leq t \leq 1\frac{1}{6}$ and $1\frac{5}{6} \leq t \leq 3\frac{1}{6}$ and $3\frac{5}{6} \leq t \leq 4$.

Dangerous from 7 p.m. to 8:10 p.m., 8:50 p.m. to 10:10 p.m., and 10:50 p.m. to 11 p.m.

**f** The minimum noise level of the concert is 90 dB.

A conversation between 60 and 65 dB cannot be heard.

**g** $$\frac{d}{dt}(10\sin(\pi t) + 100) = 10\pi\cos(\pi t)$$

**h** General solution

$$10\pi\cos(\pi t) = 0$$
$$\Rightarrow \cos(\pi t) = 0$$
$$\Rightarrow \pi t = \frac{\pi}{2}, \frac{3\pi}{2}, \frac{5\pi}{2}, \frac{7\pi}{2}\ldots$$
$$t = \frac{1}{2}, \frac{3}{2}, \frac{5}{2}, \frac{7}{2}\ldots$$
$$t = \frac{1}{2} + n, n \in Z$$



SOLUTIONS – CHAPTER 3

**i** Stationary points at $t = \frac{1}{2} + n,\ n \in Z$

1st stationary point is a maximum

at $t = \frac{1}{2}, f\left(\frac{1}{2}\right) = 110$

The 3rd stationary point is a maximum at

$t = \frac{1}{2} + 2 = \frac{5}{2}, f\left(\frac{5}{2}\right) = 110$, etc.

Given that 110 < 140, permanent hearing loss will not happen.

## Question 3

**a** $v(t) = 3t^3 - t - 2$

```
rFactor(3t^3-t-2)
                    3·(t^2+t+2/3)·(t-1)
```

$v(t) = (3t^2 + 3t + 2)(t - 1) = 0$ for $t$-intercepts.

A quadratic factor is $3t^2 + 3t + 2$.

Discriminant $= 9 - 4 \times 3 \times 2 = -15$, so there are no real factors.

**b** Stops at $v(t) = 0$, therefore at $t = 1$.

**c**

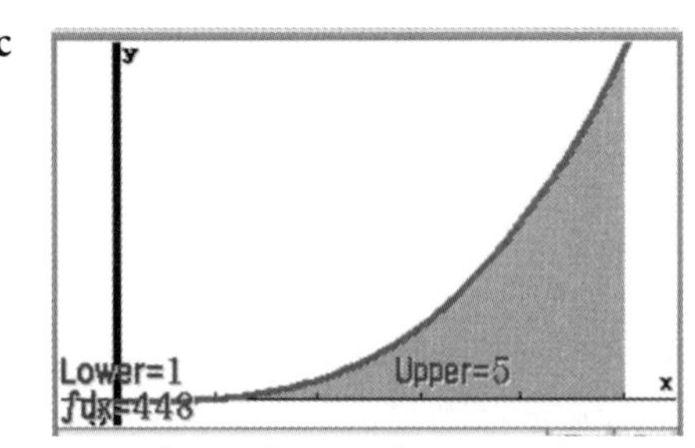


Area above the axis from $t = 1$ to $t = 5$ is 448 square metres.

Include area below the axis.

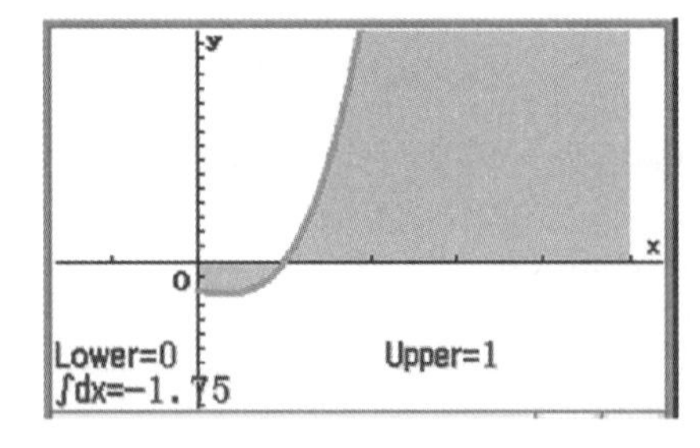


Total distance $= 1.75 + 448 = 449.75$ m

## Question 4

**a** $f(x) = x^3 + ax^2 + bx + c$

$f'(x) = 3x^2 + 2ax + b$

$f(1) = 300$ and $f'(1) = 0$

Solving simultaneously,

```
define f(x)=x^3+ax^2+bx+c
                                    done
Define g(x)=d/dx(f(x))
                                    done
{f(1)=300
{g(1)=0    |a,b
              {a=c-302, b=-2·c+601}
```

$a = c - 302,\ b = -2c + 601$

**b** $a = b$ using $a = c - 302$ and $b = -2c + 601$ gives

$c - 302 = -2c + 601$

$\therefore c = 301$

**c** When $c = 301$, $b = a = -1$

$f(x) = x^3 + ax^2 + bx + c$
$f(x) = x^3 - x^2 - x + 301$
$f'(x) = 3x^2 - 2x - 1 = (3x + 1)(x - 1)$

$f'(x) = 0$ for $x = -\frac{1}{3}, x = 1$

For the domain $x \geq 1$, the only stationary point is at $x = 1$, the endpoint of the domain.

Graph is an increasing function so max also occurs at the end of the domain.

**d** There are no stationary points within the domain.

So the endpoints are the maximum and minimum.

Since $f(x)$ is an increasing function over the domain, we have

minimum at $(1, 300)$

maximum at $(30, 26\,371)$

So on the 1st day of the 30-day month, the profit is at the minimum of \$300.

On the 30th day, the profit is at the maximum of \$26 371.

**e**

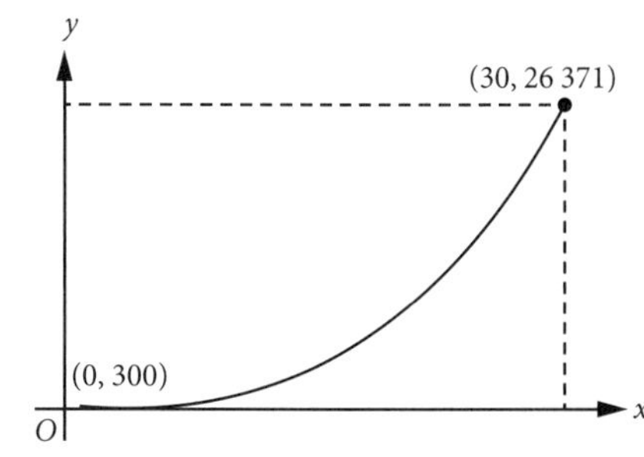


**f**

```
{f(1)=300
{g(1)=0   |a, b
          {a=c−302, b=−2·c+601}
solve(c−302=2(−2·c+601), c)
                    {c=300.8}
```

$c = 300.8$

**g** The function now is

$f_1(x) = x^3 - x^2 - x + 300.8$

Since $f_1(x)$ is an increasing function over the interval, we have

minimum at $(1, 299.8)$

maximum at $(30, 26\,370.8)$

So on the 1st day of the 30-day month, the profit is at the minimum of \$299.8.

On the 30th day, the profit is at the maximum of \$26 370.80.

The profit is reduced at both endpoints by $0.2 = 20$ cents.

The profit will be lower for every day of the month by 20 cents.

### Question 5

**a** gradient $= \frac{3}{12} = \frac{1}{4}$

**b** gradient $= \frac{4}{12} = \frac{1}{3}$

Consider the diagram below, where a 12 : 12 pitch forms an angle of 45°.

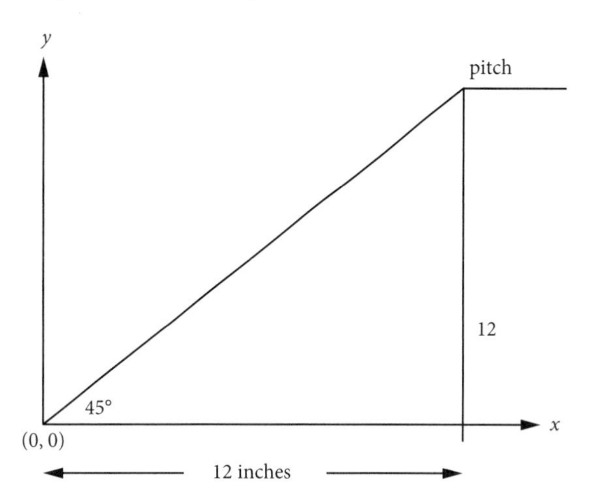


A pitch of 3 : 12 has an angle of 14°.

A pitch of 4 : 12 has an angle of 18.5°.

**c** $\tan(14°) = \frac{3}{12}$

**d** $\tan(18.5°) = \frac{4}{12}$

**e** pitch of 4 : 12

gradient $= \frac{1}{3}$

The equation of the line is $y = \frac{1}{3}x$.

**f** pitch of 3 : 12

gradient $= \frac{1}{4}$

To find the equation of line at $(2, 10)$,

$y - y_1 = m(x - x_1)$

$y - 10 = \frac{1}{4}(x - 2)$

$4y - 40 = x - 2$

$4y = x + 38$

**g** $\theta = \tan^{-1}\left(\frac{x}{12}\right)$

```
d/dx(tan⁻¹(x/12))
                                12
                             x²+144
```

$\frac{d}{dx}(\theta) = \frac{12}{x^2 + 144}$

Solve $\frac{d}{dx}(\theta) = \frac{12}{x^2 + 144} = 0$ for the stationary point.

There is no solution for $x$.

The angle will be at maximum at the end of the domain, that is, for the maximum magnitude of the vertical height $x$.

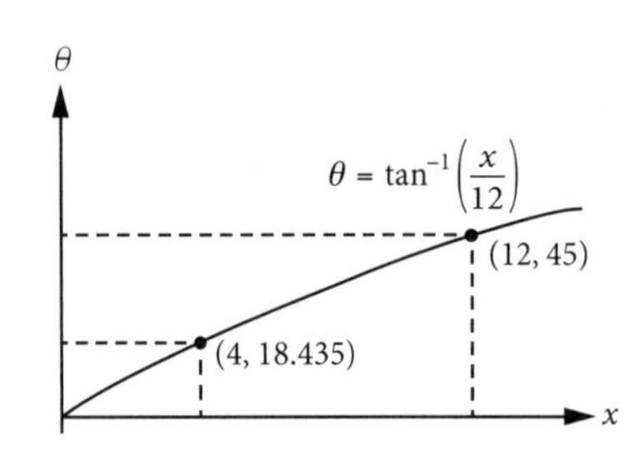


Graph of $\theta = \tan^{-1}\left(\frac{x}{12}\right)$

**Question 6**

**a**

$$f(x) = g(x)$$
$$x + 2 = 2x^2 + ax + 4$$
$$\Rightarrow 2x^2 + ax + 4 - x - 2 = 0$$
$$\Rightarrow 2x^2 + x(a - 1) + 2 = 0$$
$$\Delta = (a - 1)^2 - 16 = 0 \text{ for one solution}$$
$$(a - 1)^2 = 16$$
$$a - 1 = \pm 4$$
$$a = 5, -3$$

**b** We need $\Delta > 0$ for two solutions

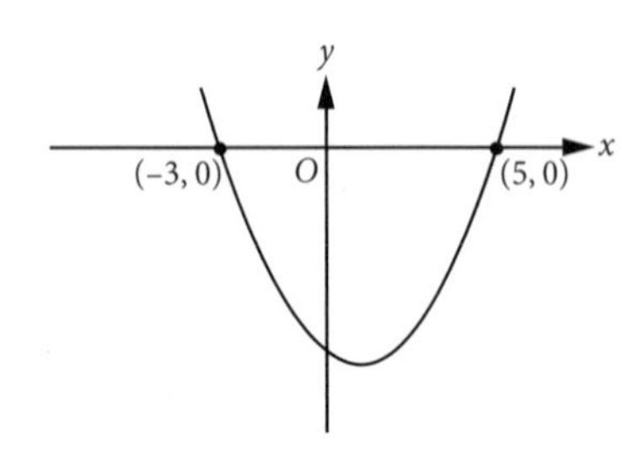


$a > 5, a < -3$

**c** We need $\Delta < 0$ for no solutions

$-3 < a < 5$

**d** For exactly one solution, $a = 5, -3$

Since $a > 0$, $a = 5$

$h(x) = f(x) - g(x) = x + 2 - (2x^2 + ax + 4)$, where $a = 5$

$\Rightarrow x + 2 - 2x^2 - 5x - 4$

$\Rightarrow -2x^2 - 4x - 2$

$h(x) = -2x^2 - 4x - 2 = -2(x^2 + 2x + 1)$

So $h(x) = -2(x + 1)^2$.

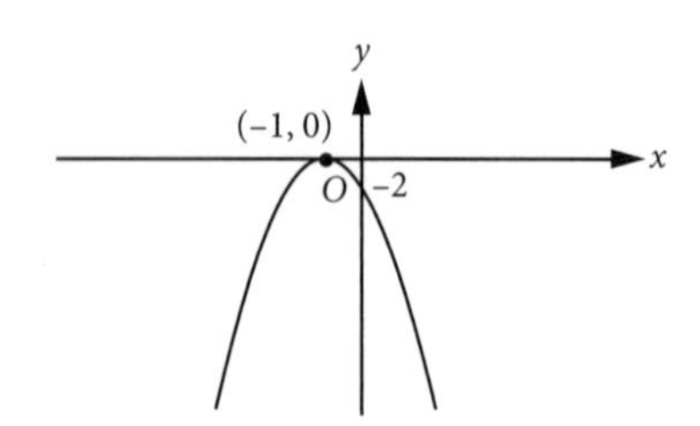


**e** $h(x) = -2(x + 1)^2$

$h'(x) = -4(x + 1) = 0$ for stationary point

This gives $x = -1$.

Coordinates are $(-1, 0)$.

**f** $h_T(x) = -h(2x - 3) + 7$

$h_T(x) = 2(2x - 2)^2 + 7$

**g**

| $h(x) = -2(x + 1)^2$ | |
|---|---|
| Dilate from the $y$-axis by 0.5 units | $y = -2(2x + 1)^2$ |
| Reflect over the $x$-axis | $y = 2(2x + 1)^2$ |
| Translate in +ve direction of the $x$-axis by 1.5 units | $y = 2(2(x - 1.5) + 1)^2$ $\Rightarrow y = 2((2x - 3) + 1)^2$ $\Rightarrow y = 2(2x - 2)^2$ |
| Translate in +ve direction of the $y$-axis by 7 units | $y = 2(2x - 2)^2 + 7$ so $h_T(x) = 2(2x - 2)^2 + 7$ |

**Question 7**

**a**

$$f(x) = (x - 1)(x^2 - k)$$
$$f(2x - 1) = (2x - 1 - 1)[(2x - 1)^2 - k]$$
$$-f(2x - 1) = -(2x - 2)[(2x - 1)^2 - k]$$
$$-f(2x - 1) + 3 = -(2x - 2)[(2x - 1)^2 - k] + 3$$

```
define f(x)=(x-1)(x²-k)
                                           done
expand(-f(2·x-1)+3)
                 -8·x³+16·x²+2·k·x-10·x-2·k+5
```

$g(x) = -f(2x - 1) + 3$

$= -8x^3 + 16x^2 + x(2k - 10) + 5 - 2k$

**b** $g(x) = -8x^3 + 16x^2 + x(2k - 10) + 5 - 2k$

$g'(x) = -24x^2 + 32x + (2k - 10)$

$g'(1) = -24 + 32 + (2k - 10)$

gradient $= 2k - 2$

When $x = 1, y = 3$

$y - y_1 = m(x - x_1)$
$y - 3 = (2k - 2)(x - 1)$

gives

$y = (2k - 2)x + 5 - 2k.$

**c** When $x = 1$, perpendicular gradient $= -\dfrac{1}{2k - 2}$

$y - y_1 = m(x - x_1)$

$y - 3 = -\dfrac{1}{2k - 2}(x - 1)$

gives $y = -\dfrac{x}{2k - 2} + \dfrac{1}{2k - 2} + 3$

```
define f(x)=(x−1)(x²−k)
                                done
Define g(x)=−f(2·x−1)+3
                                done
normal(g(x),x,1)
                 −x/(2·k−2)+1/(2·k−2)+3
```

**d** The equation of the tangent at $x = 1$ to $f(x) = (x - 1)(x^2 - k)$ is

$y = -x(k - 1) + k + 1$

```
define f(x)=(x−1)(x²−k)
                                done
tanLine(f(x),x,1)
                        −x·(k−1)+k−1
solve(−x·(k−1)+k−1=(2·k−2)·x+5−2·k,k)
                        {k=(x−2)/(x−1)}
```

Equate the 2 tangents and solve for $k$.

$(2k - 2)x + 5 - 2k = -x(k - 1) + k - 1$
$(2k - 2)x + 5 - 2k = -xk + x + k - 1$
$2kx + xk - 3k = 2x - 5 + x - 1$
$3k(x - 1) = 3x - 6$
$k = \dfrac{3(x - 2)}{3(x - 1)}$
$k = \dfrac{x - 2}{x - 1}$

For $x = 1$ this is not possible.

**Question 8**

**a** $h(x) = -x^2 + 10x + 1$

$h(0) = 1$

The ball leaves Paddy's hand at a height of 1 metre.

**b** $h(x) = -x^2 + 10x + 1$

To find the maximum, solve

$h'(x) = -2x + 10 = 0$

This gives $x = 5$.

$h(5) = 26$

$\therefore$ the maximum height is 26 metres.

**c**

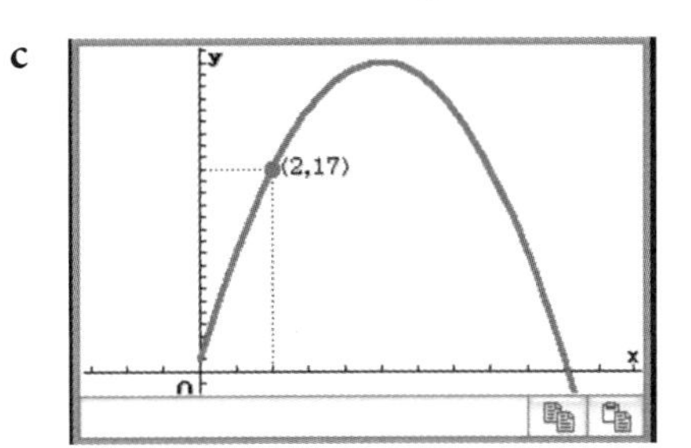


Path of the ball goes through $(2, 17)$.

Fence has a height of 2 metres.

$17 > 2$

$\therefore$ the ball won't hit the fence.

**d** Solve $h(x) = -x^2 + 10x + 1 = 1.5$.

```
define h(x)=−x²+10x+1|x≥0
                                done
solve(h(x)=1.5,x)
        {x=(−7·√2)/2+5, x=(7·√2)/2+5}
solve(h(x)=1.5,x)
{x=0.05025253169,x=9.949▶
```

Imogen gets hit after the ball goes over the fence, so take the larger $x$ value.

$x = 9.95$

Imogen is 9.95 metres from Paddy.

**e** Distance from $(30, 0)$ to $(0, 1)$

$= \sqrt{1^2 + 30^2} = \sqrt{901}$ m.

**f** Horizontal distance to $(30, 0)$ is 30 metres.

**g** $h(x) = -x^2 + 10x + 1$

So $h(2) = 17$

$h'(x) = -2x + 10$
$h'(2) = 6$

The equation of the line is

$$y - y_1 = m(x - x_1)$$
$$y - 17 = 6(x - 2)$$
$$y = 6x + 5$$

**h** **i** area $= \int_0^2 (6x + 5) - (-x^2 + 10x + 1)\,dx$

**ii** area $= \frac{8}{3}$ sq. units

### Question 9

**a** $\sin(2x) = 0$ for $x \in [0, 2\pi]$

$2x = 0, \pi, 2\pi, 3\pi, 4\pi$

$x = 0, \frac{\pi}{2}, \pi, \frac{3\pi}{2}, 2\pi.$

**b** $f(x) = 0$ for $f(x) = 3e^{\frac{x}{10}} g(x) = 0$

Solve $3e^{\frac{x}{10}} \sin(2x) = 0$

Either $3e^{\frac{x}{10}} = 0$ or $\sin(2x) = 0$

$3e^{\frac{x}{10}} = 0$ gives no solution

Answers: $x = 0, \frac{\pi}{2}, \pi, \frac{3\pi}{2}, 2\pi.$

**c** $f(x) = 3e^{\frac{x}{10}} \sin(2x)$

To find the stationary points, solve

$f'(x) = \frac{3}{10} e^{\frac{x}{10}} \sin(2x) + 6e^{\frac{x}{10}} \cos(2x) = 0$

$e^{\frac{x}{10}} (0.3 \sin(2x) + 6 \cos(2x)) = 0$

$e^{\frac{x}{10}} = 0$ has no solution

$0.3 \sin(2x) = -6 \cos(2x)$

$\Rightarrow \tan(2x) = -20$

$\Rightarrow x = 0.81, x = 2.38, x = 3.95, x = 5.52$

Coordinates of the stationary points are $(0.81, 3.25)$, $(2.38, -3.80)$, $(3.95, 4.45)$, $(5.52, -5.21)$.

**d** $x$-intercepts: $x = 0, \frac{\pi}{2}, \pi, \frac{3\pi}{2}, 2\pi.$

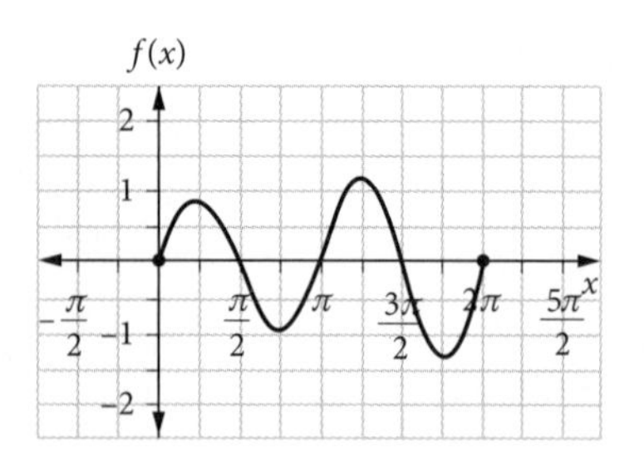


**e** $A\,(a, f(a)) = \left(\frac{\pi}{4}, f\left(\frac{\pi}{4}\right)\right)$

$B\,(b, f(b)) = \left(\frac{5\pi}{4}, f\left(\frac{5\pi}{4}\right)\right)$

gradient $= m = \frac{f(b) - f(a)}{b - a}$

$$m = \frac{f\left(\frac{5\pi}{4}\right) - f\left(\frac{\pi}{4}\right)}{\frac{5\pi}{4} - \frac{\pi}{4}} = \frac{f\left(\frac{5\pi}{4}\right) - f\left(\frac{\pi}{4}\right)}{\pi}$$

**f** $n = \frac{f\left(\frac{7\pi}{4}\right) - f\left(\frac{3\pi}{4}\right)}{\pi}$

### Question 10

**a** Factorise $p(x) = a(bx^2 - cx)^2$
$= a(bx^2 - cx)(bx^2 - cx)$
$= ax^2(bx - c)(bx - c)$
$= ax^2(bx - c)^2$

Let $ax^2(bx - c)^2 = 0$

$\Rightarrow x = 0, x = \frac{c}{b}$

**b** $p(x) = ax^2(bx - c)^2$

$p'(x) = 2ax(bx - c)(2bx - c) = 0$ for stationary points.

This gives $x = 0, x = \frac{c}{2b}$ and $x = \frac{c}{b}$.

**c** The stationary points are at $x = 0, x = \frac{c}{2b}$ and $x = \frac{c}{b}$.

$p'\left(\frac{10}{3}\right) = 0$ gives the 2nd stationary point $x = \frac{c}{b}$.

So $p'\left(\frac{c}{b}\right) = 0$ and thus $\frac{10}{3} = \frac{c}{b}$.

**d** $a = 0.1, b = 3, c = 10$

**e** The safety officer sits at $(1,0)$.

The rollercoaster goes through $(1,4.9)$.

At point $(1,4.9)$, the gradient is 5.6.

$5.6 > 5$

So the rollercoaster will be shut down.

### Question 11

**a** $f(x) = 3x^2 - x^3$

$f'(x) = 6x - 3x^2$

For stationary points, solve for $f'(x) = 0$.

$6x - 3x^2 = 3x(2 - x) = 0$

This gives $x = 0$, $x = 2$.

The coordinates are $(0,0)$ and $(2,4)$.

**b**

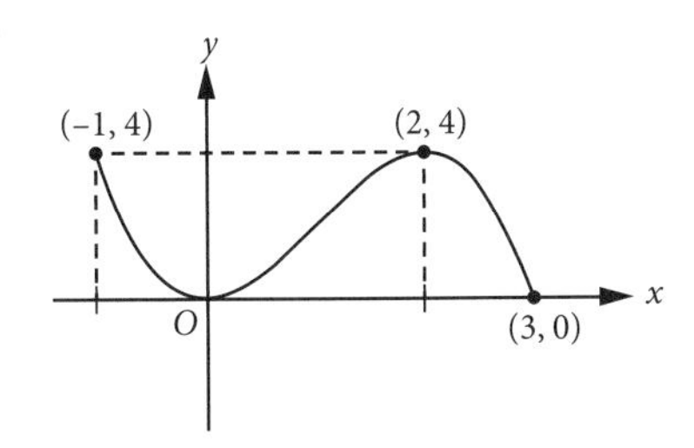


**c** Max = 4 at $x = -1$ and $x = 2$.

**d** Min = 0 at $x = 0$ and $x = 3$.

**e** Strictly increasing for $x \in [0,2]$.

**f** Use $y - y_1 = m(x - x_1)$

$f'(x) = 6x - 3x^2$

$f'\left(\frac{7}{4}\right) = \frac{21}{16}$

Also, $f\left(\frac{7}{4}\right) = \frac{245}{64}$

so $y - \frac{245}{64} = \frac{21}{16}\left(x - \frac{7}{4}\right)$.

define f(x)=3x²−x³ | −1≤x≤3

done

$\text{tanLine}\left(f(x), x, \frac{7}{4}\right)$

$\frac{21\cdot x}{16} + \frac{49}{32}$

The equation of this tangent is $y = \frac{21}{16}x + \frac{49}{32}$.

**g i**

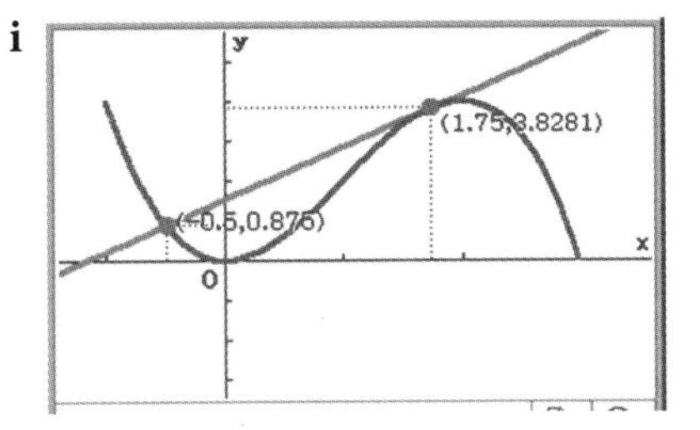


The tangent hits the curve again at point $\left(-\frac{1}{2}, \frac{7}{8}\right)$.

$$\text{area} = \int_{-\frac{1}{2}}^{\frac{7}{4}} \left(\frac{21x}{16} + \frac{49}{32}\right) - (3x^2 - x^3)\,dx$$

**ii** $\text{area} = \int_{-\frac{1}{2}}^{\frac{7}{4}} \left(\frac{21x}{16} + \frac{49}{32}\right) - (3x^2 - x^3)\,dx$

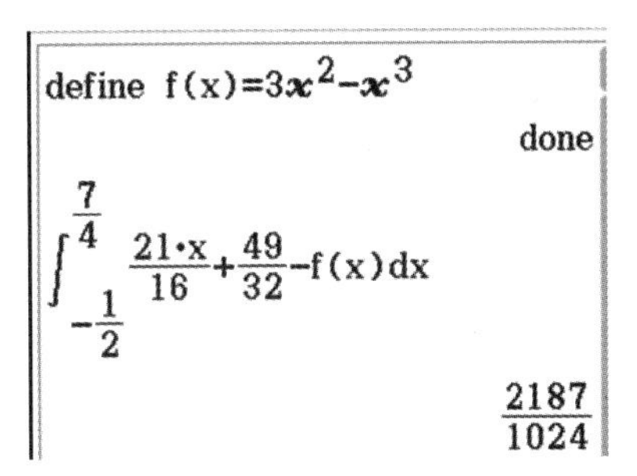
define f(x)=3x²−x³

done

$\int_{-\frac{1}{2}}^{\frac{7}{4}} \frac{21\cdot x}{16} + \frac{49}{32} - f(x)\,dx$

$\frac{2187}{1024}$

$\text{area} = \frac{2187}{1024}$ sq. units

**h** $\text{area} = 1 \times \left(\sum f(y)\right)$

$= 1(f(0) + f(1) + f(2) + f(3))$

$= 1(1 + 0.879 + 0.732 + 0.532)$

$= 3.14$ sq. units

**i** area by integration

$= \int_2^3 (3x^2 - x^3)\,dx = 2.75$ sq. units

ratio = 2.75 : 3.14

= 275 : 314

### Question 12

**a** $\text{period} = \frac{2\pi}{\frac{\pi}{6}} = 12$

amplitude = 5000

**b** Maximum = 15 000, at $t = 6$

Maximum occurs in June.

**c** Minimum = 5000, at $t = 0$ and $t = 12$

Minimum occurs in January and December.

**d** Sketch the derivative graph $p'(t)$.

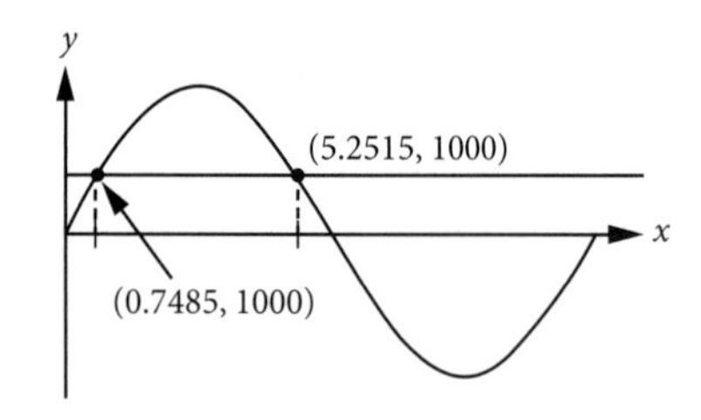


Solve for $p'(t) > 1000$:

$p'(t) > 1000$ for $0.7485 < t < 5.2515$

$\Rightarrow t \in (0.75, 5.25)$

**e**

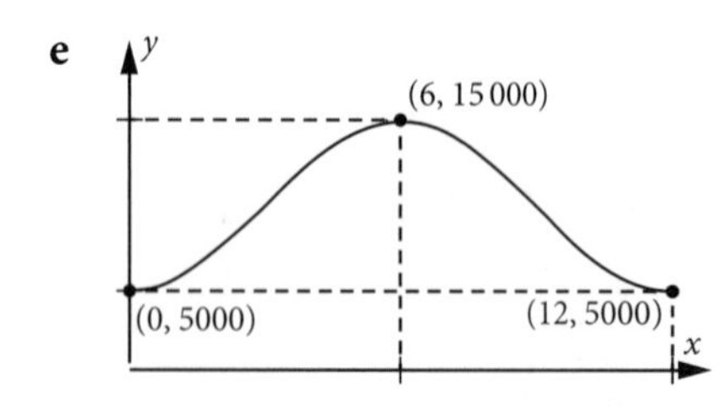


**f** $y(t) = 5000 \sin\left(\frac{\pi t}{4}\right) + 80\,000$

$y'(t) = 1250\pi \cos\left(\frac{\pi t}{4}\right)$

Sketch the derivative graph $y'(t)$.

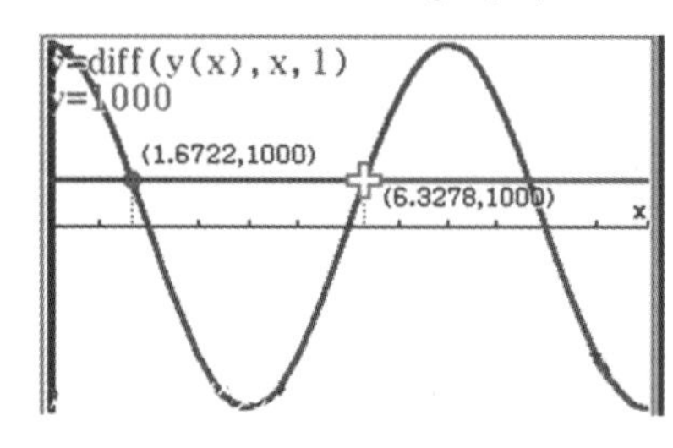


Solve for $y'(t) > 1000$:

$y'(t) > 1000$ for $0 < t < 1.6722$,
$6.3278 < t < 9.6722$

$\Rightarrow t \in (0, 1.67) \cup (6.33, 9.67)$

**Ratio comparison**

$p'(t) > 1000 \Rightarrow t \in (0.75, 5.25)$

$y'(t) > 1000 \Rightarrow t \in (0, 1.67) \cup (6.33, 9.67)$

$\text{Ratio} = \frac{p'(t)}{y'(t)} = \frac{5.25 - 0.75}{1.67 + 9.67 - 6.33} = \frac{150}{167}$

Ratio of yellow-billed loon : common loon
= 167 : 150

**Question 13**

**a** $\text{Speed} = \frac{\text{distance}}{\text{time}}$, so $\text{time} = \frac{\text{distance}}{\text{speed}}$

$T(x) = \frac{PC}{6} + \frac{CC'}{6} + \frac{C'D}{k}$

$T(x) = \frac{\sqrt{x^2 + 2000^2}}{6} + \frac{1000 - \sqrt{x}}{6} + \frac{8000 - (1000 - \sqrt{x})}{k}$

$T(x) = \frac{\sqrt{x^2 + 2000^2}}{6} + \frac{1000 - \sqrt{x}}{6} + \frac{7000 + \sqrt{x}}{k}$

**b** $T(x) = \frac{\sqrt{x^2 + 2000^2}}{6} + \frac{1000 - \sqrt{x}}{6} + \frac{7000 + \sqrt{x}}{k}$

Let $k = 10$.

So at $k = 10$:

$T'(x) = \frac{5x^{\frac{3}{2}} - \sqrt{x^2 + 2000^2}}{30\sqrt{x(x^2 + 2000^2)}}$

Solve $\frac{5x^{\frac{3}{2}} - \sqrt{x^2 + 2000^2}}{30\sqrt{x(x^2 + 2000^2)}} = 0$ for the stationary point.

```
solve(d/dx(t(x))=0,x)
                {x=54.30168894}
```

The solution for $\frac{5x^{\frac{3}{2}} - \sqrt{x^2 + 2000^2}}{30\sqrt{x(x^2 + 2000^2)}} = 0$ is $x = 54.3017$.

Test using a variety of methods if this answer is max or min. The graph shows it is a local min.

So the minimum time = $T(54.3017) = 1199.63$

Answer = 1200 seconds (to the nearest second)

**c** $x$ must not be zero so $x \in (0, \infty)$

and $1000 - \sqrt{x} > 0$.

Answer: $x \in (0, 1000^2)$

**d** A diagram will help.

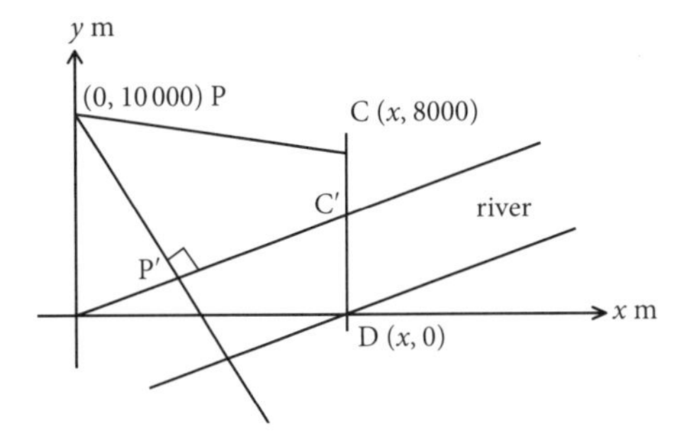


Let the point on the river bank be P′:

Speed $= \frac{\text{dist}}{\text{time}}$, so time $= \frac{\text{dist}}{\text{speed}}$.

$T(x) = \frac{PP'}{6}$

The line from the origin to C′ has the equation $y = mx$,

where $m = \frac{8000 - (1000 - \sqrt{x})}{x} = \frac{7000 + \sqrt{x}}{x}$.

So the line from the origin to C′ has the equation $y = \frac{7000 + \sqrt{x}}{x}x$.

Perpendicular gradient $= \frac{-x}{7000 + \sqrt{x}}$

To find equation of line PP′,

$y = mx + c$ where $c = 10\,000$.

Equation of line PP′ is

$y = \frac{-x}{7000 + \sqrt{x}}x + 10\,000$.

Now, equate to find the coordinates of P′.

$\frac{-x}{7000 + \sqrt{x}}x + 10\,000 = \frac{7000 + \sqrt{x}}{x}x$

Solve for $x$: $\frac{-x}{7000 + \sqrt{x}}x + 10\,000 = 7000 + \sqrt{x}$

```
solve(-x/(7000+√x)·x+10000=7000+√x, x)
                                  {x=4552.533099}
```

$x = 4552.533$

coordinates of P′ = (4552.533, 7067.472)

distance PP′

$= \sqrt{4552.533^2 + (10000 - 7067.472)^2}$

= 5415.282 metres

$T(x) = \frac{5415.282}{6} = 902.546\,97$

Time from P to P′ = 903 seconds

# UNIT 4

## Chapter 4

### Area of Study 4: Data analysis, probability and statistics

### Short-answer questions

**Question 1**

IQ scores are normally distributed, mean = 100, standard deviation = 15.

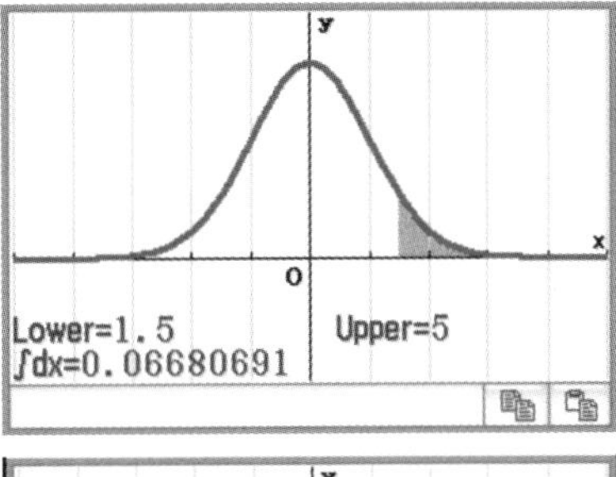


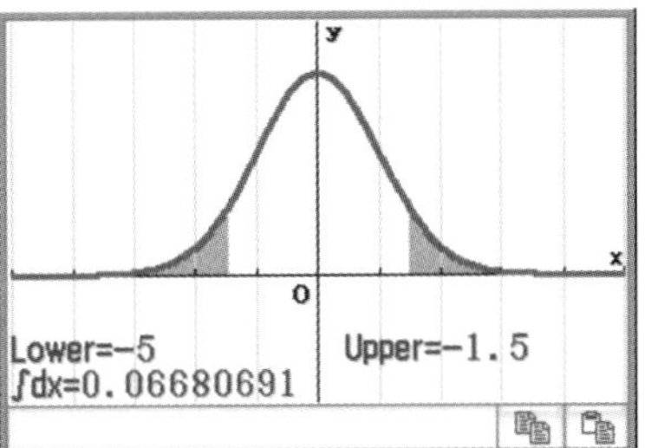


Normal distribution, mean = 100, standard deviation = 15.

$\Pr(Z > 1.5) = 0.0668$ also equals

$\Pr(Z < -1.5) = 0.0668$

Use $z = \frac{x - \mu}{\sigma}$:

$-1.5 = \frac{x - 100}{15}$

$\therefore x = 77.5$

**Question 2**

Normal variable with $\mu = 10$, $\sigma = 5$, when $X = 18$.

Using $z = \frac{x - \mu}{\sigma}$ gives $z = \frac{18 - 10}{5} = \frac{8}{5} = 1.6$

$\therefore z = 1.6$

**Question 3**

mean $= (1 \times 0.5) + (3 \times 0.3) + (5 \times 0.2) = 2.4$

**Question 4**

**a** mean

$= (-2 \times 0.4) + (-1 \times 0.3) + (0 \times 0.1) + (1 \times 0.2)$

$= -0.9$

**b** $\Pr(Y \leq 0) = 1 - \Pr(X = 1)$

$= 1 - 0.2$

$= 0.8$

**Question 5**

If $E(X) = 2$, state the value of $E(3X - 2)$.

Use the rule $E(aX + b) = aE(X) + b$.

$E(X) = 2$, so

$E(3X - 2) = 3E(X) - 2$

$= 3 \times 2 - 2$

$= 4$

**Question 6**

The mean is 1, so we get

$(0 \times a) + (1 \times b) + (2 \times 0.1) + (3 \times 0.1) = 1$

$\Rightarrow b = 0.5$

Also, $\Sigma x\Pr(X = x) = 1$

so $a + b + 0.1 + 0.1 = 1$

$a + 0.5 + 0.1 + 0.1 = 1$

$a = 0.3$

**Question 7**

Use $\text{Bi}(n, p)$, where $p = 0.2$, $n = 500$:

$E(X) = 0.2 \times 500 = 100$

100 are expected to catch the common cold.

**Question 8**

Use $\text{Bi}(n, p)$, where $p = 0.2$, $n = ?$ and mean $= 12$.

**a** mean $= np$

$12 = 0.2n$ so $n = 60$ trials

**b** $\text{Var}(X) = npq$

$= 60 \times 0.2 \times 0.8$

$= 9.6$

**c** $\Pr(X = 1) = {}^{60}C_1(0.8)^{59}(0.2)^1$

$= 60(0.8)^{59}(0.2)^1$

**Question 9**

**a**

| $x$ | 0 | 1 | 2 |
|---|---|---|---|
| $\Pr(X = x)$ | ${}^2C_0(0.5)^2(0.5)^0$ | ${}^2C_1(0.5)^1(0.5)^1$ | ${}^2C_2(0.5)^0(0.5)^2$ |
| $\Pr(X = x)$ | $\frac{1}{4}$ | $\frac{1}{2}$ | $\frac{1}{4}$ |

**b** $\Pr(X < 2) = \frac{1}{4} + \frac{1}{2} = \frac{3}{4}$

**Question 10**

Sam (S), Rain (R)

$\Pr(S) = 0.8$, $\Pr(R) = 0.2$, $\Pr(R|S) = ?$

$\Pr(S|R) = 0.5$, giving $\Pr(S|R) = 0.5 = \dfrac{\Pr(S \cap R)}{\Pr(R)}$

So $0.5 \times \Pr(R) = \Pr(S \cap R)$

and $\Pr(S \cap R) = 0.5 \times 0.2 = 0.1$.

$\Pr(R|S) = \dfrac{\Pr(S \cap R)}{\Pr(S)} = \dfrac{0.1}{0.8} = \dfrac{1}{8}$

**Question 11**

$\text{Bi}(n, p)$ where $n = 4$, $p = \frac{1}{5}$

**a** $\Pr(X = 4) = \left(\frac{1}{5}\right)^4 = \frac{1}{625}$

**b** $\Pr(X = 0) + \Pr(X = 1) = \left(\frac{4}{5}\right)^4 + {}^4C_1\left(\frac{4}{5}\right)^3\left(\frac{1}{5}\right)^1$

$= \frac{512}{625}$

**c** $\Pr(X = 3) + \Pr(X = 4)$

$= {}^4C_3\left(\frac{4}{5}\right)^1\left(\frac{1}{5}\right)^3 + \left(\frac{1}{5}\right)^4 = \frac{17}{625}$

### Question 12

normal distribution, $\mu = 18$ and $\sigma = 2.7$.

95% of the values lie within $2\sigma$ of the mean.

$$\begin{aligned}\text{range} &= \mu - 2\sigma \le X \le \mu + 2\sigma \\ &= 18 - 2(2.7) \le X \le 18 + 2(2.7) \\ &= 12.6 \le X \le 23.4\end{aligned}$$

### Question 13

normal distribution

Saturday: $x = 90$, mean = 80, standard deviation = 5

Sunday: $x = 60$, mean = 50, standard deviation = 6

Use $z = \dfrac{x - \mu}{\sigma}$ to standardise.

Saturday: $z = \dfrac{90 - 80}{5} = \dfrac{10}{5} = 2$

Sunday: $z = \dfrac{60 - 50}{6} = \dfrac{5}{3} \approx 1.67$

The store does better on Saturday because $2 > 1.67$.

### Question 14

**a** Graph of $f(x)$, shaded where $\Pr\left(X < \frac{1}{2}\right)$

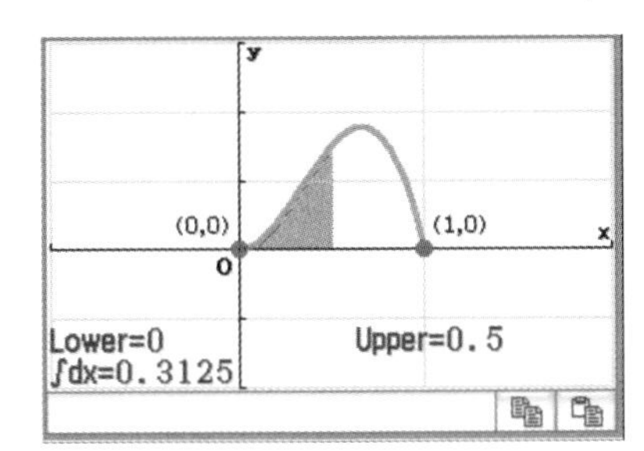


**b** 
$$\begin{aligned}\Pr\left(X < \frac{1}{2}\right) &= \int_0^{\frac{1}{2}} 12x^2(1 - x)\,dx \\ &= \int_0^{\frac{1}{2}} 12x^2 - 12x^3\,dx \\ &= \left[\frac{12x^3}{3} - \frac{12x^4}{4}\right]_0^{\frac{1}{2}} \\ &= \left[4x^3 - 3x^4\right]_0^{\frac{1}{2}} \\ &= \frac{1}{2} - \frac{3}{16} = \frac{5}{16}\end{aligned}$$

### Question 15

**a** First, test $\int_1^2 (2 - x)\,dx = \frac{1}{2}$

A PDF area = 1 so $\int_0^1 kx\,dx$ must also $= \frac{1}{2}$.

So $k = 1$.

**b** From the graph, mean = 1

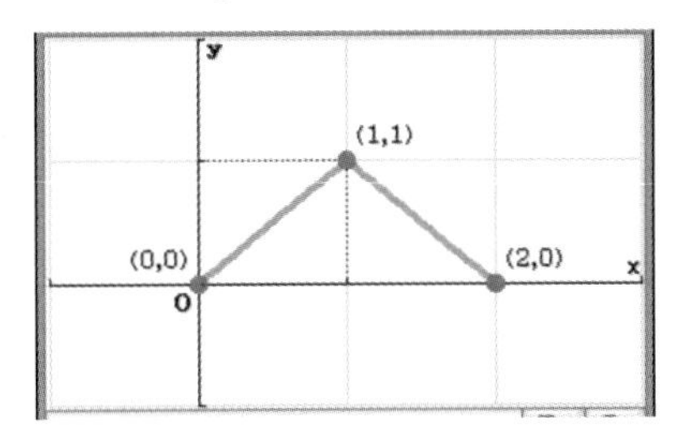


### Question 16

For a PDF $\int_0^1 (2k \sin(\pi x))\,dx = 1$

$$\therefore \int_0^1 (2k \sin(\pi x))\,dx = 2k\left[-\frac{1}{\pi}\cos(\pi x)\right]_0^1 = 1$$

$$\therefore -\frac{2k}{\pi}[\cos(\pi) - (0)] = 1$$

$$\therefore -\frac{2k}{\pi}[-1 - 1] = 1$$

$$\Rightarrow k = \frac{\pi}{4}$$

### Question 17

**a** The population is all the people waiting in a queue at the MCG on a particular Saturday afternoon.

**b** There are three population parameters: the minimum total, the average and the maximum they intended to pay for tickets.

**c** There are three sample statistics: the minimum ($10), the average ($50), and the maximum ($100).

**Question 18**

**a**

| $x$ | 0 | 1 | 2 | 3 | 4 |
|---|---|---|---|---|---|
| $\hat{p} = \frac{x}{4}$ | $\frac{0}{4} = 0$ | $\frac{1}{4}$ | $\frac{2}{4} = \frac{1}{2}$ | $\frac{3}{4}$ | $\frac{4}{4} = 1$ |
| $\Pr(X = x)$ | ${}^4C_0 0.7^0 0.3^4$ | ${}^4C_1 0.7^1 0.3^3$ | ${}^4C_2 0.7^2 0.3^2$ | ${}^4C_3 0.7^3 0.3^1$ | ${}^4C_4 0.7^4 0.3^0$ |

**b** $E(\hat{p}) = E\left(\frac{x}{n}\right)$

Using $E(aX + b) = aE(X) + b$

$E(\hat{p}) = \frac{1}{n}E(x) = \frac{1}{n}(np) = p$

**Question 19**

Success is catching a cold.

Use the formula $\hat{p} = \frac{x}{n} = \frac{20}{100} = 0.2$

The sample proportion for those who caught colds is 0.2.

**Question 20**

Since $\hat{p} = 0.5$ and $n = 100$,

$\sqrt{\frac{\hat{p}(1-\hat{p})}{n}} = \sqrt{\frac{0.5 \times 0.5}{100}} = 0.05$

Using a 95% confidence interval,

$$\begin{aligned}95\%\text{ CI} &= \left(\hat{p} - z\sqrt{\frac{\hat{p}(1-\hat{p})}{n}}, \hat{p} + z\sqrt{\frac{\hat{p}(1-\hat{p})}{n}}\right)\\ &= (0.5 - 1.96 \times 0.05, 0.5 + 1.96 \times 0.05)\\ &= (0.402, 0.598)\end{aligned}$$

## Multiple-choice questions

**1** D

The sample space of the outcomes of a fair coin being tossed twice is {HH, HT, TH, TT}. A tree diagram may be helpful.

**2** B

There are 6 possibilities when a fair die is thrown.

The sample space of the outcomes is {1, 2, 3, 4, 5, 6}.

**3** D

The height of footballers in a team is **not** discrete because heights can vary in infinitely small intervals. It is not countable.

**4** B

The shoe size of students in your class is discrete because the number of shoe sizes is countable.

**5** E

Use the rule $\text{Var}(aX + b) = a^2\,\text{Var}(X)$

$\text{Var}(X) = 2$, so

$\text{Var}(2X + 1) = 2^2\,\text{Var}(X) = 4 \times 2 = 8.$

**6** D

Use the rule $\text{Var}(aX + b) = a^2\,\text{Var}(X)$

$\text{Var}(X) = 4$, so

$\text{Var}(-X) = (-1)^2\,\text{Var}(X) = 1 \times 4 = 4.$

**7** A

Use the rule $E(aX + b) = aE(X) + b$

$E(X) = 1.1$, so

$$\begin{aligned}E(-X - 2) &= (-1)E(X) - 2\\ &= -1 \times 1.1 - 2\\ &= -3.1\end{aligned}$$

**8** A

When tossing four coins, there is only one branch of a tree diagram where we get TTTT. So the probability of getting four tails is $\frac{1}{2} \times \frac{1}{2} \times \frac{1}{2} \times \frac{1}{2} = \frac{1}{16}$.

**9** D

The graph is skewed to the right, so it is skewed positively. It is therefore incorrect to say that the graph is skewed negatively.

**10** B

This is a binomial distribution, $\text{Bi}(n, p)$ where $n = 60$ and $p = \frac{1}{6}$ (for rolling a six).

Using the formula $\text{E}(X) = np$, we get

$\text{E}(X) = 60 \times \frac{1}{6} = 10.$

The expected number of sixes is 10.

**11** D

It is not true to say that binomial experiments form a continuous probability distribution.

A binomial experiment is discrete with $n$ independent trials and 'success' and 'failure'.

**12** A

$\Pr(X = x) = {}^5C_x(0.8)^{5-x}(0.2)^x$
where $x = 0, 1 \ldots 5$.

Using the formula

$\Pr(X = x) = {}^nC_x(q)^{n-x}(p)^x$,

the probability of success is 0.2.

**13** B

mean of $X = (-1 \times 0.3) + (0 \times 0.1) + (1 \times 0.3) + (2 \times 0.2) + (3 \times 0.1) = 0.7$

**14** E

The expected value of $X^2$:

$$\begin{aligned}\text{E}(X^2) &= ((-1)^2 \times 0.3) + (0^2 \times 0.1) + (1^2 \times 0.3) \\ &\quad + (2^2 \times 0.2) + (3^2 \times 0.1) \\ &= 2.3\end{aligned}$$

**15** D

$$\begin{aligned}\text{variance of } X &= \text{E}(X^2) - (\text{E}(X))^2 \\ &= 2.3 - (0.7)^2 \\ &= 1.81\end{aligned}$$

**16** D

mean $= (-1 \times 0.3) + (1 \times 0.3) + (2 \times 0.2) + (3 \times 0.1) = 0.7$

**17** D

| $x$ | $-1$ | $0$ | $1$ |
|---|---|---|---|
| $\Pr(X = x)$ | $2a$ | $0.5a$ | $a$ |

We need the value of $a$ and the mean.

For a discrete random variable table, $\Sigma x \Pr(X = x) = 1$.

Solving $2a + \frac{a}{2} + a = 1$ gives $a = \frac{2}{7}$.

$$\text{E}(X) = \left(-1 \times \frac{4}{7}\right) + \left(0 \times \frac{1}{7}\right) + \left(1 \times \frac{2}{7}\right) = -\frac{2}{7}$$

$a = \frac{2}{7}, -\frac{2}{7}$

**18** E

$$\begin{aligned}\text{E}(X^2) &= \text{Expected value of } X^2 \\ &= (1^2 \times 0.5) + (2^2 \times 0.2) + (3^2 \times 0.1) \\ &\quad + (4^2 \times 0.2) \\ &= 5.4\end{aligned}$$

**19** C

mean of $X =$
$(1 \times 0.5) + (2 \times 0.2) + (3 \times 0.1) + (4 \times 0.2) = 2$

**20** D

A binomial trial is 'Drawing six cards from a deck of cards, with replacement, and recording the number of picture cards' because we can have picture cards or not picture cards.

**21** D

Use $\text{Bi}(n, p)$ with $n = 40$, $p = \frac{3}{4}$ and mean $= np$ and $\text{Var}(X) = npq$.

mean $= 40 \times \frac{3}{4} = 30$

$\text{Var}(X) = 40 \times \frac{3}{4} \times \frac{1}{4} = \frac{15}{2}$

So mean = 30, standard deviation $= \sqrt{\frac{15}{2}}$.

**22** B

$\Pr(X = x) = {}^4C_x(0.3)^x(0.7)^{4-x}$

$$\begin{aligned}\Pr(X \geq 3) &= \Pr(X = 3) + \Pr(X = 4) \\ &= {}^4C_3(0.3)^3(0.7)^1 + {}^4C_4(0.3)^4(0.7)^0 \\ &= 4(0.3)^3(0.7) + (0.3)^4\end{aligned}$$

SOLUTIONS – CHAPTER 4

**23** D

$\text{Var}(X) = 4$ and $Y = 2X + 3$

Use the rule $\text{Var}(aX + b) = a^2 \text{Var}(X)$

$\text{Var}(Y) = \text{Var}(2X + 3) = 2^2 \text{Var}(X) = 4 \times 4 = 16$

**24** C

If $E(X) = 4$ and $Y = 2X + 3$, find $E(Y) = E(2X + 3)$.

Use the rule $E(aX + b) = a\,E(X) + b$

$$\begin{aligned} E(2X + 3) &= 2\,E(X) + 3 \\ &= 2 \times 4 + 3 \\ &= 11 \end{aligned}$$

**25** C

$E((X - 1)^2) = ((-1)^2 \times 0.2) + ((1)^2 \times 0.1) + ((2)^2 \times 0.5) + ((4)^2 \times 0.2) = 5.5$

**26** A

Two dice are rolled.

Draw a grid representing the 36 possible outcomes.

| | | | | | |
|---|---|---|---|---|---|
| 1,1 | 1,2 | 1,3 | 1,4 | 1,5 | 1,6 |
| 2,1 | 2,2 | 2,3 | 2,4 | 2,5 | 2,6 |
| 3,1 | 3,2 | 3,3 | 3,4 | 3,,5 | 3,6 |
| 4,1 | 4,2 | 4,3 | 4,4 | 4,5 | 4,6 |
| 5,1 | 5,2 | 5,3 | 5,4 | 5,5 | 5,6 |
| 6,1 | 6,2 | 6,3 | 6,4 | 6,5 | 6,6 |

Consider the sums. For example, there is only 1 case where the sum = 2.

Now consider the pattern in the diagonals.

Set up a discrete distribution table.

| $x$ | 2 | 3 | 4 | 5 | 6 | 7 | 8 | 9 | 10 | 11 | 12 |
|---|---|---|---|---|---|---|---|---|---|---|---|
| $\Pr(X = x)$ | $\frac{1}{36}$ | $\frac{2}{36}$ | $\frac{3}{36}$ | $\frac{4}{36}$ | $\frac{5}{36}$ | $\frac{6}{36}$ | $\frac{5}{36}$ | $\frac{4}{36}$ | $\frac{3}{36}$ | $\frac{2}{36}$ | $\frac{1}{36}$ |

**27** A

For a PDF, we need $\int_1^3 k(x - 1)^2\,dx = 1$

Solve for $k$.

```
solve(∫[1,3] k·(x−1)^2 dx=1, k)
                              {k=3/8}
```

$k = \frac{3}{8}$

**28** E

$\text{Var}(X) = 6$ and $E(X) = 2$

Use $\text{Var}(X) = E(X^2) - (E(X))^2$

So $6 = E(X^2) - (2)^2$, giving $E(X^2) = 10$.

**29** D

$\Pr(A) = 0.7$

$\Pr(C) = 0.5$

Because the events are independent,
$\Pr(A \cap C) = 0.7 \times 0.5 = 0.35$ so A is correct.

Additionally $\Pr(A \cup C) = \Pr(A) + \Pr(C) - \Pr(A \cap C) = 0.7 + 0.5 - 0.35 = 0.85$,
so C is correct.

$\Pr(A \mid C) = P(A \cap C) / P(C) = 0.35/0.5 = 0.7$,
so E is correct.

It would be incorrect to say that they are mutually exclusive events because $\Pr(A \cap C) \neq 0$.

**30** B

Use $\text{Bi}(n, p)$ with mean $= np = 20$ and $\text{Var}(X) = npq = 10$.

Solve for $n$ and $p$.

$$\begin{cases} np=20 \\ np(1-p)=10 \end{cases} \Bigg|_{n,\,p}$$

$$\left\{n=40,\ p=\frac{1}{2}\right\}$$

$p = \dfrac{1}{2}$, $n = 40$

**31** E

This is without replacement. We need the first marble to be a red and the 2nd blue and vice versa.

A tree diagram for the event is

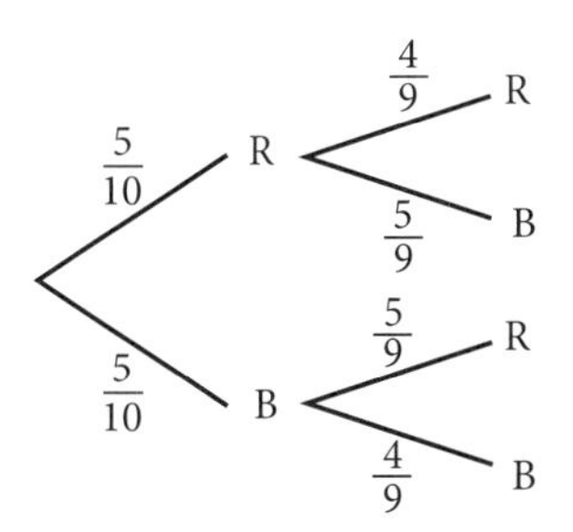


The probability of the 1st marble being red and the 2nd blue $= \dfrac{5}{10} \times \dfrac{5}{9} \times \dfrac{25}{90}$.

The probability of the 1st marble being blue and the 2nd red $= \dfrac{5}{10} \times \dfrac{5}{9} \times \dfrac{25}{90}$.

The probability of selecting one of each colour $= \dfrac{25}{90} + \dfrac{25}{90} = \dfrac{5}{9}$.

**32** A

Pr (Sue will score one run each time she bats) $= 0.7$

Sue has six shots; we need to find the probability of at least two runs.

Use $\text{Bi}(n, p)$ with $n = 6$, $p = 0.7$, $q = 0.3$

$$\begin{aligned} \Pr(X \geq 2) &= \Pr(X = 2) + \Pr(X = 3) + \Pr(X = 4) + \Pr(X = 5) + \Pr(X = 6) \\ &= 1 - (\Pr(X = 0) + \Pr(X = 1)) \\ &= 1 - {}^6C_0(0.3)^6(0.7)^0 - {}^6C_1(0.3)^5(0.7)^1 \\ &= 1 - (0.3)^6 - 6(0.3)^5(0.7)^1 \end{aligned}$$

**33** B

$\Pr(X = x) = {}^3C_x(0.2)^{3-x}(0.8)^x$

where $x = 0, 1, 2, 3$

with $n = 3$, $p = 0.8$, $q = 0.2$.

The probability distribution is

| $x$ | 0 | 1 | 2 | 3 |
|---|---|---|---|---|
| $\Pr(X = x)$ | $(0.2)^3$ | $3(0.2)^2(0.8)$ | $3(0.2)(0.8)^2$ | $(0.8)^3$ |

**34** D

Using $\text{Bi}(n, p)$ with mean $= np = 3$ and $\text{Var}(X) = npq = 1.5^2$, solve for $n$ and $p$.

$$\begin{cases} np=3 \\ np(1-p)=\left(\frac{3}{2}\right)^2 \end{cases} \Bigg|_{n,\,p}$$

$$\left\{n=12,\ p=\frac{1}{4}\right\}$$

$n = 12$, $p = \dfrac{1}{4}$

**35** C

Out of 20 questions, Amy usually answers 15 correctly.

Use $\text{Bi}(n, p)$ with $n = 20$, $p = \dfrac{15}{20} = \dfrac{3}{4}$.

We need $\Pr(X = 18) = 0.0669$.

$$\text{binomialPDf}\left(18, 20, \frac{3}{4}\right)$$

$$0.0669478076$$

**36** E

$f(x) = 1.5(1 - x^2), 0 \le x \le 1$

$$\int_0^{0.3} 1.5\cdot(1-x^2)dx \quad \frac{873}{2000}$$

$\Pr(X \le 0.3) = \frac{873}{2000}$

**37** B

$f(x) = 1.5(x - 1)^2, 0 \le x \le 2$

$$\int_0^{2} 1.5x\cdot(x-1)^2dx \quad 1$$

mean = 1

**38** B

$$\int_0^{1} x^2dx+\int_1^{2} x(2-x)dx \quad 1$$

mean = 1

**39** A

$$\int_0^{1} x^3dx+\int_1^{2} x^2(2-x)dx-1^2 \quad \frac{1}{6}$$

variance $= \frac{1}{6}$

**40** E

For $\mu = 10$ and $\sigma = \frac{1}{3}$, the graph of the distribution is

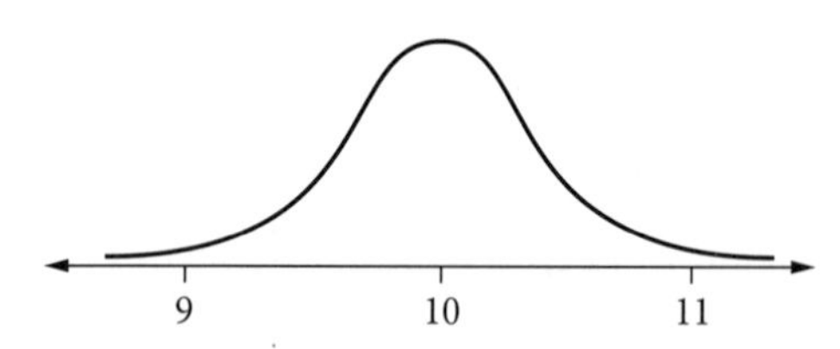


As approximately 99.7% of the graph lies between 3 standard deviations of the mean,

$3 \times \frac{1}{3}$ gives 1 unit either side of 10.

**41** E

The students' heights are normally distributed with a mean of 150, and standard deviation of 8.

16 is 2 standard deviations.

95% lies within 2 sd of the mean.

So $\Pr(X < 166) = 50\% + 47.5\% = 97.5\%$.

$\Pr(X < 166) = 0.975$

**42** E

```
normCDf(-10,1.2,1,0)
                0.8849303298
```

$\Pr(Z < 1.2) = 0.88$

**43** E

$\Pr(Z \le k) = 0.8$ and $Z$ is a standard normal variable.

```
invNormCDf("L",0.8,1,0)
                0.8416212336
```

$k = 0.842$

**44** C

normal distribution, mean = 15, standard deviation = 2

$\Pr(X < m) = 0.48$

```
invNormCDf("L",0.48,2,15)
                14.89969283
```

$m = 14.9$

**45** D

normal distribution, mean = 100, standard deviation = 15.

```
normCDf(102,200,15,100)
                0.4469648834
```

$\Pr(X > 102) = 0.447$

**46** C

normal distribution, mean = 100, standard deviation = ?

$\Pr(X < 95)$ = unacceptable = 0.03

Thus, $z = -1.88079$.

Using $z = \frac{x-\mu}{\sigma}$ gives $-1.88079 = \frac{95-100}{\sigma}$.

Solving for $\sigma$,

$\sigma = 2.658$

**47** C

$\hat{p} = \frac{X}{20}$ where $n = 20$.

To find the probability that the sample proportion is equal to the population proportion (0.7),

$\Pr(\hat{p} = 0.7) = \Pr(X = 14)$

and $\Pr(X = 14) = 0.1916$

**48** B

$\hat{p} = \frac{X}{20}$ where $n = 20$

We need 1 SD of population proportion.

$$\text{SD}(\hat{p}) = \sqrt{\frac{p(1-p)}{n}} = \sqrt{\frac{0.7(1-0.7)}{20}} = 0.1025$$

$$\sqrt{\frac{0.7(1-0.7)}{20}}$$
0.1024695077

This gives

$0.7 - 0.1025 < \hat{p} < 0.7 + 0.1025$
$= 0.5975 < \hat{p} < 0.8025$
$= 0.5975 \times 20 < X < 0.8025 \times 20$
$= \Pr(11.95 < X < 16.05)$

Because discrete $= \Pr(12 < X < 16) = 0.7796$.

```
binomialCDf(12,16,20,0.7)
                0.7795817326
```

**49** B

For a sample size of 2000, the 95% CI is (0.579, 0.621).

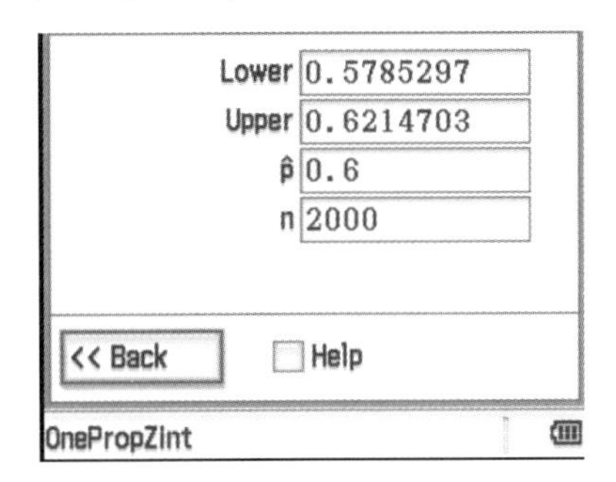


**50** E

Check $np$ and $nq$.

$np = 1000 \times 0.6 = 600$

$nq = 1000 \times 0.4 = 400$

Since $np > 5$ and $nq > 5$, the normal distribution can be used.

Find the mean and standard deviation.

$\mu = p = 0.6$

$$\text{SD} = \sqrt{\frac{p(1-p)}{n}} = \sqrt{\frac{pq}{n}} = \sqrt{\frac{0.6 \times 0.4}{100}} = 0.0155$$

Find the proportion of the standard normal distribution

```
normCDf(0.58,0.62,0.0155,0.6)
                0.80306132
```

The probability that 58% to 62% of families from a sample of 1000 own a piano is about 0.8031.

# Extended-answer questions

### Question 1

**a** **i** $\Pr(X = 3) = {}^3C_3 p^3 (1 - p)^0$

**ii** $\Pr(X = 0) = {}^3C_0 p^0 (1 - p)^3$

**b** **i** $p = \frac{1}{2}$

**ii** The binomial distribution is symmetric, so with three independent trials $X = 0$ and $X = 3$ are at opposite ends and are therefore the same probability.

**c**

| $x$ | 0 | 1 | 2 | 3 |
|---|---|---|---|---|
| $\Pr(X = x)$ | ${}^3C_0 p^0 (1-p)^3$ | ${}^3C_1 p^1 (1-p)^2$ | ${}^3C_2 p^2 (1-p)^1$ | ${}^3C_3 p^3 (1-p)^0$ |

**d** mean = 1.5

**e**

| $y$ | 0 | 1 | 2 | 3 |
|---|---|---|---|---|
| $\Pr(Y = y)$ | 0.0911 | 0.3341 | 0.4084 | 0.1664 |

**f** **i** mean = 1.650, variance = 0.742

**ii** sd = 0.862

$\Pr(1 \le X \le 2) = 0.743$

**iii** $\Pr(X >= 2) = \Pr(X = 2) + \Pr(X = 3) = 0.575$

**g** 3 trials

### Question 2

**a** $M \sim N(68, 64)$, $\Pr(60 < M < 90) = 0.838$

**b** **i** $\Pr(H|S) = \frac{\Pr(H \cap S)}{PR(S)} = \frac{0.09}{0.29} = 0.310$

**ii** No, the events are not independent.

$$\Pr(H|S) = \frac{\Pr(H \cap S)}{\Pr(S)} \neq \frac{\Pr(H) \times \Pr(S)}{\Pr(S)} = \Pr(H),$$

$0.310\ldots \neq 0.1587$

$\Pr(H \cap S) \neq \Pr(H) \times \Pr(S)$

$0.09 \neq 0.1587 \times 0.29 = 0.046\ldots$

**c** **i** $X \sim \text{Bi}(16, 0.1587)$, $\Pr(X = 1) = 0.190$

**ii** $\Pr(\hat{P} > 0.1) = \Pr(X > 1.6) = \Pr(X \ge 2) = 0.747$

**iii** $\Pr\left(\hat{P}_n > \frac{1}{n}\right) > 0.99,$

$\Pr(X > 1) = \Pr(X \ge 2) > 0.99,$

$1 - (\Pr(X = 0) + \Pr(X = 1)) > 0.99$

$\Pr(X = 0) + \Pr(X = 1) < 0.01,$

$(1 - 0.1587)^n + \binom{n}{1} 0.1587(1 - 0.1587)^{n-1} < 0.01$, $n = 38.925\ldots$, $n = 39$

**d** **i** $\hat{p} = \frac{0.102 + 0.145}{2} = 0.1235$

**ii** The 95% confidence interval for Statsville, $(0.102, 0.145)$, does not contain the Mathsland proportion, which is 0.1587.

**e** $\int_0^{\infty} (t \times M(t))\, dt = 44.6$

**f** $\int_0^{15} (M(t))\, dt = 0.0266$

**g** $0.05 \times \frac{1}{7} + x \times \frac{6}{7} = 0.0266\ldots$, $x = 0.0227$

### Question 3

**a** **i** $X \sim \text{Bi}\left(20, \frac{5}{8}\right)$, $\Pr(X \geq 10)$

$= 0.915\,29\ldots$

$= 0.9153$

**ii** $\Pr(X > 15 | X \geq 10)$

$= \dfrac{\Pr(X > 15)}{\Pr(X \geq 10)}$

$= \dfrac{0.079\,041\ldots}{0.915\,292\ldots}$

$= 0.086$

**b** $S'SSS' + S'SS'S + S'S'SS = \dfrac{3}{32} + \dfrac{1}{16} + \dfrac{3}{16}$

$= \dfrac{11}{32}$

$= 0.343\,75$

### Question 4

**a** $X \sim \text{Bi}(22, 0.1)$, $\Pr(X \geq 1) = 1 - \Pr(X = 0)$

$= 1 - 0.9^{22} \geq 0.9015$

**b** $\Pr(X < 5 | X \geq 1) = \dfrac{\Pr(X \geq 1 \cap X < 5)}{\Pr(X \geq 1)}$

$= \dfrac{\Pr(X \geq 1 \cap X \leq 4)}{\Pr(X \geq 1)}$

$= \dfrac{0.839\,389\ldots}{0.901\,522\ldots}$

$= 0.9311$

**c** **i** $\text{E}(X) = \int_0^{210} (x \times f(x))dx = 170.01$

**ii** $f$ is not a probability density function as $\int_0^{210} f(x)\,dx \neq 1$. It is a close approximation such that student calculations $\int_0^m f(x)\,dx = \frac{1}{2}$ and $\int_m^{210} f(x)\,dx = \frac{1}{2}$ both yielded values $m = 176$, correct to the nearest integer.

### Question 5

**a** **i** $\Pr(X > 7) = \int_7^8 f(x)\,dx = \dfrac{11}{16}$ or $0.6875$

**ii** $Y \sim \text{Bi}\left(3, \frac{11}{16}\right)$, $\Pr(Y = 1) = 3 \times \dfrac{11}{16} \times \left(\dfrac{5}{16}\right)^2$

$= \dfrac{825}{4096}$

**b** $\text{E}(X) = \int_6^8 (x \times f(x))dx = \dfrac{36}{5} = 7.2$

**c** oranges $\sim N(74, 9^2)$

$\Pr(O < 85 | O > 74)$

$= \dfrac{\Pr(O < 85 \cap O > 74)}{\Pr(O > 74)}$

$= \dfrac{\Pr(74 < O < 85)}{\Pr(O > 74)}$

$= \dfrac{0.389\,18\ldots}{0.5}$

$= 0.7783\ldots$

$= 0.778$

**d** **i** lemons $\sim \text{Bi}(4, 0.03)$,

$\Pr(\text{lemons} \geq 1)$

$= 1 - \Pr(\text{lemons} = 0)$

$= 1 - (0.97)^4 = 0.1147$

**ii** $\Pr(X \geq 1) > 0.5$, $1 - (0.97)^n > 0.5$

$n > 22.7566\ldots$, $n = 23$

### Question 6

**a** **i** $\dfrac{1}{64} = 0.015625$

**ii** Binomial $n = 20$, $p = \dfrac{1}{4}$

$\Pr(X \geq 10) = 0.0139$

**iii** $n \times \dfrac{1}{4} \times \dfrac{3}{4} = \dfrac{75}{16}$, $n = 25$

**b** Two cases $CCCC$, $ICCC$,

$\dfrac{1}{3} \times \left(\dfrac{3}{4}\right)^3 + \dfrac{2}{3} \times \dfrac{1}{3} \times \left(\dfrac{3}{4}\right)^2$, $\dfrac{17}{64} = 0.265\,625$

**c** $25p^{24}(1 - p) + p^{25} = 6p^{25}$, $25p^{24} - 30p^{25} = 0$,

$5p^{24}(5 - 6p) = 0$, $p > 0$, $p = \dfrac{5}{6}$

**d** $Y$ is binomial with $p = \dfrac{5}{6}$, $n = 25$.

$\Pr\left(Z \geq \dfrac{25 - a}{b}\right) = \Pr(Y \geq 22)$

$= 0.381\,566\ldots$

$\Pr\left(Z \geq \dfrac{20 - a}{b}\right) = \Pr(Y \geq 18)$

$= 0.955\,268\ldots$

$\dfrac{20 - a}{b} = -1.698\,23\ldots$

$\dfrac{25 - a}{b} = 0.30137\ldots$

$a = 24.246$, $b = 2.500$

### Question 7

**a** 1 minute

**b** $\Pr(T \le 3 \mid T > 0) = \dfrac{\Pr(0 < T \le 3)}{\Pr(T > 0)}$

$= \dfrac{0.273\,37\ldots}{0.5}$

$= 0.547$

**c** $k$ can be found by a translation of 1.5 units in the direction of the negative $t$-axis, use symmetry to find second value for $k$ or solve

$$\int_{-4.5}^{0.5}\left(\frac{1}{4\sqrt{2\pi}}e^{-\frac{1}{2}\left(\frac{t-k}{4}\right)^2}\right)dt = 0.4648 \text{ for } k,$$

$k = -1.5$ or $k = -2.5$

**d** Let $X \sim \text{Bi}(8, 0.85)$, $\Pr(X \le 3) = 0.003$

**e** **i** $\Pr(\text{at least one late}) = 1 - 0.85^n$

**ii** $1 - 0.85^n > 0.95$, $n > 18.43\ldots$, $n = 19$

**f** $0.85(1 - y) + xy = 0.75$,

$y = -\dfrac{0.1}{x - 0.85} = \dfrac{2}{17 - 20x}$, $0.3 \le x \le 0.7$

Minimum value for $y$ is $\dfrac{2}{11}$ or $0.\overline{18}$ and

maximum value for $y$ is $\dfrac{2}{3}$ or $0.\dot{6}$.

### Question 8

**a** $X \sim N(67, 1)$, $\Pr(X < 68.5) = 0.9332$

**b** $X \sim N(67, 1)$, $\Pr(65.6 < 68.4) = 0.8385$

**c** $\dfrac{0.838\,48\ldots}{0.933\,19} = 0.8985$

**d** $\Pr(X < 68.4) = 0.995$, when $x = 68.4$, $z = 2.5758$

So $2.5758 = \dfrac{68.4 - 67}{\sigma}$, $\sigma = 0.54$ mm.

### Question 9

**a** $X \sim N(14, 4^2)$, $\Pr(X > x) = 0.1$,

$x = 19.1$ cm or 191 mm

**b** $\Pr(X < 9) = 0.105\,65\ldots$

$0.105\,65\ldots \times 2000 = 211$ basil plants

**c** $E(X) = \displaystyle\int_0^{50}\left(x \times \frac{\pi}{100}\sin\left(\frac{\pi x}{50}\right)\right)dx = 25$ cm

**d** Solve $\displaystyle\int_0^{a}\left(\frac{\pi}{100}\sin\left(\frac{\pi x}{50}\right)\right)dx = 0.15$ for $a$ where $a$ is the maximum height of a coriander plant.

$a = 12.7$ cm or 127 mm

**e** $Y \sim \text{Bi}(n, 0.2)$, $\Pr(Y \ge 1) > 0.95$.

Solve $0.8^n < 0.05$ for $n$.

$n = 14$ tomato plants

**f** **i** $0.7 \times 0.7 + 0.3Y(1 - p) = 0.79 - 0.3p$

Probability that the third pot is smooth $= 0.79 - 0.3p$

**ii** Solve $0.79 - 0.3p = 0.61$, $p = 0.6$

### Question 10

**a** $(0.0169, 0.1431)$

**b** mean $= \displaystyle\int_{290}^{330}\left(\frac{3w}{640000}(330 - w)^2(w - 290)\right)dw$

$= 306$ grams

**c** $\displaystyle\int_{290}^{306}\left(\frac{3}{640000}(330 - w)^2(w - 290)\right)dw = 0.5248$

### Question 11

**a** **i** $X \sim N(3, 0.64)$, $\Pr(3 \le X \le 5) = 0.4938$

**ii** $\Pr(3 \le Y \le 5)$, $\Pr(3 \le Y \le 4) + \Pr(4 < Y \le 5)$

$= \displaystyle\int_3^4\left(\frac{y}{16}\right)dy + \int_4^5(0.25e^{-0.5(y-4)})\,dy$

$= 0.4155$

or

$\Pr(3 \le Y \le 5) = \displaystyle\int_3^5(f(y))\,dy$

$= 0.4155$

**b** $\displaystyle\int_0^4\left(\frac{y^2}{16}\right)dy + \int_4^{\infty}(0.25ye^{-0.5(y-4)})\,dy = 4.333$

**c** **i** $\displaystyle\int_0^4\left(\frac{y}{16}\right)dy = \frac{1}{2}$ or $\displaystyle\int_{-\infty}^{m}(f(y)) = \frac{1}{2}$

Upper 50% of $Y$ exists above 4.

**ii** $\displaystyle\int_4^a 0.25e^{-0.5(y-4)}\,dy = 0.2$

or $\displaystyle\int_a^{\infty} 0.25e^{-0.5(y-4)}\,dy = 0.3$

or $\displaystyle\int_{-\infty}^{a}(f(y))\,dy = 0.7$

$a = 5.02$

**d** $W \sim \text{Bi}\left(10, \dfrac{9}{32}\right)$ or $\dbinom{10}{4}\left(\dfrac{9}{32}\right)^4\left(\dfrac{23}{32}\right)^6$

$\Pr(W = 4) = 0.1812$

**e** $\Pr(X > 3) = 0.5, \Pr(Y > 3) = \frac{23}{32}$

$\Pr(\text{longer than 3}) = 0.5 \times 0.5 + 0.5 \times \frac{23}{32}$

$= 0.609\,375\ldots$

$\Pr(\text{machine } A \mid \text{longer than 3}) = \frac{0.5 \times 0.5}{0.609\,375}$

$= 0.4103$

### Question 12

**a**

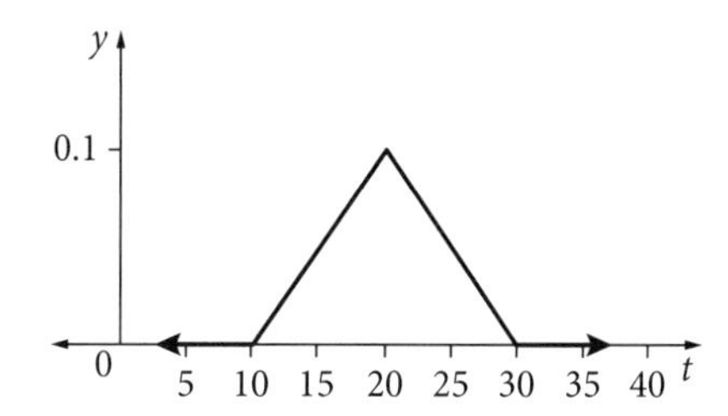


**b** $\Pr(f(t) < 25) = 1 - \frac{1}{100}\int_{25}^{30}(30 - t)\,dt = \frac{7}{8}$

**c** $\frac{\Pr(T \le 15) \cap \Pr(T \le 25)}{\Pr(T \le 25)} = \frac{\frac{1}{8}}{\frac{7}{8}} = \frac{1}{7}$

**d** $X \sim \text{Bi}\left(6, \frac{7}{8}\right), \Pr(X > 4) \approx 0.9709$

**e** $\binom{6}{3}p^3(1 - p)^3 + \binom{6}{4}p^4(1 - p)^2$

$= 5p^3(1 - p)^2(4 - p)$

**f** **i** Solve $Q'(p) = 0$ for $p$.

$p = 2 - \sqrt{2}, Q = 20(17 - 12\sqrt{2})$

**ii** $\frac{1}{100}\int_{b}^{30}(30 - t)\,dt = 1 - (2 - \sqrt{2}), b \approx 20.9$

### Question 13

**a**

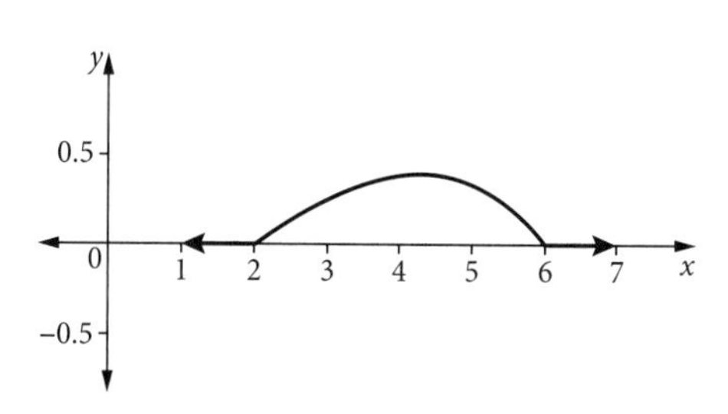


Maximum is $(4.31, 0.38)$.

**b** $\int_{2}^{3} f(x)\,dx = 0.1211$

**c** $\int_{2}^{6} xf(x)\,dx = 4.1333$

### Question 14

**a**

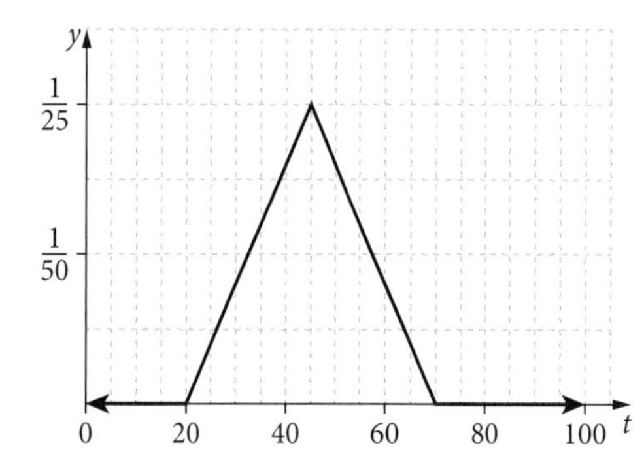


**b** $\int_{25}^{55}(f(t))\,dt = \frac{4}{5}$

**c** $\Pr(T \le 25 \mid T \le 55) = \frac{\Pr(T \le 25)}{\Pr(T \le 55)}$

$= \frac{\int_{20}^{25}(f(t))\,dt}{\int_{20}^{55}(f(t))\,dt}$

$= \frac{1}{41}$

**d** $\int_{a}^{70}(f(t))\,dt = 0.7$

or $\int_{20}^{a}(f(t))\,dt = 0.3$

or $\int_{a}^{45}(f(t))\,dt = 0.2$

$a = 39.3649$

**e** **i** $X \sim \text{Bi}\left(7, \frac{8}{25}\right), \Pr(X > 3) = 0.1534$

**ii** $\Pr(X \ge 2 \mid X \ge 1) = \frac{\Pr(X \ge 2)}{\Pr(X \ge 1)}$

$= \frac{0.7113\ldots}{0.93277\ldots}$

$= 0.7626$

**f** $q(p) = \binom{7}{2}p^2(1 - p)^5 + \binom{7}{3}p^3(1 - p)^4$

$= 7p^2(p - 1)^4(2p + 3)$

**g** **i** Solve $q'(p) = 0$ for $p$

$p = 0.3539, q = 0.5665$

**ii** $\int_{d}^{70} f(t)\,dt = 0.353\,88\ldots d$

$= 48.9676\ldots$

$= 49$ minutes

### Question 15

**a** $\Pr(L < 0) = 0.0062$

**b** $\Pr(L > 15) = 0.1056$

**c** **i**

| $c$ | 0 | 100 | 200 |
|---|---|---|---|
| $\Pr(C = c)$ | 0.006 | 0.888 | 0.106 |

**ii** $E(C) = 100 \times 0.8881... + 200 \times 0.1062...$
$= \$110$

**iii** $\text{sd}(C) =$
$\sqrt{100^2 \times 0.881... + 200^2 \times 0.1062... - (109.94...)^2}$
$= \$32$

**d** **i** $(0.030, 0.056)$

**ii** The proportion of concerts that begin more than 15 minutes late for the Mathsland Concert Hall is outside the sample CI.

**e** $E(M) = \int_0^{\infty} (x \times f(x))\,dx = 2$

**f** **i** $\int_{15}^{\infty} (f(x))\,dx = \frac{4}{289}$

**ii** $\left(\frac{285}{289}\right)^9 \times \frac{4}{289} = 0.0122$

**iii** $\Pr(M < 20 | M > 15) = \left(\frac{\Pr(15 < M < 20)}{\Pr(M > 15)}\right)$
$= 0.403$

**Question 16**

**a** $X_A \sim N\left(11, \left(\frac{1}{4}\right)^2\right), \Pr(X_A > 10.5) = 0.977$

**b** $E(X_B) = \int_0^{12} (xf(x))\,dx = 7.75$ hours

**c** $E(X_B) = \sqrt{\int_0^{12} x^2 f(x)\,dx - \left(\int_0^{12} xf(x)\,dx\right)^2}$
$= 2.31$ hours

**d** $\Pr(X_B > 10.5) = \int_{10.5}^{12} f(x)\,dx = 0.1134$

**e** $\Pr(\text{boxes mislabelled}) = \Pr(A \cap (X_A < 10.5)) + \Pr(B \cap (X_B < 10.5))$
$= 0.5 \times 0.0228 + 0.5 \times 0.1134$
$= 0.068$

**f** $\Pr(B | \text{mislabelled}) = \frac{(B \cap \text{mislabelled})}{\Pr(\text{mislabelled})}$
$= \frac{0.5 \times 0.1134}{0.0681}$
$= 0.833$

**g** $X_1 \sim \text{Bi}(26, 0.05)$ or $1 - 0.95^{26}$,
$\Pr(X_1 \geq 1) = 0.7365$

**h** $X_2 \sim \text{Bi}(100, 0.05)$,
$\Pr(\hat{P}_A > 0.04 | \hat{P}_A < 0.08) = \frac{\Pr(5 \leq X_1 \leq 7)}{\Pr(X_1 \leq 7)}$
$= \frac{0.4361}{0.8720}$
$= 0.5000$

**i** A 90% confidence interval for the population proportion from this sample is $(0.02, 0.10)$.

**Question 17**

**a** 0.106

**b** 10.7

**c** $E(\hat{P}) = 0.08 = \frac{2}{25}, \text{sd}(\hat{P}) = \frac{\sqrt{46}}{125}$

**d** $X \sim \text{Bi}(25, 0.08), \Pr(X > 2.5), \Pr(X \geq 3) = 0.323$

**e** 50

**f** $\sigma = \sqrt{\int_0^{50} (x^2 f(x))\,dx - \left(\int_0^{50} xf(x)\,dx\right)^2} = 10.3$

**Question 18**

**a** $\int_0^5 \left(\frac{4x}{625}(5x^3 - x^4)\right) dx = \frac{10}{3}$

**b** $80\int_2^5 \left(\frac{4}{625}(5x^3 - x^4)\right) dx = 73$ butterflies

**c** $\Pr(X \geq 4 | X \geq 2) = \frac{\int_4^5 \left(\frac{4}{625}(5x^3 - x^4)\right) dx}{\int_2^5 \left(\frac{4}{625}(5x^3 - x^4)\right) dx}$
$= 0.2878$

**d** $X \sim N(14.1, 2.1^2), \Pr(16 \leq X \leq 18) = 0.1512$

**e** 10.6, correct to one decimal place

**f** **i** $X \sim \text{Bi}(36, 0.0527), \Pr(X \geq 3) = 0.2947$

**ii** $\Pr(6 \leq X \leq 36) = 0.010659...$,
$\Pr(7 \leq X \leq 36) = 0.002436...$, smallest value of $n = 7$.

**iii** $E(\hat{P}) = 0.0527$

$\text{sd}(\hat{P}) = \sqrt{\frac{0.0527(1 - 0.0527)}{36}} = 0.0372$

**iv** $\Pr(0.01546... < \hat{P} < 0.08993...)$
$= \Pr(0.55659... < X < 3.237...)$
$= \Pr(1 \leq X \leq 3) = 0.7380$

**g** proportion = 0.055, $0.055 + 1.96\sqrt{\dfrac{0.055 \times 0.945}{n}}$

$= 0.0866$

or $0.055 - 1.96\sqrt{\dfrac{0.055 \times 0.945}{n}} = 0.0234$

or $p - 1.96\sqrt{\dfrac{p(1-p)}{n}} = 0.0234\ldots$ [1]

and $p + 1.96\sqrt{\dfrac{p(1-p)}{n}} = 0.0866\ldots$ [2]

$n = 200$

# Mathematical Methods formulas

## Mensuration

| | | | |
|---|---|---|---|
| area of a trapezium | $\frac{1}{2}(a+b)h$ | volume of a pyramid | $\frac{1}{3}Ah$ |
| curved surface area of a cylinder | $2\pi rh$ | volume of a sphere | $\frac{4}{3}\pi r^3$ |
| volume of a cylinder | $\pi r^2 h$ | area of a triangle | $\frac{1}{2}bc\sin(A)$ |
| volume of a cone | $\frac{1}{3}\pi r^2 h$ | | |

## Calculus

| | | | |
|---|---|---|---|
| $\frac{d}{dx}\left(x^n\right) = nx^{n-1}$ | | $\int x^n dx = \frac{1}{n+1}x^{n+1} + c,\ n \neq -1$ | |
| $\frac{d}{dx}\left((ax+b)^n\right) = an(ax+b)^{n-1}$ | | $\int (ax+b)^n dx = \frac{1}{a(n+1)}(ax+b)^{n+1} + c, n \neq -1$ | |
| $\frac{d}{dx}\left(e^{ax}\right) = ae^{ax}$ | | $\int e^{ax} dx = \frac{1}{a}e^{ax} + c$ | |
| $\frac{d}{dx}\left(\log_e(x)\right) = \frac{1}{x}$ | | $\int \frac{1}{x} dx = \log_e(x) + c,\ x > 0$ | |
| $\frac{d}{dx}\left(\sin(ax)\right) = a\cos(ax)$ | | $\int \sin(ax)dx = -\frac{1}{a}\cos(ax) + c$ | |
| $\frac{d}{dx}\left(\cos(ax)\right) = -a\sin(ax)$ | | $\int \cos(ax)dx = \frac{1}{a}\sin(ax) + c$ | |
| $\frac{d}{dx}\left(\tan(ax)\right) = \frac{a}{\cos^2(ax)} = a\sec^2(ax)$ | | | |
| product rule | $\frac{d}{dx}(uv) = u\frac{dv}{dx} + v\frac{du}{dx}$ | quotient rule | $\frac{d}{dx}\left(\frac{u}{v}\right) = \frac{v\frac{du}{dx} - u\frac{dv}{dx}}{v^2}$ |
| chain rule | $\frac{dy}{dx} = \frac{dy}{du}\frac{du}{dx}$ | Newton's method | $x_{n+1} = x_n - \frac{f(x_n)}{f'(x_n)}$ |
| trapezium rule approximation | $Area \approx \frac{x_n - x_0}{2n}\left[f(x_0) + 2f(x_1) + 2f(x_2) + \ldots + 2f(x_{n-2}) + 2f(x_{n-1}) + f(x_n)\right]$ | | |

## Probability

| | | | |
|---|---|---|---|
| $\Pr(A) = 1 - \Pr(A')$ | | $\Pr(A \cup B) = \Pr(A) + \Pr(B) - \Pr(A \cap B)$ | |
| $\Pr(A \mid B) = \frac{\Pr(A \cap B)}{\Pr(B)}$ | | | |
| mean | $\mu = E(X)$ | variance | $\text{var}(X) = \sigma^2 = E((X-\mu)^2) = E(X^2) - \mu^2$ |
| binomial coefficient | $\binom{n}{x} = \frac{n!}{x!(n-x)!}$ | | |

| Probability distribution | | Mean | Variance |
|---|---|---|---|
| discrete | $\Pr(X = x) = p(x)$ | $\mu = \sum x\, p(x)$ | $\sigma^2 = \sum (x-\mu)^2 p(x)$ |
| binomial | $\Pr(X = x) = \binom{n}{x} p^x (1-p)^{n-x}$ | $\mu = np$ | $\sigma^2 = np(1-p)$ |
| continuous | $\Pr(a < X < b) = \int_a^b f(x)dx$ | $\mu = \int_{-\infty}^{\infty} x\, f(x)dx$ | $\sigma^2 = \int_{-\infty}^{\infty} (x-\mu)^2 f(x)dx$ |

## Sample proportions

| | | | |
|---|---|---|---|
| $\hat{P} = \frac{X}{n}$ | | mean | $E(\hat{P}) = p$ |
| standard deviation | $\text{sd}(\hat{P}) = \sqrt{\frac{p(1-p)}{n}}$ | approximate confidence interval | $\left(\hat{p} - z\sqrt{\frac{\hat{p}(1-\hat{p})}{n}},\ \hat{p} + z\sqrt{\frac{\hat{p}(1-\hat{p})}{n}}\right)$ |



9780170465366